Cyber Security Innovation
for the Digital Economy
A Case Study of the Russian Federation

RIVER PUBLISHERS SERIES IN SECURITY AND DIGITAL FORENSICS

Series Editors:

William J. Buchanan
Edinburgh Napier University, UK

Anand R. Prasad
NEC, Japan

Indexing: All books published in this series are submitted to the Web of Science Book Citation Index (BkCI), to CrossRef and to Google Scholar.

The "River Publishers Series in Security and Digital Forensics" is a series of comprehensive academic and professional books which focus on the theory and applications of Cyber Security, including Data Security, Mobile and Network Security, Cryptography and Digital Forensics. Topics in Prevention and Threat Management are also included in the scope of the book series, as are general business Standards in this domain.

Books published in the series include research monographs, edited volumes, handbooks and textbooks. The books provide professionals, researchers, educators, and advanced students in the field with an invaluable insight into the latest research and developments.

Topics covered in the series include, but are by no means restricted to the following:

- Cyber Security
- Digital Forensics
- Cryptography
- Blockchain
- IoT Security
- Network Security
- Mobile Security
- Data and App Security
- Threat Management
- Standardization
- Privacy
- Software Security
- Hardware Security

For a list of other books in this series, visit www.riverpublishers.com

Cyber Security Innovation
for the Digital Economy
A Case Study of the Russian Federation

Sergei Petrenko

Innopolis University
Russia

LONDON AND NEW YORK

Published 2018 by River Publishers
River Publishers
Alsbjergvej 10, 9260 Gistrup, Denmark
www.riverpublishers.com

Distributed exclusively by Routledge
4 Park Square, Milton Park, Abingdon, Oxon OX14 4RN
605 Third Avenue, New York, NY 10017, USA

First issued in paperback 2023

Cyber Security Innovation for the Digital Economy A Case Study of the Russian Federation / by Sergei Petrenko.

Routledge is an imprint of the Taylor & Francis Group, an informa business

Publisher's Note
The publisher has gone to great lengths to ensure the quality of this reprint but points out that some imperfections in the original copies may be apparent.

ISBN 13: 978-87-7022-921-0 (pbk)
ISBN 13: 978-87-70220-22-4 (hbk)
ISBN 13: 978-1-00-333778-2 (ebk)

While every effort is made to provide dependable information, the publisher, authors, and editors cannot be held responsible for any errors or omissions.

Contents

Foreword   ix

Preface   xv

Acknowledgements   xvii

List of Figures   xix

List of Tables   xxvii

List of Abbreviations   xxix

Introduction   1

1   Relevance of Cyber Security Innovations   7
    1.1   Digital Transformation of State and Society   7
          1.1.1   State Program "Digital Economy"   8
          1.1.2   Main Information Infrastructure Development
                  Objectives   28
          1.1.3   Implementation of the Long-Term Evolution (LTE)
                  Technology   39
          1.1.4   IIoT/IoT Technologies Development   53
    1.2   Typical Cyber Security Threats   68
          1.2.1   Possible Scenarios of Cyber-Attack on the Information
                  Infrastructure   68
          1.2.2   Threats and Consequences of Wireless LAN IEEE
                  802.1x Implementation   106
          1.2.3   Cyber Security Threats of Corporate Digital and
                  IP-ATX (Private Automatic Telephone Exchanges) .   117
          1.2.4   Threats and Security Profile of the Mobile Operating
                  System OS Sailfish and Tizen   123

1.3 Cyber Security Threats Monitoring Necessity 136
    1.3.1 Cyber Security Incidents Factual Account 136
    1.3.2 Need for Joint Initiatives of Society and States . . . 144
    1.3.3 Capture the Flag Competition on Vulnerability Detection 148
    1.3.4 Security Operations Center (SOC) Key Role 150

2 MSSP/MDR National Operator Development   **163**
2.1 Ultimate Opportunity of National MSSP/MDR Operators . . 163
    2.1.1 Relevance of MSSP/MDR Cyber Security Services 164
    2.1.2 MSSP/MDR Best Organization Practice 167
    2.1.3 Sample of MSSP by AT&T 171
    2.1.4 Sample of MSSP Model by UBIqube 173
    2.1.5 Feasible Technical Solutions 174
2.2 Possible Ways of Providing Cyber Security Services 181
    2.2.1 Typical MSSR/MDR Services 184
    2.2.2 IS Sourcing Model Analysis 192
    2.2.3 The IS Sourcing Practice 193
    2.2.4 Sample SLA Content for the Provision of Cyber Security Services 200
    2.2.5 Best Practices for Providing Cyber Security Service 204
2.3 Development of National MSSP/MDR Based on Big Data . 208
    2.3.1 Big Data Processing Requirements Analysis 208
    2.3.2 Best Big Data Processing Practice 211
    2.3.3 MSSP/MDR Subsystem Functionality for Big Data Processing 215
    2.3.4 Sensor Cloud Architecture Advantages 219
2.4 New Methods of Cyber Security Knowledge Management . 221
    2.4.1 Possible State of the Art 221
    2.4.2 Cyber Security MDM Principles 228
    2.4.3 MDM Cyber Security System Example 230

3 Innovative Methods for Detecting Anomalies   **237**
3.1 Justification of a New Method for Detecting Anomalies . . . 237
    3.1.1 Analysis of the Existing Approaches to the TCP/IP Network Abnormal Functioning Detection 237
    3.1.2 Possible Statement of the Detecting Anomalies' Problem 242
    3.1.3 Definition of New Informative Features 244
    3.1.4 Detection of Anomalies Based on Dimensions . . . 248

      3.1.5   Investigation of Properties of Invariants of
             Dimension 252
  3.2 The Main Provisions of the New Method for Detecting
      Anomalies 261
      3.2.1   The Main Hypotheses for Detecting Anomalies . . 261
      3.2.2   Control of Semantic Correctness Criteria 268
      3.2.3   Sufficient Condition for the Criteria Fulfillment . . 273
      3.2.4   Implementation of the New Method of TCP
             Transport Layer Protocol 277
  3.3 Startup of Anomaly Detection Based on Dimensions 283
      3.3.1   Possible Architecture Solutions 283
      3.3.2   Features of the Transfer and Control Criteria 287
      3.3.3   Experiment Results 290
      3.3.4   Trends and Development Prospects 293
  3.4 New Method of Analytical Verification 299
      3.4.1   Data Processing Model on the Example of Oracle
             Solution 299
      3.4.2   Marked Data Visualization 301
      3.4.3   Formalization of HTTP and SQL * Net Protocols . 304
      3.4.4   Presentation of the Transport Layer Protocol (TCP) 309
      3.4.5   Presentation of the Networking Layer IP 313
      3.4.6   Control of the Platform Semantic Correctness . . . 314
      3.4.7   Platform Semantic Correctness Control 317
      3.4.8   Verification of Applied Queries 323
      3.4.9   TCP Verification 326
      3.4.10 IP Verification 327

4 Development of Cyber Security Technologies   **335**
  4.1 Cyber Security R&D Best Practice 335
      4.1.1   Cyber Security R&D Importance 336
      4.1.2   Cyber Security Project Management 340
      4.1.3   New Cyber Security Problems Statement 344
  4.2 Development of the Cyber Security Requirements in Terms
      of GOST R IEC 61508 346
      4.2.1   Analysis of the Cyber Security Requirements 346
      4.2.2   Need for GOST R IEC 61508 Development 351
      4.2.3   Method for Anomaly Detection in the CF CPCS
             Behavior 355
  4.3 Creation of New Cyber Security Ontologies 362

4.3.1   Analysis of New Requirements of Cyber Security .   362
4.3.2   Known Cyber Security Ontologies   365
4.3.3   Proposed Cyber Security Ontology   373
4.3.4   Ontology Structure Example   379
4.4   Development of Cyber Security Platforms   385
4.4.1   Principles of Designing Special Computing Systems   386
4.4.2   Feasible Computing Classifications   390
4.4.3   Characteristics of the Known Computing Systems .   395
4.4.4   Development of the Supercomputer Technologies .   398
4.5   Security Software Development based on Agile
Methodology   406
4.5.1   Main Ideas and Principles of the Agile
Methodology   406
4.5.2   Best Practices of Agile Methodology   407
4.5.3   Adapting Agile for Secure Application
Development   409
4.6   Development of BI-platforms for Cyber Security Predictive
Analytics   414
4.6.1   BI-security Platform Requirements   415
4.6.2   BI Security Platform Startup   419
4.6.3   Expected Results   426

Conclusion   **433**

References   **441**

Index   **455**

About the Author   **457**

Foreword

Rector of Innopolis University*
Alexander Tormasov

Dear Readers!

The book presents valuable experience and results of organizing and conducting innovative R & D in the Information Security Center of the Innopolis University, which allowed creating on the basis of modern Industry 4.0 technologies a number of safe and trusted technological platforms for the digital enterprises of the domestic economy.

The modern level of development in information and communication technologies (ICT) now makes it possible to take industrial production and scientific research in information security to a fundamentally higher plane, but the effectiveness of such a transition directly depends on the availability of highly qualified specialists. Every year, about 5,000 Russian specialists graduate in the field of information security, whereas the actual industrial demand is estimated at 21,000 per year through 2020. For this reason, the Russian Ministry of Education and Science, along with executive governmental bodies, has created a high-level training program, which they continually develop, for state information security employees. This initiative includes 170 universities, 40 institutions of continuing education, and 50 schools of secondary vocational training. In evaluating the universities' performance over 30 academic disciplines, information security has scored the highest for three consecutive years on the Russian Unified State Examination Единый Государственный Экзамен. In addition, employee training subsystems operating in the framework of the Russian Federal Security Service, the Russian Ministry of Defense, the Russian Federal Protective Service, Russian Federal Service for Technical and Export Control, and the Russian Emergencies Ministry of Emergency Situations are similar to the general system for training information security specialists at the Russian Ministry of Education and Science, which trains personnel according to the concrete needs of individual departments.

Yet, there remains the well-known problem that the vast majority of educational programs in information security struggle to keep pace with the rapid development in the ICT sphere, where significant changes occur every 6 months. As a result, existing curricula and programs do not properly train graduates for the practical reality of what it means to efficiently solve modern information security problems. For this reason, graduates often find themselves lacking the actual skills in demand on the job market. In order to ensure that education in this field truly satisfies modern industrial demands, Innopolis University students and course participants complete actual information security tasks for commercial companies as well as governmental bodies (e.g., for the university's over 100 industrial partners). Also, Innopolis University students participate in domestic and international computer security competitions, e.g., the game Capture the Flag (CTF), considered to be among the most authoritatives in the world.

Currently, Innopolis University trains information security specialists in "Computer Science and Engineering" (MA program in Secure Systems and Network Design) The program is based on the University of Amsterdam's "System and Network Engineering" program with its focus on information security. In 2013, it was ranked as the best MA program for IT in the Netherlands (Keuzegids Masters 2013), and in 2015 it won the award for best educational program (Keuzegids Masters 2015). The University of Amsterdam is one of Innopolis University's partners and is included in the Top 50 universities of the world (QS World university rankings, 2014/2015). An essential feature of this program is that Innopolis University students take part in relevant research and scientific-technical projects from the beginning of their studies. In solving computer security tasks, students have access to the scientific-technical potential of 3 institutes, 13 research laboratories, and 3 research centers engaged in advanced IT research and development at Innopolis University. This partnership also extends to Innopolis University's academic faculty, both pedagogic and research-oriented, which numbers more than 100 world-class specialists.

The information security education at Innopolis University meets the core curriculum requirements set out in the State Educational Standards for Higher Professional Education 075 5000 "Information Security" in the following degrees: "Computer Security", "Organization and Technology of Information Security", "Complex Software Security", "Complex Information Security of Automated Systems", and "Information Security of Telecommunication Systems". At the same time, high priority is given to practical security issues of high industrial relevance; however, given the relative novelty of these needs, they remain insufficiently addressed in the curricula of most Russian universities and programs. These issues include the following:

- Computer Emergency Response Team (CERT) based on groundbreaking cognitive technologies;
- Trusted cognitive supercomputer and ultra-high performance technologies;
- Adaptive security architecture technologies;
- Intelligent technologies for ensuring information security based on Big Data and stream processing (Big Data + ETL);
- Trusted device mesh technology and advanced system architecture;
- Software-defined networks technology (SDN) and network functions virtualization (NFV);
- Hardware security module technology (HSM);
- Trusted "cloud" and "foggy" computing, virtual domains;
- Secure mobile technologies of 4G +, 5G and 6G generations;
- Organization and delivery of national and international cyber-training sessions;
- Technologies for automated situation and opponent behavior modeling (WarGaming);
- Technologies for dynamic analysis of program code and analytical verification;
- Quantum technologies for data transmission, etc.

The current edition of the "**Cyber Security Innovation for the Digital Economy: A Case Study of the Russian Federation**" was written by Sergei Petrenko, Prof. Dr.-Ing., Head of the Information Security Center at Innopolis University. The work of this author has significantly contributed to the creation of a national training system for highly qualified employees in the field of computer and data security technologies. This book sets out a notion of responsibility in training highly qualified specialists at the international level and in establishing a solid scientific foundation, which is prerequisite for any effective application of cyber security technologies for the Digital Economy.

Rector of the Innopolis University,
Dr. Sci. in Physics and Mathematics,
Professor Alexander Tormasov

Deputy Director General SAP CIS
Dmitry Shepelyavyi

Dear Readers!

The concept of the "Digital Economy" over the past few years has become one of the key vectors of Russian development. Actively going on the process of creating the following required conditions: technological, legal, financial, and managerial, to ensure the maximum effectiveness of the state program. Implementation of the focal points of the Digital Economy program is directly related to the introduction of technological innovations, such as Industrial Internet and Internet of Things, Big Data, cloud technologies, cyber security technologies, etc. From the commercialization point of view, especially on entering the world market, the transition to innovative technologies in production and business processes is becoming one of the significant and competitive advantages for the country. Implementation of technologies always considers new products, services, jobs, and also increases the efficiency of manufacture and business in general.

In addition, the technological innovations also devoted to a completely different decision-making speed, which is critical not only for business and production, but also for all government levels. The modern Urbanism trend evidences that according to the data, provided by UNICEF, 75% of the world's population will be concentrated in cities by 2050. Naturally, this will seriously complicate the infrastructure of cities and, consequently, their management. There is a need to create highly intelligent systems for collecting, processing, and analyzing data that will increase the megapolis manageability and the operation efficiency of urban infrastructure, as well as the speed of response to emergency and critical damages.

The executive bodies already need to have constant operational monitoring of the socio-economic and socio-political situation, in this regard, the creation of situational centers of various directions becomes increasingly important. The effectiveness of the situational center depends on the combined efforts of a large number of specialists and on the capabilities of the applied technical solutions, since the processing and analyzing the data sets with a complex structure should be carried out in real time. The coordination of large urban projects is a good example. Such projects as "smart airport" and "smart stadium", where a centralized solution manages the smart sensor ecosystem, with the capabilities of predictive analytics directs the flows of people, ensures the safety and smooth operation of many internal structures. The concept of "smart stadium" has already been implemented in Germany-the "Allianz Arena" in Munich.

In most cases, such projects are initiated by the state, but they cannot be implemented without the support of the business community and developers, or without the involvement of global expertise. Situational centers and corresponding technological platforms of digital enterprises are already successfully functioning in many countries of the world and solve urgent problems of megacities. For example, the four largest cities in China (Beijing, Kunming, Chongqing, and Tianjin), where the traffic congestion was the case of everyday life, implemented a unique solution that combined cloud computing and mobile applications. It helped in real time to monitor the traffic situation and redistribute the flow of public transport and taxis in the best possible way.

Developments in the Digital Enterprise Platforms are useful not only for urban infrastructures, but they can also be used in production, in the process of transition to a digital model. New technologies in this scenario are applied at all stages of the product value chain: design, production, supply chain, maintenance, and repair. Automation, hyperconnection, and artificial intelligence can significantly reduce the duration of the production cycle and costs.

To summarize, it should be noted that the digital economy is now coming to the forefront as one of the country's potential competitive advantages: the commodity-based economy is gradually losing the ground, it is necessary to seek the new opportunities for the development, the digital economy is one of such options. While we have a lot of separate projects in different industries, the processes of digitalization are at absolutely different rates and the task of the state, the business community, and experts is to create conditions for the most productive development.

The foregoing raises the urgency of the presented monograph "**Cyber Security Innovation for the Digital Economy: A Case Study of the Russian Federation**". I consider that this book will be a very valuable tool for the development and formation of highly qualified specialists of a new class in the field of information technology and cyber security, following the over-the-top direction "Digital Economy" of the Russian Federation.

Deputy Director General SAP CIS
Dmitry Shepelyavyi

Preface

This scientific monograph considers possible solutions to the relatively new scientific-technical problem of developing innovative solutions in the field of cyber security for the Digital Economy. The solutions proposed are based on the results of exploratory studies conducted by the author in the areas of Big Data acquisition, cognitive information technologies (cogno-technologies), new methods of analytical verification of digital ecosystems on the basis of similarity invariants and dimensions, and "computational cognitivism", involving a number of existing models and methods.

In practice, this allowed to successfully create new entities-the required safe and trusted digital ecosystems-on the basis of the development of digital and cyber security technologies, and the resulting changes in their behavioral preferences. Here, the ecosystem is understood as a certain system of organizations, created around a certain technological platform that use its services to make the best offers to customers and access to them to meet the ultimate needs of clients-legal entities and individuals. The basis of such ecosystems is a certain technological platform, created on advanced innovative developments, including the open interfaces and code, machine learning, cloud technologies, Big Data collection and processing, artificial intelligence technologies, etc. The mentioned technological platform allows creating the best offer for the client both from own goods and services and from the offers of external service providers in real time.

The book is designed for undergraduate and post-graduate students, for engineers in related fields as well as managers of corporate and state structures, chief information officers (CIO), chief information security officers (CISO), architects, and research engineers in the field of information security.

Acknowledgements

The author would like to thank Professor Alexander Tormasov (Innopolis University) and Deputy Director General SAP CIS Dmitry Shepelyavyi for the foreword and support.

Author sincerely thanks Prof. Alexander Lomako and Prof. Igor Sheremet (Russian Foundation for Basic Research, RFBR) for valuable advice and their comments on the manuscript, the elimination of which contributed to improving its quality.

The author would like to thank Prof. Alexander Lomako and Dr. Alexey Markov (Bauman Moscow State Technical University) for the positive review and semantic editing of the monograph.

The author thanks his friends and colleagues: Kirill Semenikhin, Iskander Bariev and Zurab Otarashvili (Innopolis University) for their support and attention to the work.

The author expresses special gratitude to Nikolai Anatolyevich Nikiforov – Minister of Informatization and Communication of Russian Federation, Roman Shayhutdinov – Deputy Prime Minister of the Republic of Tatarstan, Minister of Informatization and Communication of the Republic of Tatarstan, Igor Kaliayev – Academician of the Russian Academy of Sciences (RAS), Alexander Smirnov – President of the National Institute for Global Security Studies.

I would also like to thank Khismatullina Elvira for translation the original text into English language as well as Mark de Jong – Publisher at River Publishers for providing us this opportunity of the book publication and Junko Nagajima – Production coordinator who tirelessly worked through several iterations of corrections for assembling the diverse contributions into a homogeneous final version.

This work was supported by the Russian Foundation for Basic Research Grant (No. 18-47-160011 and No. 16-29-04268 ofi_m) and Grant of the President of the Russian Federation (NSh-6831.2016.8).

Sergei Petrenko
(s.petrenko@rambler.ru)

List of Figures

Figure 1.1   Goals and objectives of the Russian Federation "Digital Economy" Program.   10

Figure 1.2   Damage assessment of cyber security incidents for the global business.   12

Figure 1.3   Infrastructure and connection evaluations in Russia.   16

Figure 1.4   The proposed governance structure of the Digital Economy of the Russian Federation.   17

Figure 1.5   Growth of the information security threats.   19

Figure 1.6   Development directions of service and content providers.   29

Figure 1.7   National information and communication infrastructure.   30

Figure 1.8   Perspective infrastructure of the communication operator.   32

Figure 1.9   Project dynamics of "Electronic Government". . . .   33

Figure 1.10   Public service portal infrastructure.   33

Figure 1.11   Typical services of the national cloud platform. . .   35

Figure 1.12   O.7 Housing system structure.   35

Figure 1.13   O7.112 system structure.   36

Figure 1.14   National cloud platform services.   36

Figure 1.15   The forthcoming of the wireless multimedia LTE networks.   40

Figure 1.16   From 1G to 5G networks.   40

Figure 1.17   Evolution of the LTE specifications.   42

Figure 1.18   System Architecture Evolution (SAE).   43

Figure 1.19   Differences between 2G, 3G, and 4G security mechanisms.   44

Figure 1.20   The security architecture of 4G LTE networks. . . .   45

Figure 1.21   Structure and content of a typical LTE network. . .   47

Figure 1.22   A feasible flowchart of the LTE network.   51

Figure 1.23   Possible composition of a complex IIoT/IoT solution.   54

Figure 1.24   Block diagram of the IIoT/IoT-solutions Control
Center.   55

Figure 1.25   Evolution of communication standards
for IIoT/IoT.   61

Figure 1.26   Comparison of known communication technologies
for IIoT/IoT.   61

Figure 1.27   Telematic module scheme in operation in IoT
network.   62

Figure 1.28   General scheme of operation in the IoT
environment.   63

Figure 1.29   Telematic module scheme for the "smart parking". .   63

Figure 1.30   Telematic module scheme for the connection of
sensors for opening doors and windows.   63

Figure 1.31   Telematic module scheme for smoke sensor.   64

Figure 1.32   Telematic module scheme for CO, CO_2, CH_4, NO_2
gas concentration control.   64

Figure 1.33   Telematic module scheme in operation in GPS/
GLONASS systems.   65

Figure 1.34   Telematic module scheme for encryption work
defined in the standard GOST.   65

Figure 1.35   Chip scheme with encryption support defined in the
standard GOST.   65

Figure 1.36   NB-IoT for nonstandard bandwidth.   66

Figure 1.37   NB-IoT for nonstandard frequency bands.   68

Figure 1.38   ARP attack schema.   72

Figure 1.39   Fake router forcing by ICMP "Redirect".   74

Figure 1.40   Forcing a fake X RIP router to intercept traffic
between P and Q networks.   75

Figure 1.41   Impersonation without feedback.   78

Figure 1.42   The attack scheme of a TCP connection impersonation
without feedback.   79

Figure 1.43   Actions of the malicious intermediary in the
desynchronized connection between the A and B
nodes (L1, L2–the amount of data in the forwarded
segments).   82

Figure 1.44   Early TCP connection desynchronization (ACK
storm segments are not shown).   83

Figure 1.45   TCP connection desynchronization with zero data
(ACK storm segments are not shown).   85

Figure 1.46   Tunneling through the filtering router.   88

Figure 1.47   Acknowledgments splitting.   93

Figure 1.48   False acknowledgment duplicates.   95

Figure 1.49   Interaction of wireless traffic protection
subsystems.   107

Figure 1.50   The keying scheme.   109

Figure 1.51   Wireless technology classification.   110

Figure 1.52   Example of an enterprise wireless network.   110

Figure 1.53   Example of a distributed wireless network.   111

Figure 1.54   Enterprise networking WLAN.   111

Figure 1.55   IP connection establishment.   122

Figure 1.56   History of the Tizen OS and Sailfish OS
development.   125

Figure 1.57   Original architecture of the Tizen OS kernel.   125

Figure 1.58   Requirements for the Russian Tizen OS Profile.   . .   129

Figure 1.59   Feasible Tizen Security Profile architecture.   132

Figure 1.60   Feasible scenarios of cyber-attacks to a typical
subway management system.   137

Figure 1.61   Feasible scenarios of cyber-attacks on the Moscow
subway Wi-Fi.   139

Figure 1.62   Feasible scenarios of cyber-attack on the typical
subway supervisory system.   142

Figure 1.63   An example of a prospective early warning
cyber-attacks center.   143

Figure 1.64   Example of localization and adaptation of SAP
solutions.   151

Figure 1.65   SOC perspective structure.   153

Figure 1.66   The "magic" SIEM Gartner quadrant, 2008.   154

Figure 1.67   The "magic" SIEM Gartner quadrant, 2018.   155

Figure 1.68   Big Data on cyber security processing typical
scheme.   156

Figure 1.69   Analysis cognitive technologies of Big Data on
cyber security.   157

Figure 1.70   SAP ETD data gathering on application level.   . . .   157

Figure 1.71   Integration scheme of SAP ETD with HP
ArcSight.   158

Figure 1.72   SAP ETD Architecture basis.   158

Figure 1.73   SAP ETD Typical interfaces.   159

Figure 1.74   Threat detection with SAP.   160

Figure 2.1   Dynamics of the modern cyber-attack complexity growth. 164

Figure 2.2   Losses from security incidents. 165

Figure 2.3   Example of a corporate e-mail protection service. . 166

Figure 2.4   Example of a cyber-attack (DoS/DdoS) protection service. 166

Figure 2.5   Example of a Web-filtering service. 166

Figure 2.6   Feasible portfolio of security services. 167

Figure 2.7   MSSP market dynamics in North America (2007–2015). 168

Figure 2.8   Typical MSSP models. 169

Figure 2.9   Traditional CPE model MSSP. 169

Figure 2.10   Perspective cloud MSSP model. 170

Figure 2.11   Dynamics of MSSP development. 170

Figure 2.12   Virtual security domaining (VDOM). 171

Figure 2.13   MSSP model by AT&T Company. 171

Figure 2.14   System appearance of the UBIqube MSSP control system. 173

Figure 2.15   Example of UTM solutions. 175

Figure 2.16   Typical components of the Fortinet network security platform. 177

Figure 2.17   Feasible MSSP scheme using Fortinet solutions. . . 177

Figure 2.18   Feasible scheme for balancing MSSP performance using Fortinet solutions. 178

Figure 2.19   Typical set of company IS policies and regulations. 182

Figure 2.20   IS service structure and composition. 184

Figure 2.21   IS services classification example. 190

Figure 2.22   Possible IS sourcing models. 191

Figure 2.23   The step sequence to design the IS service management model. 193

Figure 2.24   The life cycle of designing the service model of the IS service management system. 194

Figure 2.25   The IS sourcing strategy space. 194

Figure 2.26   The resources usage algorithm. 195

Figure 2.27   The IS management system maturity levels. 197

Figure 2.28   Interaction models with IS service providers. 198

Figure 2.29   Relationships based on the contract. 199

Figure 2.30   Possible scheme to provide IS services. 200

Figure 2.31   Approach to the IS sourcing concept implementation strategy.   204

Figure 2.32   The IS sourcing concept implementation. Project time frame.   205

Figure 2.33   The IS services package development.   206

Figure 2.34   Management development of staff and resources. .   207

Figure 2.35   National monitoring MSSP/MDR platform.   209

Figure 2.36   National MSSP/MDR interaction organization example.   210

Figure 2.37   The MSSP/MDR monitoring platform implementation example.   211

Figure 2.38   Technological reserve for the MSSP/MDR modernization.   212

Figure 2.39   Example of Cybersecurity Reporting Using Big Data Technologies.   216

Figure 2.40   MSSP/MDR report examples on the digital enterprises.   217

Figure 2.41   Sensor Cloud architecture.   219

Figure 2.42   The MDM place and role in IT.   222

Figure 2.43   The "magic" MDM Gartner quadrant.   223

Figure 2.44   The Gartner "Hype Cycle" for information technologies.   224

Figure 2.45   Artificial cognitive a, b agents architecture.   228

Figure 2.46   Information interaction scheme of the cognitive agent with the environment.   228

Figure 2.47   The cognitive warning system architecture.   231

Figure 2.48   The "generalized data model" scheme.   233

Figure 2.49   Fault tolerance scheme.   233

Figure 2.50   Possible switching schemes.   234

Figure 3.1   The graph family representative is the $\omega(s_X)$ pattern.   253

Figure 3.2   The g_{AK} graph is an example of the dimensional system pattern with insignificant elements.   254

Figure 3.3   The $R(g_{AK})$ graph.   256

Figure 3.4   An example graph that is a result of adding the dimensional graphs of the source station and the destination station.   260

Figure 3.5   The g_X graph.   272

Figure 3.6   The $\psi(g_X)$ hypergraph.   272

Figure 3.7   Combining the control graphs of some TCP implementations. 278

Figure 3.8   Aggregation of the control graphs of some IP implementations. 281

Figure 3.9   Placement of an SDAO sensor on a broadcast network segment. 284

Figure 3.10   Placement of SDAO sensors at receiving stations. . 284

Figure 3.11   Dedicated additional information channel. 285

Figure 3.12   Associated additional information channel. 286

Figure 3.13   Statistics console of the test SDAO. 291

Figure 3.14   Situation of detecting a test SDAO incident. 292

Figure 3.15   Feasible control systems for invariants in the three-tier CS architecture. 295

Figure 3.16   ERP Oracle E-Business Suite data processing schema. 301

Figure 3.17   Thin client program control graph. 302

Figure 3.18   The information graph of the data processing model. 304

Figure 3.19   Protocols informational graph within the model framework under consideration. 310

Figure 3.20   The TCP information graph within the entered notation scope. 312

Figure 3.21   The IP information graph in one IP packet transmission. 314

Figure 3.22   A system state graph example for a number of fragments equal to two. 315

Figure 3.23   Sequence of operations performed by the application program. 317

Figure 3.24   SQL clause SELECT. 318

Figure 3.25   SQL clause INSERT. 319

Figure 3.26   SQL clause UPDATE. 319

Figure 3.27   SQL clause DELETE. 319

Figure 4.1   The role and place of DARPA in US R&D. 341

Figure 4.2   DARPA performance. 342

Figure 4.3   Project management scheme in DARPA. 343

Figure 4.4   Standard development scheme. 348

Figure 4.5   Standard evolution coordination. 348

Figure 4.6   GOST R IEC 61508 applications. 349

Figure 4.7   The general idea of risk management in IEC 61508. 350

Figure 4.8   Correlation of basic safety standards concepts. . . . 352

Figure 4.9   Probability of a dangerous malfunction for an hour. 352

Figure 4.10   Correlation of safety standards concepts. 352

Figure 4.11   A safety system's life cycle in accordance with IEC 61508. 353

Figure 4.12   V-model of software development for security systems. 354

Figure 4.13   Spiral security software development model. 354

Figure 4.14   Smart Grid cyber security challenges. 364

Figure 4.15   Standard means of Smart Grid cyber protection. . . 364

Figure 4.16   Regulatory requirements. 372

Figure 4.17   The immunity formation scheme. 381

Figure 4.18   The class structure of cyber security ontology. . . . 382

Figure 4.19   Interaction scheme between implementations. . . . 383

Figure 4.20   Self-recovering Smart Grid. 385

Figure 4.21   von Neumann Architecture. 388

Figure 4.22   Classification of multiprocessor computing systems (MIMD). 391

Figure 4.23   The architecture of multiprocessor systems with shared memory: systems with uniform (a) and nonuniform (b) memory access. 391

Figure 4.24   The architecture of multiprocessor systems with distributed memory. 392

Figure 4.25   An example of a hierarchical organization's communication environment of a cluster CS. 394

Figure 4.26   An example of a spatially distributed multicluster computing system. 395

Figure 4.27   Sunway TaihuLight supercomputer architecture, China. 400

Figure 4.28   Sunway TaihuLight supercomputer software stack architecture, China. 401

Figure 4.29   Development of the Digital Economy applications based on Agile. 407

Figure 4.30   The need to take into account the security requirements. 409

Figure 4.31   Practice of Microsoft SDL. 410

Figure 4.32   The role and place of the GOST R 56939 recommendations. 411

Figure 4.33   The main stages in the methodology for measures
              justification.   413
Figure 4.34   The system of Russian standards in the field of
              secure software and their interconnections.   413
Figure 4.35   Typical BI security platform content.   420
Figure 4.36   Typical data loading agents.   420
Figure 4.37   Functional connector diagram.   421
Figure 4.38   A feasible data model of the BI security platform. .   422
Figure 4.39   Feasible data types of the BI security platform.   . .   423
Figure 4.40   The computing cluster's structure of the BI security
              platform.   423
Figure 4.41   The structure of the BI security platform gateway. .   424
Figure 4.42   An example of a computing algorithm.   425
Figure 4.43   Data processing.   425
Figure 4.44   Analytical core's composition of the BI security
              platform.   427
Figure 4.45   Adapting data models.   427
Figure 4.46   Representation of calculation results.   428
Figure 4.47   Algorithm for detecting security incidents.   428

List of Tables

Table 1.1	Investment evaluation in the Digital Economy development in Russia	17
Table 1.2	Main objectives of the "Digital Economy" program of the Russian Federation in "Information Security"	20
Table 1.3	Evolution of wireless standards	41
Table 1.4	Comparative characteristics of the feasible options for establishing a trusted LTE network	52
Table 1.5	Cost estimates of NB-Fi and NB-IoT ownership . . .	62
Table 1.6	Wireless network security standards Master key source subsystem	107
Table 2.1	Criteria description for the IS services classification .	188
Table 2.2	Distinguished ways of ontology construction	226
Table 3.1	Informative features selection impact on the abnormal functioning detection system characteristics	242
Table 3.2	σ/σ' equivalent graph examples	259
Table 3.3	Matrix for a set of several TCP implementations . .	279
Table 3.4	Matrix R for a set of several TCP implementations .	280
Table 3.5	S matrix for a set of several IP implementations . . .	282
Table 3.6	R matrix for a set of several IP implementations . . .	282
Table 3.7	Details characteristics table	321
Table 3.8	Table of constants	321
Table 3.9	The A_c equations system representation	322
Table 3.10	The A_E dimensional equations system representation	322
Table 3.11	The A_E matrix representation	323
Table 3.12	Relationships between application-layer protocol dimensions	324
Table 3.13	Matrix A_{L5}	327
Table 3.14	Matrix B_{L5}	327
Table 3.15	Matrix A_{L3}	328
Table 3.16	Matrix B_{L3}	328
Table 4.1	GOST R IEC 61508 development prerequisites . . .	347

Table 4.2   Correspondence of the safety standards concepts . . 351

Table 4.3   CF CPCS complexity factors and the resulting difficulties   359

Table 4.4   Cyber security risk assessment example 366

Table 4.5   Known risk assessment techniques 368

Table 4.6   Correspondence between UML and IDEF5 graphical elements   374

Table 4.7   Correspondence between OWL tags and UML constructs 374

Table 4.8   M. J. Flynn computer architectures classification . . 390

Table 4.9   Accounting for security requirements and software functionality   411

Table 4.10   Typical formats of data loading agents 421

List of Abbreviations

AC&S	Access Control and Security
ACL	Access Control List
AuthN/AuthZ	Authentication/Authorization
AP	Access Point
APT	Advanced Persistent Threat attacks
ARP	Address Resolution Protocol
CAIDA	Cooperative Association for Internet Data Analysis
CDC U.S.	Centers for Disease Control and Prevention
CEP	Complex Event Processing
CERT	Computer Emergency Response Team
CIA	Confidentiality, Integrity, and Availability
CIAC	Computer Incident Advisory Capability
CISO	Chief information security officers
CLI	Command-Line Interface
CMC	Crisis Management Center
CMVP	Cryptographic Module Validation Program
CNCI	Comprehensive National Cybersecurity Initiative
COM	Component Object Model
CPU	Central Processing Unit
CSA	Cloud Security Alliance
CSA BDWG	Cloud Security Alliance Big Data Working Group
CSIRT	Computer Security Incident Response Team
CSIS	Center for Strategic and International Studies
CSP	Cloud Service Provider
CSRC	Computer Security Resource Center
CSV	Comma Separated Values
CVE	Common Vulnerabilities and Exposures
CTF	"Capture the flag" game
CSIRT	Computer Security Incident Response Team
DARPA	Defense Advanced Research Projects Agency's
DDoS	Distributed Denial of Service
DHCP	Dynamic Host Configuration Protocol

DISA	Defense Information Systems Agency
DLL	Dynamic Link Library
DMZ	Demilitarized Zone
DNS	Domain Name System
DOD U.S.	Department of Defense
DoS	Denial of Service
DS	Distribution System
DShield	Distributed Intrusion Detection System
EICAR	European Institute for Computer Antivirus Research
ESP	Encapsulating Security Payload
EU	European Union
FIPS	Federal Information Processing Standards
FISMA	Federal Information Security Management Act
FPLG	field programmable logic devices
FSTEC	Federal Service for Technical and Export Control
FTP	File Transfer Protocol
GIG	Global Information Grid
GHz	Gigahertz
GPS	Global Positioning System
GRC	Governance, Risk Management, and Compliance
GUI	Graphical User Interface
JSOC	Joint Special Operations Command
JTF-GNO	Joint Tactical Force for Global Network Operations
JFCCNW	Joint Functional Component Command for Network Warfare
HAP	High Assurance Platform
HSM	Hardware Security Module technology
HTTP	Hypertext Transfer Protocol
HTTPS	Hypertext Transfer Protocol over SSL
ICT	Information and Computer Technology
ICMP	Internet Control Message Protocol
IDPS	Intrusion Detection and Prevention System
IDS	Intrusion Detection System
IEEE	Institute of Electrical and Electronics Engineers
IETF	Internet Engineering Task Force
IGMP	Internet Group Management Protocol
IM	Instant Messaging
IMAP	Internet Message Access Protocol
IP	Internet Protocol

IPIB	Institute of Information Security Problems
IPS	Intrusion Prevention System
IPsec	Internet Protocol Security
IRC	Internet Relay Chat
ISC	Internet Storm Center
IoT	Internet of Things
IT	Information Technology
ITL	Information Technology Laboratory
LAN	Local Area Network
MAC	Media Access Control
MDR	Managed Detection and Response Services
MIFT	Moscow Institute of Physics and Technology
MSSP	Managed Security Service Provider
M2M	Machine to Machine
NBA	Network Behavior Analysis
NCSD	National Cyber Security Division
NBAD	Network Behavior Anomaly Detection
NFAT	Network Forensic Analysis Tool
NFS	Network File System
NFV	Network functions virtualization
NIC	Network Interface Card
NIEM	National Information Exchange Model
NIST	National Institute of Standards and Technology
NSA	National Security Agency
NTP	Network Time Protocol
NVD	National Vulnerability Database
OMB	Office of Management and Budget
OS	Operating System
OSS	Operations Systems Support
PaaS	Platform as a Service
PKI	Public Key Infrastructure
PDA	Personal Digital Assistant
PoE	Power over Ethernet
POP	Post Office Protocol
RF	Radio Frequency
RFC	Request for Comment
ROM	Read-Only Memory
RPC	Remote Procedure Call
SAML	Security Assertion Markup Language

SATOSA	System of traffic analysis and network attack detection
SDN	Software-defined networks technology
SEM	Security Event Management
SIEM	Security Information and Event Management
SIM	Security Information Management
SIP	Session Initiation Protocol
SLA	Service-Level Lgreement
SMB	Server Message Block
SMTP	Simple Mail Transfer Protocol
SNMP	Simple Network Management Protocol
SOC	Security Operations Center
SOPCA	Prevention and Cyber Security Incident Response
SPOCA	Computer Attack Detection and Prevention System
SP	Special Publication
SSH	Secure Shell
SSID	Service Set Identifier
SSL	Secure Sockets Layer
STA	Station
STS	Security Token Service
TCP	Transmission Control Protocol
TCP/IP	Transmission Control Protocol/Internet Protocol
TFTP	Trivial File Transfer Protocol
TLS	Transport Layer Security
TTL	Time to Live
USE	Unified State Examination
UDP	User Datagram Protocol
USB	Universal Serial Bus
US-CERT	United States Computer Emergency Readiness Team
VLAN	Virtual Local Area Network
VM	Virtual Machine
VPN	Virtual Private Network
WEP	Wired Equivalent Privacy
WLAN	Wireless Local Area Network
WPA	Wi-Fi Protected Access
WVE	Wireless Vulnerabilities and Exploits
XACML	eXtensible Access Control Markup Language
XML	Extensible Markup Language

Introduction

Nowadays, a number of high-tech companies from the USA, the EU, China, and Russia (Facebook, Google, Amazon, Apple, Alibaba, Tencent, Rostelecom, MTS, Sberbank, etc.) have already made significant progress in building promising digital ecosystems of the future. The emergence of these ecosystems led to a radical change in the economic landscape: the shift of the value chain from production to distribution. The owners of the corresponding technological platforms, able to efficiently aggregate the producers of goods and services around them and those of them, based on the analysis of behavioral features to form the best offers for clients, became major actors. According to analytical forecasts, by 2025, the digital ecosystems would cover up to 30% of income in their regions through optimizing the chain of intermediaries between producers of goods/services and consumers.

The players with low capital costs, which become providers of services and the first choice for clients, began to appear in the Russian economy. Generally, they have the following advantages over traditional players: better customer orientation, higher flexibility and efficiency, radical efficiency of applied business models, integration into the value chain between the manufacturer and the customer, etc. In response, traditional players of different industries are using the opportunity to expand their business models beyond their core competencies, achieving higher rates of change, customer recognition, and higher financial results, ensuring the viability of their business.

Today, digital enterprises are the most innovative among other classical enterprises of the world economy. This is due to the fact that the use of modern and safe information technologies for these enterprises is a vital prerequisite for technological development and improvement. Here, innovative activity, on the one hand, provides a certain "window of opportunity", which is expressed in applying the bold and breakthrough technologies, unknown to most traditional (not digital) companies. For example, the technology of Industrial Internet of Things (IIoT/IoT), Big Data collection and processing, stream processing ETL data, predictive analytics,

NBIC (nano-, bio-, info-, and cogno-) technologies, cognomorphic and neuromorphic calculations of high and ultra calculations with memory, etc. On the other hand, this innovation activity carries certain challenges and risks.

Generally, innovation is an activity, aimed at improving both production and providing services, technologies, skills, and competencies, i.e., all ecosystem key components of a successful technological leader. From a functional point of view, this means being flexible and ready for change, having the adaptability and self-organization qualities, being able to develop key applications, services, and platforms of the Digital Enterprise with Agile's flexible software development methodology, having a good communication skills and working in cross-functional teams, following the cyber security rules, deal with modern Industry 4.0 technologies at the level of their developers, etc. At the same time, the majority of successful digital enterprises also aspire to become conductors (providers) of a number of nonprofile, noncore services and services that accompany the main activity, a kind of integrated aggregator and marketplace.

Internal innovations imply the project initiation within the Digital Enterprise and the formation of its own project team. This approach has become widespread in practice. For example, technology development leaders such as IBM, HP, Cisco, SAP, etc. each year spend billions of dollars on hundreds of internal projects. In a number of technologically advanced telecom operators, especially in the USA, having more than a hundred thousands of developers in their staff, they would also prefer to independently carry out the projects. However, for most domestic companies, this approach is not always economically viable. Therefore, some of the digital enterprises are moving along the path of combining internal and open innovations, successfully integrating the merits of the named approaches to the organization and conduction of the research and development. So, for example, if the operational development is required, the project would not need any significant time and money resources, then it would be more profitable for the enterprise to carry out the development independently.

If the development is already on the external market, then it is more profitable for the enterprise to purchase these developments and/or to participate in the joint development projects with an appropriate technology partner. For this, it is necessary to monitor the work results of the promising project teams around the world, mainly the publication of various innovative prototypes in the early stages of a development; be able to assess the usefulness of each elaboration in terms of the possibility of generating profit within the next – 2–3 years after its implementation; if there is a positive estimate, then there is a high probability of profit, and then initiating work with this interesting idea. Moreover, if the fundamentally new results are expected, then it may be necessary to allocate certain monetary grants, getting back a direct access to intermediate and final results of the research.

It is recommended to work with well-known companies, developing the advanced information technologies (SAP, IBM, HP, Cisco, EMC, Oracle, etc.). However, there is also a certain problem—the overestimated expectations and the overexpected scale of projects. First of all, the mentioned developers are interested in large-scale projects with long terms of implementation and budgets of several tens of millions of dollars. However, the Digital Enterprise may require solutions that are not very costly, for example, retail improvement projects, etc. In practice, most digital enterprises work with large companies on large projects (10%), medium companies with small projects (30%), and with small start-ups (60%).

For the effective, innovative work, it is essential to work with leading universities and leading advanced, innovative developments, for example, Innopolis University in Russia and the MIT University in USA. It is also desirable to establish cooperation with leading analytical companies, for example, Forester and Gartner.

The next stage is implementing innovative ideas into commercial realization, the so-called innovative-technological experiment. Here, each experiment continues for a predetermined and limited time period. Experiments are realized both within the Digital Enterprise and the companies, which work with enterprise in collaboration. Today, most Digital enterprises take the path of organizing and conducting an experiment. Therefore, first, an independent selection of innovations most interesting for the enterprise, based on the objective assessments of the state of development of a certain branch of the economy, as well as on the information-providing cyber security technologies, is carried out. There occurs the so-called "great funnel" of technological expectations and opportunities. Then, as a rule, on the basis of "brainstorming", the answers to the following burning issues are sought. What are the project risks and how to optimize them? Do the received evaluations fit into the criteria of the innovative projects of the Digital Enterprise? Will this give the company some advantages? Are any manufacturing and/or supplying technological systems required for this change? Is a Digital Enterprise able to conduct a technology transfer within 3–6 months? Is there enough money and resources for the experiment and will there be enough infrastructure and organizational capabilities to implement the plan? In the case when the majority of answers to these and other questions are positive, and also taking into account the readiness of the environment and the internal infrastructure of the Digital Enterprise to introduce new ideas, innovations are brought to the stage of the experiment.

Why is it so important to carry out experiments, and not to run projects at once? The fact is that in each project, as a rule, there are some limitations, sometimes not visible at first glance, for example, insufficient level of the chosen technology. And if these limitations are not identified and not taken into account, they will occur at the most inopportune moment and negatively affect the implementation idea. Therefore, it is important to have an ability to identify and optimize risks of innovations in a timely manner. For example, for the needs of the enterprise, it is proposed to deploy a pilot of the Private Cloud whose development environment will be provided not in the form of a set (farm) of physical servers, but in the cloud. Developers propose to use the set of Open Stack libraries from the open-source software. The only question is whether this software complies with the technological landscape of the Digital Enterprise and whether it is completely trusted and secure. What if the enterprise will receive a "Trojan horse" with the appropriate logical or digital "bombs, the so-called undeclared capabilities instead of the expected result that will negatively affect, occurring at the X hour? Therefore, it is so important to risk consciously and firstly learn how to qualitatively test the innovative ideas during the experiment.

It is important to allocate the required budgets and resources, in order to establish and conduct these experiments, as well as to involve the team of the own and external developers. It should be taken into account that, in practice, the project team, as a rule, is not universal in terms of skills and competencies, required for the experiment. Therefore, external teams and contractors with the required qualities are allowed to participate in the experiment. Such a typical project team usually includes up to 20 specialists who have all necessary power to choose a promising technology. Further, the experimental framework is established, the required resources are specified, the requirements and criteria for the success of the experiment are determined, etc. In case when developers use Agile flexible software development methodology, the relevant sections of the project activity are to be made every week, new design requirements are to be constantly taken into account, the results of the development, and, if it is necessary, should be implemented a flexible adjustment of the program of action.

On the basis of the results of such experiment, the several solutions and subsequent scenarios for the project developments are possible. For example, it is possible to ascertain the fact that the experiment failed: the technology does not work and the project needs to be closed; the recognition that the experiment was successful: the results are positive, but the Digital Enterprise is not yet ready to implement the results; and finally, it is also possible to come to the decision on the implementation of the experimental results into production. At the same time, certain limitations and obstacles can be encountered in the implementation process, for example, with the corporate framework of accepted business processes or with the limitations of the current legislation, as well as with the lack of the necessary budget for the project implementation due to delay in approval time frame and the current year budgets, etc. In order to comply with market changes and achieve the required mobility and flexibility, it will be necessary to promptly make changes in the corporate policies of the Digital Enterprise, i.e., it is necessary to correctly launch and implement experiments, as well as to apply the results in practice in a qualitative and flexible manner.

It should be noted that in practice the share of successful experiments, growing into full-scale projects of Digital Enterprises, is about 15%. An increase of this share to 100% is very unlikely, at best, the indicators may achieve 30–40%, most probably, not more than 20%.

All the forgoing, allows coming to the conclusion that the discussed scientific problem lies in the effective organization and implementation of innovative activities in the field of cyber security of the Digital Economy, which has great theoretical, scientific, and practical importance. According to the author, this monograph is the first work on the mentioned problem and may be intended for the following reader groups:

- Corporate and State CEO, responsible for the "Digital Economy" national programs in the direction of "Cyber Security of Digital Enter-prises";
- Chief information officers (CIO) and Chief information security officers (CISO), responsible for corporate information security programs and organization of the information security regime of digital enterprises;
- Constructors and research engineers responsible for the technical design of the Security Threat Monitoring Centers in the various Situation Centers and government (and corporate) segments of detection systems for the prevention of cyber-attack consequences.

This book can also be a useful training resource for undergraduate and postgraduate students in related technical fields, since these materials are largely based on the authors' teaching experience at the Moscow Institute of Physics and Technology (MIFT) and Innopolis University.

This book contains four chapters devoted to the following subjects:

- The relevance of the given scientific-technical problem of the effective organization and implementation the innovative activities in the cyber security of Digital Economy;
- Determination of the limiting capabilities of the known technologies for control and monitoring of digital enterprises cyber security by national MSSP/MDR operators;
- Search for possible scientific and technical solutions for the proper provision of cyber security of the Russian Federation Digital Economy; as well as the development of new methods for detecting anomalies and the analytical verification of the corresponding Digital Economy functioning;
- Organization of perspective research study in the area of Russian Digital Economy cyber security.

The book is written by Professor and the Director of Information Security Center of Innopolis University S.A. Petrenko.

In advance, the author would like to thank and acknowledge all readers. Anyone wishing to provide feedback or commentary may address the author directly at:

Sergei Petrenko
s.petrenko@rambler.ru.
Russia-Germany
January 2018

1

Relevance of Cyber Security Innovations

The relevance of innovation in cyber security for digital enterprises is shown. Examples of typical threats of cyber security are given. It was revealed that "classical" models, methods, and means of information security are already insufficient to provide the required cyber security. It is necessary to develop fundamentally new models, methods, and tools based on such quantitative and qualitative patterns of information confrontation that will not only detect, but also pre-warn, as well as preempt group and mass cyber-attacks by intruders. The importance of the development of corporate monitoring centers for cyber-security threats (SOC) is shown on the basis of modern Industry 4.0 technologies. Including technologies for collecting and processing large volumes of structured and unstructured information from various Internet/Intranet and IoT/IIoT sources (Big Data and Big Data Analytics), cognomorphic and neural-like software engineering and work technology, trusted cloud and fog technologies, new LTE and 5G communication technologies.

1.1 Digital Transformation of State and Society

This chapter reveals the concept of the digital economy and assesses the readiness of Russia for the transition to a digital economy. The main goals and objectives of the state program "Digital Economy" established by the Government of the Russian Federation are set out, and its key indicators are given. The importance of the objective accomplishment of the IT product import substitution for the program implementation is shown. The priorities of the program in the area of information security are listed.

1.1.1 State Program "Digital Economy"

Historical Background

On December 1, 2016, in the annual message to the Federal Assembly, Russian President V. Putin emphasized the importance of the Russian digital economy formation to secure the technological and economic sovereignty[1] of the State under global digitalization and the improvement of Russian economic sector efficiency by the state-of-the-art information technologies[2]. On July 28, 2017, the State Program "Digital Economy"[3] was approved by the Russian Government #1632-p. The measures of the program, aimed to create the required legal, technical, organizational, and financial terms to develop and integrate the Russian digital economy in the digital economy space, starting with member states of the EEU, APEC, SCO, and BRICS, were defined. It is essential that the solution of the set objectives and achieving of the given program goals under the increase of threats to information security directly depend on a proper solution of the import substitution task and the implementation of the primary national information technologies, including information security technologies.

However, we shall return a few years back. Since the beginning of the new millennium in a number of high technological countries (in the leading countries of the European Union, as well as in Israel, Japan, South Korea, Brazil, Canada, China, Iceland, Mexico, Australia, New Zealand, Switzerland, Norway, Turkey, Russia, and the USA), the national programs under the name of "Information Society" have been successfully implemented. The series of the key documents have been prepared worldwide, including the Okinawa Charter on Global Information Society (established on Okinawa island on July 22, 2000)[4], the Declaration of Principles "Shaping Information Societies is a global objective in a new millennium" (Geneva, 2003), the Action Plan (Geneva, 2003), Tunis Commitment (Tunis, 2005)[5], etc.

In the Russian Federation, the following documents, considering the national interests and the requirements of national security and state sovereignty, were developed [1–12]:

---

[1]http://kremlin.ru/events/president/news/53379/.
[2]http://minsvyaz.ru/ru/activity/directions/779/.
[3]http://static.government.ru/media/files/.
[4]http://kremlin.ru/supplement/3170/.
[5]http://www.gov.karelia.ru/Leader/Inform/Egov/tunis.html.

- "Information Society" State Program (2011–2020) of the Russian Federation (approved by the decision of the Russian Government on April 15, 2014, #313)[6];
- Information Security Doctrine of Russian Federation (approved by the presidential decree of the Russian Federation of December 5, 2016, #646)[7];
- Information Society Development Strategy of the Russian Federation 2017–2030 (approved by the presidential decree of the Russian Federation on May 9, 2017, #203)[8];
- Indicators Profile and Implementation Plan of the Information Society Development Strategy of the Russian Federation 2017–2030 (developed by the Russian Federation Government for a compulsory implementation by all Authorities)[9];
- "Digital Economy" Program of the Russian Federation (approved by the Russian Government order #1632-ð of July 28, 2017).

In the "Digital Economy" Program of the Russian Federation (Figure 1.1), the digital economy is defined as "economic activity, where the key production factor is information in digital form that promotes the information society formation, considering the needs of citizens and society in receiving proper and authentic data, the development of the Russian Federation information structure, design and implementation of the Russian information and communication technologies (ICT), as well as the formation of the new technological base for social and economic spheres". Evaluating the digital economy contribution to GDP, The Boston Consulting Group[10], a known international consulting company, shares interesting data. For instance, the GDP contribution of the G-20 countries digital economy of 2010 is about 2.3 USD trillion or 4.1%, and at the moment is 5.5% of total GDP of those countries (in the USA – 5.4%). Yet the leader of the digital economy share in GDP is Great Britain – 12.4%. The Russian digital economy share of GDP is two to three times less than in the G-20 countries, in 2011 it was 1.6%, in 2015 – 2.1%, and in 2016 – 2.8% or 75 USD billion. Analysts argue that the digital economy can be the driver of the economic and technological

---

[6]http://minsvyaz.ru/ru/activity/programs/1/.

[7]http://www.scrf.gov.ru/security/information/document5/.

[8]http://www.kremlin.ru/acts/bank/41919/.

[9]http://www.kremlin.ru/acts/bank/41919/.

[10]https://www.bcg.com/ru-ru/default.aspx.

Figure 1.1   Goals and objectives of the Russian Federation "Digital Economy" Program.

development under the low growth rate of the global economy in general [11–21].

It should be noted, that the digital economy definition is closely associated with the known concepts as technological (digital) revolution and industrial revolution (the Fourth Industrial Revolution) [6–11, 13, 22] and requires the common use of Big Data, Industrial Internet of Things (IIoT), Internet of Things (IoT), cloud and fog computing, quantum and NBIC technologies (nano-, bioinformation and cognitive technologies), additive manufacturing, robotization, etc. In other words, digital economy acts as an economy of the new technological wave, and it is directly related to the common use of the information and communication technologies and the so-called "cross-cutting" information technologies and information security technologies. The use of the state-of-the-art information and communication technologies provides the Russian economy with new opportunities to perform commercial transactions on the international markets generally without any middleman in the online mode, including the use of "cryptocurrency". Moreover, the antagonistic relationships between the seller and the customer of the mass industrial production era gave place to the consistent interaction and cooperation, resulting in a personalized product or service that best meets the customer needs. Bringing the product closer to the end user and the transaction transparency raises the international competition to a fundamentally new level. At the same time, the product competitive advantage (goods or services) becomes the key criterion to market demand [18, 19, 23–26].

It should be noted that for the majority of the high technological countries (G-20), the digital economy has already passed the development stage of the ICT infrastructure, and the mentioned countries are now at the next stage, called the API-economy. The basis of the API-economy are technologies of the software build automation by request, with the required set of standard software modules, responsible for the performance of the standard and highly demanded functions. Achieved technological advantage provided the G-20 countries to take the leading positions on the global market by reindustrialization based on the advanced IT and the implementation of the ICT international standard system, which allows institutionalizing the proprietary rights on ICT and secondary innovative products based on them.

However, the "simplified" concept of the digital economy accelerated development based on the creation of the proper ICT infrastructure and national legal basis reformation are generally offered to developing countries including Russia. At the same time, the removal of the international integration barriers in respect of the new business forms development based on ICT is declared at the large scale. In fact, in the absence of the competitive environment and equal access to ICT, it is suggested to take the responsibility for the significant part of the financial costs to create the national digital economy infrastructure that meets the requirements of digitalization of leading countries, as well as to "willingly" create the welcoming environment for transnational cyber corporations (TNC) to penetrate national markets and to take over the national market players. Moreover, the control over TNC and ICT international standard system allows the USA and other leading countries in digitalization to establish rules and to control the digital economy development, including the creation of technical barriers within the technical regulation and certification on the compliance with international standard requirements. It is obvious that the implementation of such digital economy concept in "accelerated and limited variant" is not acceptable for Russia, as it has a strategically significant threat of the technological sovereignty loss and strengthening the long-term status of "technological periphery" [1–5, 10, 13, 14–18, 23].

Therefore, the importance of the complete national digital economy formation in the Russian Federation is proved by the need to develop new forms of the long-term cooperation in different integration formats, starting with the EEU, APEC, SCO, and BRICS, with the perspective to implement opportunities of the supranational regulation of the global production and distribution chain under the emerging global economic order shaping, with the compulsory consideration of the state national interests and geopolitical goals, including the proper national security and economic and technological sovereignty of the Russian Federation [1–6, 11, 13–24].

Russian Readiness to the Digital Economy

To evaluate whether the state is ready for digital economy, the so-called international Networked Readiness Index can be applied, which has been introduced by the specialists from "World Economic Forum"[11]. The Index shows (Figure 1.2) how well the national economy uses the digital technologies to improve the competitive advantage and the welfare. It also indicates the factors influencing the digital economy development in general. Given Networked Readiness Index is updated annually and communicated to the public in the relevant "Global Information Technologies" annual reports. According to the 2016 report, Russia ranked 41 in the list of readiness of countries to the digital economy with the significant underrun from 10 leading countries including Singapore, Finland, Sweden, Norway, the USA, the Netherlands, Switzerland, Great Britain, Luxembourg, and Japan[12].

Figure 1.2   Damage assessment of cyber security incidents for the global business.

---

[11]https://www.weforum.org/.
[12]https://www.weforum.org/reports/the-global-information-technology-report-2016/.

At the same time, Russia showed the following results according to the other criteria: 38th place in the use of the economic and innovative results of the digital economy (Inset 1.1, Inset 1.2) and 43rd place in the investments to innovations along with infrastructure development, skills, and effective markets.

Inset 1.1

Russian Digital Economy Ecosystem
Digital economy levels
- Technological companies
- Suppliers of the technological companies
- Cost reduction/efficiency improvement by digital technologies for the companies from other economic sectors
- Cost reduction/improvement of the population live quality

RuNet influence on the economy
- 19% of GDP (dependent markets)
- 2.8% of GDP (Internet markets)
- Human resources 2.3 million employees
- Infrastructure and for RUB 2000 billion
- eCommerce RUB 1238 billion
- Digital content RUB 63 billion
- Marketing and advertisement RUB 171 billion

Regulation
1. Digital economy cross-border nature
2. Special attention to cyber security
3. Promotion and support of the directions:
_ Import substitution
_ Ensuring the equal conditions of the Internet companies' operations
_ Digital technology sector taxation reform
_ Information technologies export
_ Data access and storage infrastructure development

4. *Preventive legislation*
_ Big Data and artificial intelligence
_ Robotics
_ Independent platforms and Internet of Things
_ Block chain
5. RIA should consider plans and projected growth for 10–20 years

Inset 1.2
Russian technological and economic development scenarios What is waiting for us—two possible scenarios: • Complete digital economy, integrated in the global one; • Conservative development of certain digital directions What goals are pursued: • Smart cities and autonomous transport; • Soft-sell and effective Internet advertising; • Cyber-attack security and responsible attitude to personal data; • Region development and bridging of the digital gap; • Telemedicine and smart agriculture; • Increase of the digital literacy; • Trust mechanisms in the Internet, a public space

Another evaluation of the readiness to the building of the digital economy, the so-called international I-DESI index, is based on the more general Digital Economy and Society Index (DESI) for the European Union countries in comparison to the same indicator for countries such as Australia, Brazil, Canada, China, Iceland, Israel, Japan, South Korea, Mexico, New Zealand, Norway, Switzerland, Turkey, Russia, and the USA. I-DESI index allows evaluating the level of the digital technology implementation to the business, digital service provision to the population, and the sector development in general. At the same time, the statistical data of some credible organizations such as the Organization for Economic Co-operation and Development, Organization of the United Nations, International Telecommunication Union, etc., are considered. For example, according to the I-DESI[13], in 2016–2017, Russia dropped behind the European Union, Australia, and Canada in the digital economy development, but was ahead of China, Turkey, Brazil, and Mexico. At the same time, according to the access to the fixed broadband communication in 2016, Russia along with the USA was ahead of the European Union and other countries. In regard to human capital, Russia outran Turkey, Mexico, and Brazil, but significantly fell behind Japan, South Korea, Sweden, Finland, Great Britain, and other leading European Union countries. On the basis of the frequency of Internet use, Russia does not show high positions in comparison with the EU, the USA, New Zealand, and Australia, but goes ahead of China, Brazil, and Mexico. According to the criteria of digital technology implementation in the industrial companies,

---

[13]http://unctad.org/meetings/en/Presentation/dtl_eweek2016_AMateus_en.pdf.

Russia slightly outran Turkey, China, and Mexico, but significantly fell behind leading European Union countries[14].

According to foreign and national experts [1, 4, 5, 22], the fact that Russia underruns compared to the global leaders in the digital economy formation and development could be explained primarily by problems in the national digital economy regulatory environment, as well as by deficiently supportive environment for the business and innovation implementations, and as a consequence, the low level of digital technologies implementation in the national industrial companies. Moreover, the latter factor (in comparison with the digital technologies application in government agents and civil society) was also noted in the World Bank Report on Global Development 2016[15]. In the World Economic Forum Report on Global Competitive Advantage 2016–2017, the attention was paid to the lack of investments for the new information technologies along with the need of infrastructure development and existing market efficiency improvement[16].

The Program Primary Goals and Objectives

The primary goals and objectives of the "Digital Economy" Program of the Russian Federation are:

- Keeping the economic and technological sovereignty under global digitalization;
- Forming the new instruments for long-term international cooperation in different integrated formats, starting with the EEU, APEC, SCO, and BRICS;
- Improving competitiveness of the certain economic sectors as well as the Russian economy in general on the global market;
- Developing the digital economical ecosystem of the Russian Federation, where the efficient cooperation of business (including cross-border), scientific, and educational communities, state and civil society will be ensured;
- Creating required and sufficient conditions of institutional and infrastructural nature to remove the existing barriers and limitations for the design and development of the high-technology business, and not to

---

[14]http://unctad.org/meetings/en/Presentation/dtl_eweek2016_AMateus_en.pdf.

[15]http://pubdocs.worldbank.org/en/894401495493535366/RER-37-May26-FINAL-with-summary-RUS.pdf.

[16]http://www3.weforum.org/docs/GCR2016-2017/05FullReport/TheGlobalCompetitiveness Report2016-2017_FINAL.pdf.

allow the appearance of new barriers and limitations in traditional economic sectors as well as in new sectors and high-technological markets.

At the same time, there are three main levels of the Russian digital economy that have a significant impact on people's life and society in general (Figures 1.3 and 1.4, Table 1.1):

- Markets and economic sectors (business domain) that include the specific subject interaction (vendors and customers of products, works, and services);
- Platforms and technologies, where the competences for the development of markets and economic sectors (business domain) are formed;
- Environment that creates the conditions for the platform and technology development and for the efficient interaction of market subjects and economic sectors (business domain) and covers the regulatory environment, information infrastructure, human resourcing, and information security.

The "Digital Economy" program identifies five development areas (Inset 1.1, Inset 1.2) for the period up to 2024:

- Formation and development of the regulatory environment;
- Staff training and improving the education system;
- Formation of research competence and technical capacity;
- Information infrastructure development;
- Proper information security [9–11, 14, 15, 23, 24].

There are brief comments on the mentioned directions of digital economy development in the Russian Federation.

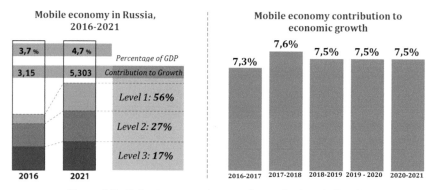

Figure 1.3   Infrastructure and connection evaluations in Russia.

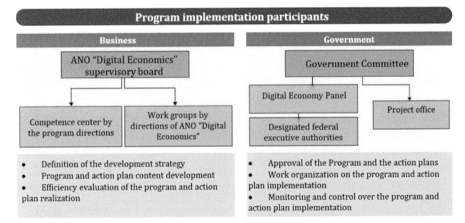

Figure 1.4 The proposed governance structure of the Digital Economy of the Russian Federation.

Table 1.1 Investment evaluation in the Digital Economy development in Russia

Investment Stage	Volume (RUB, Million)/Number of Startups (Units)			
	2013	2014	2015	2016
Pre-seed	139/44	216/135	205/125	349/132
Seed	1,451/106	1,293/71	1,508/61	1,429/64
Round A	2,099/58	2,404/29	2,635/18	4.017/33
Round B	3,112/23	3,636/18	2,519/10	1,880/7
Later rounds and purchases	26,947/20	52,336/29	27,359/11	101,014/25
Total	**33,758/261**	**59,884/29**	**34,225/225**	**108,690/261**
Without later rounds and purchases	8,811/231	7,548/253	6,867/214	7,677/236

Within the direction "Formation and development of the regulatory environment", it is supposed to create a new regulatory environment that provides a favorable legal regime for the generation and development of the state-of-the-art technologies, as well as for the implementation of economic activities connected with the digital economy application. This direction is intended to complete the following actions:

- Create a constantly active control mechanism of changes and competences (knowledge) within the digital economy regulation;
- Remove the key legal limitations and create specific legal institutions, aimed at solving the primary goals of the digital economy formation;

- Shape a complex legislative regulation of relations, emerging from the digital economy development;
- Take measures aimed at the promotion of the economic activity, connected with the modern technologies application, data collection, and storage;
- Form the digital economy development policy within the territory of the Eurasian Economic Union to harmonize approaches to the legal regulatory environment, promoting the digital economic development of the Eurasian Economic Union territory;
- Create a methodology of the competence development in the digital economy regulation.

Within the "Staff training and improving the education system" it is supposed to:

- Create required conditions to staff training for the digital economy;
- Improve an educational system that must provide the digital economy with the competent specialists;
- Create a labor market that must meet the digital economy requirements;
- Create a motivation system for skill and competence development of Russian specialists for the needs of Russian digital economy.

Within the "Formation of research competence and technical capacity", the proper support system for exploratory researches in the digital economy is created. This direction is intended to fulfill the following needs:

- Shape an institutional environment to promote research and development in the digital economy;
- Form a technical capacity in digital economy;
- Form a research competence in digital economy.
- Within the "Information infrastructure development", the following actions are suggested:
- Develop the communication networks for the collection and accredited transfer of digital economy data;
- Develop a system of the Russian centers of data processing to provide requested information services with the required quality and security;
- Implement the digital platforms to work with the digital economy data;
- Create an effective system of collection, storage, and provision of spatial data to the consumers.

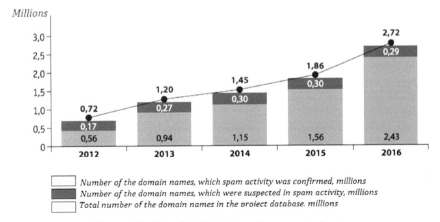

Figure 1.5 Growth of the information security threats.

Within the "Proper information security", the following actions are supposed (Figure 1.5, Inset 1.3, Table 1.2):

- Protect the digital sovereignty and consider the national interests in international information space;
- Provide an organizational and legal protection of the personal rights, business and state interests when interacting in the digital economy;
- Provide the required stability features and ICT security of the Russian Federation;
- Create conditions for Russian leading positions in export of the national services and information security technologies.

The Program Key Indicators

For the successful implementation of the "Digital Economy" Program of the Russian Federation, it was suggested to develop the relevant "Road maps" and "Primary action plans" with the deadline and executives responsible for the implementation (Inset 1.3) and to define proper sources and amount of financing. At the same time, it is supposed that the given maps and plans will be established for 3 years and, if necessary, will be modified annually.

The several key implementation indicators by 2024 were defined for the Russian "Digital Economy" program performance management (Inset 1.4).

Table 1.2   Main objectives of the "Digital Economy" program of the Russian Federation in "Information Security"

Objectives	Deadlines
5.1. Provide stability and security of the Russian national telecommunications network functioning (including the Russian Internet segment)	IV quarter of 2022
5.2. Provide controllability and reliability of the Russian Internet segment functioning	I quarter of 2020
5.3. Provide technological independence and functioning security of the hardware and the data processing infrastructure	IV quarter of 2024
5.4. Provide stability and functioning security of information systems and technologies	IV quarter of 2024
5.5. Provide legal regulations and technological tools of service functioning and data use ("Internet of Things", Big Data)	II quarter of 2020
5.6. Provide legal regulations of the M2M connectivity for cyber-physical systems	2022
5.7. Provide the legal regulations of the machine and cognitive interfaces functioning, including the Internet of Things	2022
5.8. Provide protection of personal rights, freedoms, and legal interests in digital economy	IV quarter of 2020
5.9. To create technical tools that provide secure public interactions in digital economy	I quarter of 2020
5.10. Provide business rights and legal interests protection in digital economy	2024
5.12. Create effective mechanisms of state regulation and support of information security under the integration of national economy to global economy	2024
5.13. Create a basis for building the trusted area within the EEU that provides common information security	2023
5.14. Provide participation of Russia in the preparation and implementation of international documents on information security, related to digital economy	2021

The following are the key indicators for the digital economy ecosystem in general:

- Successful operation of more than 10 leading companies (ecosystem operators), competitive on the global markets;
- Successful operation of more than 10 sectoral (industrial) digital platforms for the major key economic domains (including for the digital healthcare, digital education, and "smart cities");

- Successful operation of more than 500 small and medium businesses in the sphere of the digital technology development and digital services platforms.

Inset 1.3

The primary action plan in "Information security"

August 25 to September 29

Competence center: action plan project development

September 29

Competence center: submission of the action plan projects for approval of the work groups

September 29 to October 6

Work groups: summary formulation based on the work results of the work groups

October 6

Competence center: submission of the action plan projects and the work group summary to designated federal executive authorities and project office

October 6–13

Designated federal executive authorities: approval of the action plan project by the interested federal executive authorities

October 13

Designated federal executive authorities: submission of the approved action plan project (or with the remarks and the disagreement table) to Competence center and the project office

October 14

Competence center: submission of the action project to project office to take for the approval of the Digital Economy Panel

October 14–16

Project office: to take the action plan project or the approval of the Digital Economy Panel

(Continued)

Inset 1.3 Continued
October 16
Presidential Administration of Russia: consideration and approval (or submission for the development) of action plan projects on Digital Economy Panel
October 20–24
Designated federal executive authorities: Submission to the IT Governance Committee the action plan project approved by the Digital Economy Panel
October 24–28
Presidential Administration of Russia: consideration and approval of the action plan by the IT Government Committee

Inset 1.4
Russian Digital Economy challenges and threats
• Information, infrastructure, and population security • Confidentiality • Information space pollution • Transparency of the automate decision-making algorithms • The reconsideration necessity of the legislation and international affairs • Staff deficiency and the substitution of the traditional economy work places • Digital feudalism of the Presidential Administration of Russia: consideration and approval of the action plan by the IT Government Committee

The following are the key indicators for the human resources and education indicators:

- Graduate number of the educational organizations of higher education with the degree related to the ICT is 120,000 people per year;
- Graduate number of higher and secondary professional education with the competencies in information technologies on the worldwide average level is 800,000 people per year;
- Proportion of the population with digital skills is 40%, in regard to the formation of research competence and technical capacity;

- Number of implemented projects in the digital economy (with the volume of no less than RUB 100 million) is 30 units;
- Number of the Russian organizations participating in the large-scale projects (with the volume of RUB 3 million) implementation in focus areas of international scientific and technical cooperation in the digital economy is 10.

The following information infrastructure development key indicators:

- Proportion of households that have the broadband access to the Internet (100 Mb/s) out of total household number is 97%;
- Within all big cities (with the population of 1 million) there is a stable 5G and higher network coverage.

The following are the information security key indicators [5, 3, 12, 18–21, 25, 26]:

- Percent of the subjects using the secured information interaction standards of the state and public institutions is 75%;
- Proportion of the local network traffic on the Russian Internet segment routed through international servers is 5%.

The Importance of the Import Substitution Task

The "Digital Economy" Program identifies the following so-called basic cross-cutting (universal technologies intended for use across different Russian economic sectors) information technologies:

- Supercomputer technologies with high and ultrahigh performance (exascale);
- Technologies for the electronic component base development;
- Technologies for the hardware cyber-physical complexes creation;
- Technologies of the human–machine (cyber biological) systems development;
- Quantum technologies of the data collection, processing, and transfer;
- Connection and data transfer technologies (software-defined networks (SDN), LTE technologies, 4G, and 5G);
- Robotic technologies (drone systems, computer vision);
- Computer engineering technologies (engineering problem solutions);
- Mathematical modeling (simulation study, theory of queue and inventory);
- Technologies of the cloud and fog computing;

- Cognitive technologies (corticomorphic, neural network, and genetic technologies);
- Technologies of the Big Data capturing;
- Technologies of the virtual environment (hardware and software hypervisors, dispatchers, and planners);
- Navigation technologies;
- Geo information technologies;
- Technologies of the virtual and alternative realities;
- Information security technologies (of a new technological order);
- Distributed ledger technologies (block chain);
- Software engineering technologies (software production), etc.

Including information security technologies (in more extended view) [1–6, 11, 13–17, 19–22, 25]:

- Cognitive technologies of the cyberspace control and early computer attack warning (iSOPKA);
- Technologies of the computer engineered environment and opponent behavior prediction (WarGaming);
- Intelligent information security technologies based on the Big Data and data flow (Big Data+ETL);
- Adaptive security architectural technologies (Adaptive Security Architecture);
- Technologies of the software-defined networks (SDN) and network functions virtualization (NFV);
- Cryptographic module technologies (hardware security module–HSM);
- Security technologies of the Industrial Internet of Things and the Internet of Things (IIoT and IoT);
- Technologies of the trusted device mesh and safe system architecture (advanced system architecture);
- Technologies of the trusted cloud, fog, and virtual environments;
- Safe mobile technologies;
- Technologies of the software code dynamic analysis;
- Quantum trusted data transfer technologies, etc.

There are several competence centers in Russia specialized in the information security that can be used for the solution of the primary digital economy objectives in the direction of "information security":

- Leading industrial enterprises such as State Corporation "Rostec", JSC "Rostelecom", RTI Group, Mikron Group, JSC "MCST", JSC "Baikal

Electronic, FGUP "RFNC-VNIIEF", JSC "Concern "Avtomatika", Manufacturers Association of the radio-electronics industry, etc;
- Head academic centers and institutes: Federal Research center "Informatics and Management" of the Russian Academy of Sciences, the Institute of the Control Sciences named after V.A. Trapeznikov of the Russian Academy of Sciences, Research Institute of the Multiprocessors named after A.V. Kalyaeva, FSAEI HE Southern Federal University, The Institute for System Programming of the Russian Academy of Sciences, The Program Systems Institute of the Russian Academy of Sciences, etc;
- Leading higher education institutes and universities: Lomonosov Moscow State University, Moscow Institute of Physics and Technology, Bauman Moscow State Technical University, Saint Petersburg Electrotechnical University "LETI", ITMO University, A.F. Mozhaysky's Military-Space Academy, ANO HE "Innopolis University", etc;
- Producers of the information security tools: Echelon Company, Confident, Atlas, Penza Research Electrotechnical University, Infotecs, Elvis+, Informzashchita (Security Code), Ancud, Factor-TS, Cryptosoft, CryptoPro, Kaspersky Lab, Positive Technologies, etc.

It is essential that the given program assigns the core role of the specific national (import-substituting) information technologies and information security technologies. At the same time, it is mentioned that the list of the basic cross-cutting information technologies can be reconsidered and added in the course of the "Digital Economy" Program.

From one side, the formation and development of the national digital economy segment based on the use of primary national trusted information and communication technologies (ICT) and its further integration to the global digital economy space (starting with the countries of the EEU, APEC, SCO, and BRICS) are a "window of opportunities" to build the Russian economy in forming global economic order. From the other side, it carries significant risks and challenges to economic security and state sovereignty [1–6, 11, 13–17, 19–22, 25].

The following threats and challenges are defined in the "Digital Economy" Program of the Russian Federation:

- Problem of the human rights protection in the digital world, including identification (person correlation with his digital image), user digital data security, as well as the problem of the public digital environment assurance;

- Threats to individual, business, and the state, connected with the pattern to build complex hierarchic information and communication systems, which widely use visualization, remote (cloud) data storages, and different connection technologies and terminals;
- Increase the possibilities for external information and technical impact on information infrastructure, including critical information infrastructure;
- Global growth in cybercrime;
- Underrun from leading foreign countries in the development of the information technologies competitiveness;
- Dependence of social and economic development on the foreign countries' export policy;
- Insufficient effectiveness of the scientific researches related to the advanced information technology development, low level of national development implementation, and low level of human resourcing in information security.

In particular, it is noticed that Russia falls behind leading countries (G-20) for 5–8 years in the digitalization level. In the next 5 years, this gap may increase to 15–20 years, as there is an exponential modern ICT development. It is noted that today the information technology import to Russia is up to 90% out of 100% of the hardware used and up to 60% out of 100% of the system and application software. According to the Institute of Statistical Studies and Economy of Knowledge of the Higher School of Economics [17], the information technology export from Russia is between USD 3.5 billion and USD 5.1 billion, while their import grows to USD 11.7 billion. Such dependence of the national industry on the foreign electronic component base, computer hardware and communication tools, system and application software, puts social and economic development of Russia dependent on the geopolitical interests of a number of foreign countries and carries additional security risks, for example, in terms of address implementation and undocumented features destructive application in the supplied foreign hardware and software [1–5, 11, 13–17].

The primary objectives of the "Digital Economy" State Program of the Russian Federation in "Information Security" that require immediate solution are listed below.

The legitimization of the cryptocurrency use as a payment method in some state economy spheres. The importance of this task is explained by

---

[17]https://issek.hse.ru/data/2016/11/09/1109680675/NTI_N_27_09112016.pdf.

the fact that the cryptocurrency transactions stay out of the legal regulation and state control in the absence of the legal status. Therefore, there are opportunities to use cryptocurrencies in illegal tax avoiding, disinvestment, criminally obtained income legalization (laundering), terrorism financing, and sale of drugs, people, and weapon. It is noted that since 2014, the cryptocurrency (bitcoin) operations have been authorized for individuals in the EU countries and in Switzerland, the USA, Japan, Singapore, and China. The work on improving the control and legal regulation in this sphere is still in progress. Currently, bitcoin does not have a legal status in Russia. Moreover, the Government of the Russian Federation instructed the Ministry of Finance of the Russian Federation (Instruction No. ISH-P13-5204 of September 29, 2016) to monitor the quasi-cash circulation (including cryptocurrencies) by the end of this year, as well as to additionally analyze the risks of possible cryptocurrency use for illegal (criminal) purposes, and to submit proposals on modifications of the respective regulatory and legal documents and acts.

National block chain technology (the technology of open distributed ledgers) development and implementation. During the operation performance, every transaction is recorded in the system as a new chain element (block). At the same time, an unlimited number of participants with equal rights collect and store the information concerning the whole transaction chain. Therefore, the unauthorized data modification is impossible: the corresponding automated system will not approve, but deny the operation. Nowadays, there are all possible open resources developed on the base of the block chain technologies that record and store the information concerning the trade operations, obligations, rights, etc. The automation system tools provide the verification of the given catalogues. It is expected that under the relevant state regulation, the block chain and cryptocurrency application will exclude the possibility of illegal deals variety, in particular in the state procurements and "bring to the light" the shadow business, etc.

Currently, several international bank structures (e.g., R3 consortium–Goldman, Sachs, Barclays, Bank of America, JP Morgan, and Sberbank of Russia) are applying the block chain technology to decrease the costs and to get the competitive advantage, as well as to spread their business domain in general. In particular, the issue of the cancellation of the bank SWIFT system services and the possibility of the transfer to interbank payment system based on the block chain are under discussion. It is clear that underrun and an absence of regulations in this sphere in Russia will not only lead to partial financial isolation and inability to provide competitive banking services, but also to control loss over some financial operations in case when the leading bank structures will abandon the SWIFT services to this new advanced technology.

The development of the national cognitive hardware and software systems, including the systems created for the purposes of strategic and operational forecasting, calculation performance, securities trading, etc. For instance, at the moment, freemium client online brokers like Robinhood, Loyal3, FreeTrade, etc., allocated on the common smartphones of the end user, are widely used. Moreover, as they provide global broker services, it creates the threat of the investment outflow from national to foreign markets in cross-border cryptocurrency application. Here, the traditional financial institutions are replaced by new players with significant user population, for example, a variety of social networks and Internet aggregators like Amazon.com, Alibaba.com, Yandex, Rambler, Mail.ru, etc.

To neutralize the listed threats and challenges of the Russian digital economy, it is recommended to follow the below-listed basic information security principles [3, 5–10, 13–16, 20–22]:

- Apply the technical regulatory measures, especially in regard to imported equipment and components, hardware and software of transport, power systems and other strategic infrastructure, as well as the key objects of vital activity;
- Implement an import substitution program and prior use the national equipment and software;
- Apply only national information security technologies (control of integrity, confidentiality, authentication, and availability of data transfer and processing);
- Apply information security technologies with Russian cryptographic standards.

Only strict adherence to the above-mentioned security principles will allow timely and qualitative accomplishment of the objectives of the "Digital Economy" Russian Federation State Program. In turn, this will improve the welfare and life quality of population by increasing the availability and quality of goods and services, promoting awareness and digital literacy, improving the availability and quality of the public services for citizens, ensuring national security both inside and outside the country.

1.1.2 Main Information Infrastructure Development Objectives

Since 2015, the infrastructure models of the Digital Economy of the Russian Federation have undergone significant changes. The main reasons were the continuous improvement of the trunk of the unified national communications network, the emergence and development of the IIoT infrastructure and the Internet of Things (IoT), the creation of new wireless communication standards, as well as the emerging mass migration of domestic telecom operators from traditional infrastructure models to more perspective models of service providers. As a result, emerged new infrastructural models of the Digital Economy of the Russian Federation, realizing the great potential of special services based on effectively organized "cloud" and "fog" computing, Industrial IIoT, new Big Data and Big Data Analytics technologies, etc.

Today, a number of leading international telecom operators, such as AT&T, Deutsche Telekom, and France Telecom, as well as content providers (Figure 1.6), actively continue to form the market of the so-called content services. The main prerequisites include the technological evolution of Industry 4.0 based on Internet technologies of IIoT/IoT, the widespread penetration of broadband access, the creation of new promising standards for wireless communication, the emergence of free services, and an increase in the total number of smartphones/tablets and IIoT/IoT. For example, the following main segments of the content services market were distinguished: games and

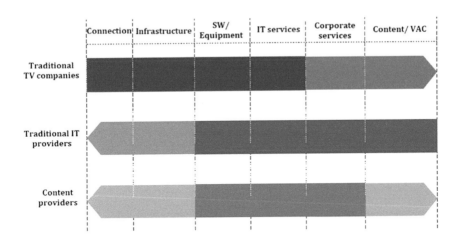

- Traditional Telecom companies offer corporate services and content
- Traditional IT providers develop infrastructure
- Content providers offer more content, develop infrastructure
- All market sectors perform service conversion structure
- All market sectors perform service conversion

Figure 1.6 Development directions of service and content providers.

applications (from US $2489 million), music applications (from US $964 million), and IPTV television (from US $1145 million). At the same time, the services of online publications (more than 100% per year), mobile TV and VoD (54% per year), and pay-TV and IPTV (18% annually) are rapidly developing [27].

Under these circumstances, PJSC "Rostelecom" (after the inclusion of the interregional connection holding Svyazinvest in April 2011) becomes the largest Russian telecommunications company and the main actor claiming leadership in a number of new segments. If earlier Rostelecom was a trunk operator for other telecommunication operators (the monopolist in the long-distance telecommunications market), after the mentioned amalgamation, the company was transformed into the largest telecommunication company with a wide range of services offered specifically for the end users in the domestic market (Figure 1.7). As a result of the RTO's merger, the Rostelecom network increased several times [28]. This propelled the company to the top in terms of coverage and mileage of the network in Russia, but, on the other hand, raised the need to integrate hundreds of thousands of kilometers of communication lines into a single system, as well as to organize its management and control.

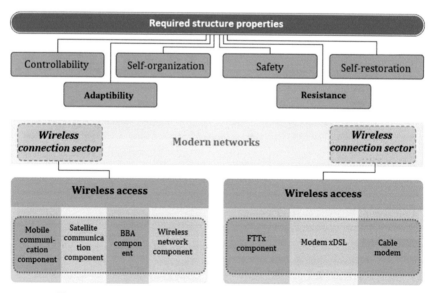

Figure 1.7   National information and communication infrastructure.

At the same time, the necessity to significantly expand the communication lines for the growing amount of Internet traffic was raised. The modern development of Rostelecom network infrastructure implied the installation of equipment with a multiply increased throughput for access networks with 100G/400G trunk, the building up of a federal mobile operator based on a multi-standard network (LTE, UMTS, EV-DO, CDMA), and the organization of optical networks (FTTH, PON). In addition, it required the organization of a single network management center, ensuring the standardized quality of provided services and improving the management efficiency, and the creation of a centralized architecture of IT solutions based on integrated ERP and OSS/BSS systems.

Another major initiative of Rostelecom ("RTComm.RU") was the development of a K-band satellite network in the framework of the project "Providing high-speed access to information networks through satellite communication systems". By the end of 2016, the mentioned network had already provided integrated coverage of the country's territory, including hard-to-reach areas of the Far North, Siberia, and the Far East and provided a single package of integrated services for 15 million or more domestic users. The project also included building corporate networks that required a satellite component and high speeds comparable to terrestrial solutions. For project implementation, it was required to build the so-called terrestrial component (more than seven base stations and corresponding communication and flight control centers), as well as a space component consisting of four satellites (Express-AM5, Express-AM6, etc.). It is significant that within the K-band framework, currently used in Russia, the basic speeds of providing Internet access were 128, 256, 512, and more kilobits. The implementation of this project allowed access at speeds of up to 15 megabits per second, which is comparable to Internet access in large cities with applied ground-based systems.

Modernization of infrastructure allowed Rostelecom to start a major project implementation to create a content management system and a national platform for cloud computing (Figure 1.8), announced within the framework of the state "Information Society" program for the period up to 2024.

It has required the deployment of a single multifunctional distributed computing platform for IaaS class services, network development of distributed data centers for "cloud" and "fog" computations, and formation of an appropriate innovation ecosystem. Today, the infrastructure of PJSC "Rostelecom" includes not only one of the world's largest trunk communication networks, but also a large network of distributed data centers. The formed

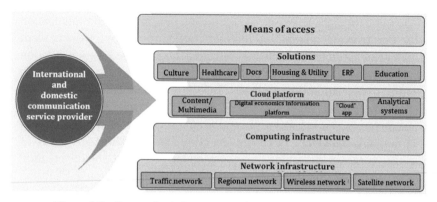

Figure 1.8   Perspective infrastructure of the communication operator.

innovation ecosystem includes more than 70 commercial partner companies, nine basic departments, and competence centers of leading domestic specialized universities. This has realized the implementation of a number of large federal projects [28], including "Electronic Government" (Unified Portal of State and Municipal Services, Inter-Agency Electronic Interaction System-SMEW), "Web-election 2018", and National Cloud Platform (healthcare services, education, security, housing and utility services, etc.). For instance, in the first quarter of 2018, the number of users on Public Services Portal exceeded 65 million, and there were more than 40 million registered personal offices. Today, the Public Service portal provides an opportunity to receive more than 6,500 electronic and 75,000 information services. The next stage in the development of the public services portal was the implementation of socially significant services, for example, a utility services payment or a doctor appointment. A simplified registration procedure (based on the full name, e-mail address, or mobile phone number) has been implemented for new services. In order to implement the above-mentioned services, the corresponding applications for mobile devices, iOS, Android, Windows, etc., have been prepared

The development of the Unified Federal Service for Electronic Registration and the Integrated Electronic Medical Card was completed. Both systems operate on the basis of the National Cloud Platform (Figure 1.9). The Ministry of Healthcare of the Russian Federation together with the created systems will be able to shortly bring all the medical institutions of the country into a single network, with a common database of medical records and with the possibility of fixing an appointment to specialists via the Internet. The O7.Medicine., service for the regional level, has been developed.

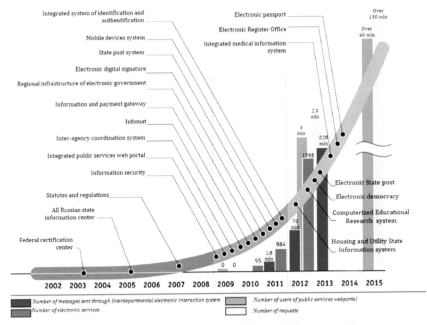

Figure 1.9 Project dynamics of "Electronic Government".

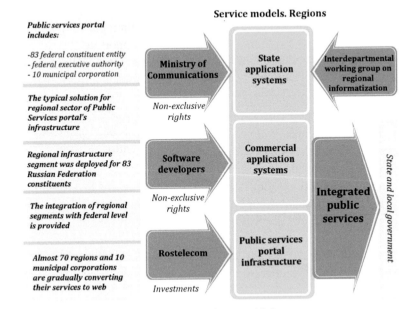

Figure 1.10 Public service portal infrastructure.

The operation of the State Information System (GIS) of housing and utility services in the regions of Russia started operating. The system is designed to support the operational access of citizens, authorities, and resource providers of information on all significant aspects of managing multi-apartment buildings and providing public services, as well as electronic document management in the area of housing and utility services.

For the further development of e-government and the provision of relevant public electronic services, the Ministry of Communications of Russia is implementing two main projects. The first is the conversion of state services into electronic form. The second is the transition of executive bodies to interagency electronic interaction to provide state and municipal services.

According to analysts, almost all federal agencies, regions, and municipalities participate in these projects, and the dynamics of the country are generally positive. At the same time, new goals and priorities have been set. By 2020, 85% of the country's population (only 45% of today's) should receive public services through a Web portal, meanwhile 90% of citizens should be satisfied with the quality of government services. The relevant primary tasks include the following services:

- Implementation of simple registration;
- Electronic appointment fixing;
- New channels for the service provision, expanding their list and providing not only state and municipal services, accounting and adaptation of the gathered user experience;
- Implementation of recommendations and requirements for the personal data security [29, 30] for the new infrastructural models of personal data operators.

December 18, 2017 The Russian Federation Government Commission, regarding the use of information technologies to improve the quality of life and business environment (Report of proceedings at a meeting on December 18, 2017, No. 2), approved the Action Plan for Information Infrastructure of the government Program "Digital Economy of the Russian Federation". It should be noted that the mentioned Program was approved by the relevant order of the Government of the Russian Federation dated July 28, 2017, No. 1632-r.

The approved Action Plan for the "Information Infrastructure" can be found at http://static.government.ru/media/files/DAMotdOImu8U89bhM7lZ8 Fs23msHtcim.pdf.

Web Portal
- Integrated public services portal
- Electronic public services provision

07. Medicine
- Integrated health care record
- Electronic reception
- Medical information system

07. Education
- Electronic school and electronic kinder garden
- Student social network, educational channel

07. Housing & Utility service
- Public funds and tariff control
- Accounting and financial assets control

07. City
- City infrastructure monitoring
- Prediction and incident prevention

07.112
-Message Reception and Processing with general number
- Interdepartmental coordination

07. Doc
- Office system automation
- Document storage

07. Business
- Automation of core business processes
- Virtual office

Figure 1.11　Typical services of the national cloud platform.

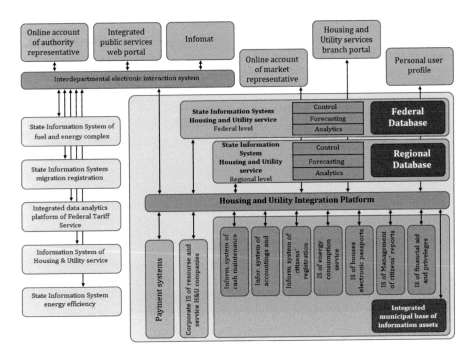

Figure 1.12　O.7 Housing system structure.

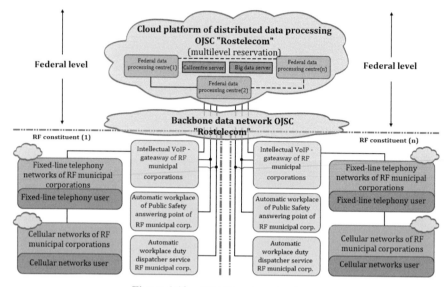

Figure 1.13   O7.112 system structure.

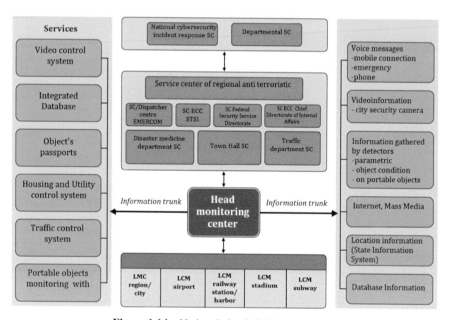

Figure 1.14   National cloud platform services.

According to this Plan, the following main tasks are defined for the development and improvement of the National Telecommunication Network (NTN) of the Russian Federation (RF) for the perspective up to 2024 for the proper provision of the "Digital Economy of the Russian Federation" program:

- Determination of the required and sufficient conditions for the development of trunk transport telecommunication network, fixed and mobile telephone communications;
- Accounting of the main trends and prospects for the development of telecommunication technologies, their advantages and disadvantages, the degree of standardization, the possibility of import substitution and certification for information security requirements, technical support, and maintenance in the following areas:
  - Technologies for building trunk transport networks, including OTN, GMPLS/ASON, photon networks, and satellite networks;
  - Technologies for building wired and wireless access networks, including GPON, GePON, WSN, Carrrier WiFi, Small Gell Network, 4G/5G, etc.;
  - Virtual and Augmented Reality technologies, virtualization and support for heterogeneous networks, Industrial Internet of Things (IIoT), and Internet of Things (IoT);
  - Technologies for the development of future packet networks of the next generation SDN and NFV;
  - Technologies for the development of convergent networks and IMS and vIMS services;
  - Technologies for the implementation of Internet services of Things (IoT), including services of Industrial Internet of Things (IIoT), inter-machine interaction (M2M), personal networks, content distribution networks, and self-organizing networks;
  - Technologies for creating specialized IT platforms for Internet Intranet and IoT/IIoT networks based on models and methods of "cloud", "edge", and "fog" computing, Big Data and Big Data Analytics, neuromorphic and cognomorphic computations, etc.
- Generation of the following proposals for the following perspective NTN target architectures up to 2024:
  - Trunk transport networks, given the introduction of the new telecommunication technologies;

- o Wired and wireless access networks;
- o Networks with the convergence of fixed, mobile and telematic services based on the virtualization technologies for the network elements and functions, cloud technologies, etc.;
- o IT platforms to provide infrastructure capabilities to service and content providers for the implementation of content-oriented services, virtual presence services, Internet services of things, etc.

- Elaboration of proposals for the construction and development of the NTN RF infrastructure with respect to the following actions:

  - o Planning, calculation, and optimization of the NTN RF networks;
  - o Organization of a unified numbering and addressing system;
  - o Ensuring the quality of providing information telecommunication services;
  - o Increasing the stability of the communication network functioning in general;
  - o Establishment of an equalized accounting and reporting time and organization of a network synchronization system;
  - o Ensuring the proper information security, including block chain and quantum technologies;
  - o Implementation of the SORM intellectual system;
  - o Monitoring and management of the network infrastructure;
  - o Management of the public communication network in emergency situations (ES), etc.

- Clarifying the course of telecommunications and information communication services in Russia, taking into account the dynamics of changes in the traffic volume and structure, as well as a predictive assessment of the communication service development;
- Selection of the perspective technologies for the service development and task solution of the NTN RF development up to 2024 to ensure the implementation of the "Digital Economy of the Russian Federation" program;
- Evaluation of the system properties of typical NTN RF networks, including manageability, sustainability, cyber security, adaptability, and self-organization, using appropriate and predetermined criteria and indicators;
- Organization of the supervision and management of the system properties of the NTN RF by taking into account the use of the so-called "end-to-end" info communication technologies of the "Digital Economy of the Russian Federation", including information security technologies;

- Generation of proposals for the development of the regulatory and legal framework for the construction and development of the NTN RF infrastructure of the new telecommunication technologies;
- Generation of proposals for enhanced training and retraining of personnel for the implementation of the main provisions of the NTN RF development, etc.

1.1.3 Implementation of the Long-Term Evolution (LTE) Technology

At present, LTE networks have become widespread in Russia and other technologically advanced countries of the world. For example, in order to hold the XXII Olympic Winter Games (2014) in Sochi, the fastest LTE-Advanced networks were created, which transferred data at 300 Mbit/s. According to the GSMA association[18], at the beginning of 2014, there were 263 LTE networks operating in the world (in February 2012, there were 49 ones) in the ranges of 700, 800, 1800 MHz and 2.3 and 2.6 GHz. More than 425 operators from 126 countries started investing in LTE networks. The investment in mobile communications in 5 years was expected rise to 1.1 trillion dollars, with the increase in the number of LTE network subscribers from 90 million (2013) to 3.7 billion (2017) and that of broadband subscribers up to 1.1 billion in the same time. Let us consider the key features of modern LTE networks and then propose the possible options for creating a trusted LTE network of the enterprise.

The relevance of LTE technology
The evolution of wireless communication technologies (Table 1.3) led to the association of two previously independent development trends in mobile systems (GSM and CDMA cellular communications) and data transfer systems of the IEEE 802 family (Wi-Fi, WiMAX) (Figure 1.15).

For this reason, the networks of the next, fourth generation (4G) are called universal mobile multimedia data networks. As a result, 3GPP Long-Term Evolution (LTE) evolved into an open international standard for cellular communication of the fourth generation (4G) [7, 14] developed and advanced by the noncommercial consortium 3GPP[19]. Currently, LTE networks are considered as the most promising (the forthcoming of 5G networks is expected in 2020) (Figure 1.16).

---

[18]http://www.gsma.com.
[19]http://www.3gpp.org/DynaReport/36-series.htm.

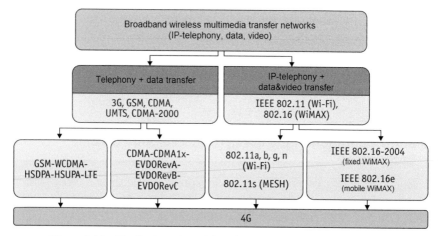

Figure 1.15   The forthcoming of the wireless multimedia LTE networks.

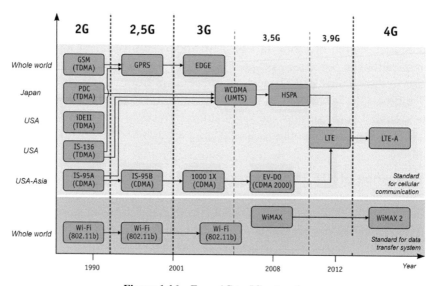

Figure 1.16   From 1G to 5G networks.

The development of the LTE technology began in 2004 [7, 14] First, the possible technologies of the physical layer of W-CDMA (used in HSPA) and OFDM were considered; these technologies provide a high-speed data transfer. As a result, the OFDM technology was chosen; in May 2006, the first specification for the radio interface for the Evolved UMTS Terrestrial Radio

Table 1.3   Evolution of wireless standards

Wireless Standards	3GPP	Quakcomm	China	IEEE	
Used by operators	AT&T, T-Mobile US, and other leading telecom carriers	Sprint, Verizon Wireless	China Mobile	Sprint	
2G: telephony + data transfer	GSM: 2G GPRS: 2,5G EDGE: 2,75G	CDMAOne			
3G:telephony + data transfer at the rate more than 200 kbit/siPhone 4/4S(at the rate more than 5 Mbit/s)	Release 4	UMTS 3G	CDMA2000 EDVO rev 0	TD-SCDMA (up to 2 Mbit/s)	Mobile WiMAX 3,9G(4 Mbit/s... EVO "4G")
	Release 5	HSDPA 3,5G(up to 21 Mbit/s)	CDMA2000 EDVO rev A(1.8–3.1 Mbit/s)		
	Release 6	HSUPA 3,5G(up to 5,8 Mbit/s)	EDVO Rev C/Ultra Mobile-Broadband cancelled:		
	Release 7	HSPA + 3,5G	Sprint moved to WiMAX,		
	Release 8/9	LTE 3,9G	Verizon moved to 3GPP LTE		
4G:IP-telephony + data transfer at the rate more than 100 Mbit/siPhone 6/7(at the rate more than 130 Mbit/s)	Release 10	LTE Advanced		TD-LTE	WiMAX 4G

Access (E-UTRA) packet data network was created in 3GPP. Its fundamental differences are listed below:

- Increase in a peak network bandwidth;
- High spectral efficiency (bit/Hz/s);

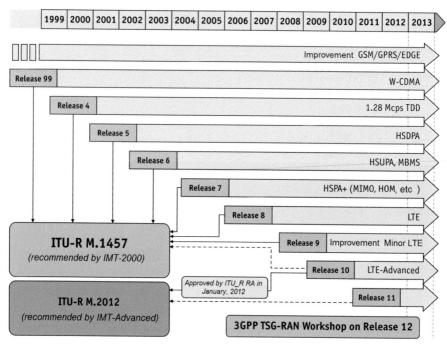

Figure 1.17   Evolution of the LTE specifications.

- Increased flexibility in spectrum management;
- Improved network architecture (SAE), that is, the radio access (EUTRAN) and the core network (Evolved Packet Core, EPC).

Further LTE specifications have evolved from the so-called 3GPP Release 7 to 3GPP Release 11 [7, 19] (Figure 1.17).

As a result, the following basic requirements for the standard of the fourth-generation (4G) mobile networks were created:

- The maximum data transfer rate in the downlink up to 1 Gbit/s, in uplink–up to 500 Mbit/s (average throughput per subscriber is three times higher than in LTE);
- Bandwidth in the downstream radio channel is 70 MHz, in the upstream– 40 MHz;
- Maximum spectrum efficiency in the downlink is 30 bit/s/Hz, in the uplink–15 bit/s/Hz (three times higher than in LTE);
- Full compatibility and interaction with LTE and other 3GPP-systems.

To fulfill the above-listed requirements [7, 14], the following innovations were proposed:

- Broader radio channels (up to 100 MHz), asymmetric separation of bandwidths between the uplink and downstream channels in the case of frequency duplex;
- FDD and TDD use;
- Use of the one set of radio-frequencies by all base stations (eNodeB) and interference coordination;
- Sharing of information concerning the transmission channel when planning access to the environment;
- Synchronous broadcast transmission;
- Better advanced systems of coding and error control (Hybrid ARQ);
- Hybrid OFDMA and SC-FDMA technology for the uplink, the use of MIMO (up to eight antennas), etc.

Key features of LTE

Let us note the following features of the new LTE technology. First, it is universal and hence a more modern network architecture (SAE–System Architecture Evolution) (Figure 1.18). SAE, proposed by the 3GPP consortium,

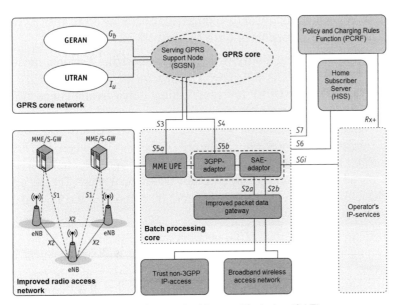

Figure 1.18 System Architecture Evolution (SAE).

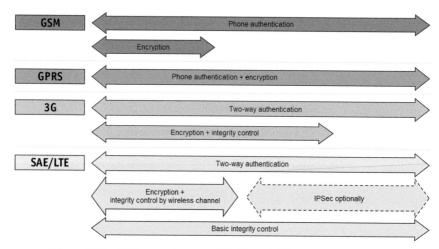

Figure 1.19   Differences between 2G, 3G, and 4G security mechanisms.

provides the required stability of the data network, providing services to IP-based subscribers when moving between disparate wireless access networks (GSM, UMTS, WCDMA, etc.) [7, 19].

In the SAE architecture, the number of typical nodes has been reduced to two: base stations (evolved NodeB, eNodeB) and access gateways (Access Gateway, AGW), which significantly reduced the operating costs of LTE/SAE networks. The core components of the SAE network include:

- Mobility Management Entity (MME) for authorizing terminal devices in land mobile networks and general management;
- User Plane Entity (UPE) for establishing downstream connections, routing, and data protected by encryption;
- 3GPP-adaptor for interaction between 2G/3G and LTE networks;
- SAE adaptor to support the continuity of the service when moving a subscriber between 3GPP and other networks (I-WLAN, etc.). Second, it is an advanced security architecture based on IPSec. The security mechanisms here are noticeably different from those in the 2G and 3G networks (Figure 1.19).

The security architecture of 4G LTE networks is disclosed in TS 33.401 [7, 14], according to which (Figure 1.20), for a secure data exchange in the LTE network, a reliable connection between the user device and the operator public land network (Public Land Mobile Network), as well as between the user device and the core network (IMS Core Network Subsystem) is required. The following LTE security issues are highlighted:

- General issues of the security architecture;
- Network layer security;
- User layer security;
- Application security;
- User security policies.

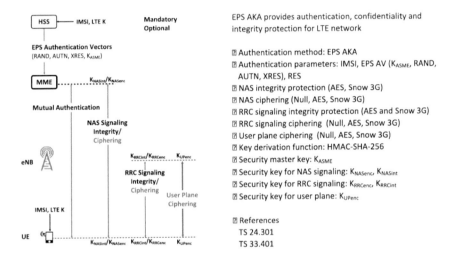

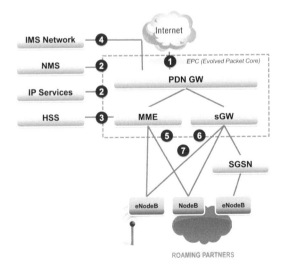

Figure 1.20   The security architecture of 4G LTE networks.

Thus, LTE technology is an evolutionary development of very successful cellular communication technologies of the second (GSP, GPRS, EDGE) and the third (UMTS, HSPA, HSPA+) generations. The main features of LTE in comparison with the technologies of previous generations are:

- Commitment to a high-speed packet data transferring;
- Commitment to integration with IP networks;
- Ensuring the quality of the transmission of video and voice traffic;
- Improved mechanisms for ensuring information security;
- Increased spectral efficiency, flexibility in the use of frequency resources (operation in frequencies from 400 MHz to 6 GHz and in the band from 1.4 to 80 MHz, frequency, and time duplexing).

Compared to third-generation technologies (UMTS, etc.), LTE provides high-speed (peak speeds of up to 150 Mbps in LTE and more than 1 Gb/s in LTE-Advanced) data transfer and low latency of real-time information transmission, such as video streams. In addition, LTE simplified the network architecture and improved the mechanisms for providing information security. Difference from WiMax technology (group of IEEE 802.16 standards) is a compatibility with other 3GPP technologies, as well as providing a substantially greater choice of user equipment. It should be recognized that in the competition of technologies, WiMax lost and gradually reduces its presence in the market. Compared with Wi-Fi technology (IEEE 802.11 group of standards), LTE provides guaranteed quality of service provision, greater spectral efficiency when servicing a large number of subscribers. An important difference is the operation of LTE systems in licensed frequencies, in contrast to Wi-Fi networks operating in unlicensed ISM bands (2.4 and 5 GHz). The operation in the licensed range can reduce the level of unintentional interferences, which is expressed in the best communication quality.

Possible options for creating a trusted LTE network

In practice, a typical LTE network consists of four main segments: subscriber equipment (SE), base stations (BS), core network, and network management system (Figure 1.21). Physically, the network of management of base stations and core network elements can match with the transport network.

Let us consider the possible options[20] of creating a trusted LTE network in practice.

---

[20]http://www.telum.ru.

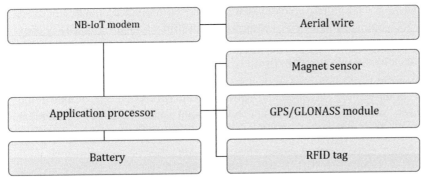

Figure 1.21   Structure and content of a typical LTE network.

Option 1. Use of commercial operator network

This is the simplest option, which implies the use of a commercial mobile operator network, that is, the base stations, the core of the network, and the network management system (Figure 1.21) are the property of the communication operator and are completely under his control. Here it is necessary to logically separate the traffic of the trusted network subscribers from the traffic of the commercial operator subscribers by creating a virtual operator (MVNO). Mandatory is an additional protection of the transferred information, for example, by means of VPN tunnels between the subscriber equipment.

In this version, the reliability and security of the solution corresponds to some basic level of the commercial cellular network, which is significantly lower than the level, required for the professional communication systems. Bandwidth is low, since the network resources are shared with commercial subscribers. The level of communication services accessibility corresponds to the level offered by commercial operators in large buildings. In accordance with the requirements for cellular communication networks, the system controllability from the operator side is high; however, it is limited by the enterprise. This provides mobility of subscribers within the network of a commercial operator, including LTE roaming zones.

If necessary, it can be achieved with a reference to 2G (GSM, EDGE) and 3G (UMTS, HSPA, etc.) communication networks. Compatibility with a wide range of subscriber equipment and integration with global networks is ensured, but integration with local enterprise systems is difficult.

The LTE system deployment based on the commercial operator network does not require the use of specialized equipment. The solution deployment consists in the selection of the operator-provider of communication services and the organization of a virtual operator based on it, as well as the organization of secure communication channels over the operator network.

In general, the considered option is characterized by rather small expenses, but at the same time, by the minimal values of the controllability, stability, and security of the LTE network.

Option 2. Base station of external location

This option involves the deployment of the own LTE network, independent from commercial operators. We need to deploy a core network, base stations (one or more), and network management systems. A license for a bandwidth of 20 MHz or higher is required. The core of the LTE network and the network management system should be deployed on the server/hardware/enterprise data center.

This option involves placing one LTE base station on the street, outside the monitoring area. To cover a relatively small area, a base station of reduced power (Microcell) can be applied. A base station may have one or more sectors located to provide the desired coverage. The details depend on the building geometry, the materials used, etc., and radio planning is necessary. As a side effect, coverage of a certain area outside the facility will be ensured. For the installation of the antenna-feeder path, BS can be used a special mast and existing constructions (pipes, etc.).

In this version, communication services are provided as needed. Since the system is fully controlled, it is possible to introduce additional traffic classes and prioritize subscribers (it requires additional scientific and technical support).

The solution reliability is determined by the reliability of the used equipment (base station, core network, communication channel between them, network management system). Failure of any network element results in the failure of the entire system. It is recommended to reserve network elements. The solution sustainability is achieved by adapting the operating frequency to the interference situation using a network management system. Ensuring the solution security is aggravated by a lack of trusted equipment of macro-base stations LTE class, as well as a high signal level outside the monitored object. Bandwidth depends on the number of sectors, but is estimated at a level of 100–300 Mbit/s, which is much higher than in option 1.

Ensuring a high level of accessibility of indoor telecommunication services is aggravated due to the radio signal attenuation in the walls. System controllability is achieved by using a network management system. It provides mobility of subscribers within the coverage area of the base station, compatibility with LTE subscriber equipment operating in the selected frequency range, and integration of local telephone and IP networks of the enterprise (it requires additional scientific and technical support). In general, the considered option is characterized by a higher cost than option 1, as it involves the own LTE network establishment. In this case, it becomes possible to provide the required values of controllability, stability, and security. At the same time, the failure of a single base station results in the denial of a service for all subscribers.

Option 3. Distributed antenna system inside the building

Unlike option 2, it involves placing the LTE base station in the communication room/operating room/data center inside the enterprise building (the monitoring object). A distributed antenna system (DAS) is installed to distribute the radio signal on the object. The solution allows achieving a high degree of coverage area (accessibility of communication services). It is possible to use a passive or active antenna system of different manufacturers. The number of installed base station sectors must be determined based on the peak system load.

Communication services are provided as needed. Since the system is fully controlled, it is possible to introduce additional traffic classes and prioritize subscribers (it requires additional scientific and technical support). The solution reliability is ensured by the reliability of its components. It is recommended to use the redundancy of the network core and network management systems. Also, a reserve base station can be used, but this will significantly affect the solution cost.

Since the system is completely centralized, and only passive microwave elements are distributed across the building, the failure of the base station or core results in the denial of the entire system. Ensuring the solution stability is achieved by adapting its operation modes to a jamming environment (it requires additional scientific and technical support). Ensuring the solution security is complicated by the lack of trusted equipment of the macro-base stations LTE class. At the same time, almost complete localization of the signal at the monitoring site is possible.

The bandwidth is directly related to the number of sectors connected to the distributed antenna system (usually one to three). Thanks to the placement of the distributed antenna system elements, the level of accessibility of the communication services (coverage) can be made arbitrarily close to 100%. A high potential service availability level is the main advantage of distributed antenna systems. System controllability is achieved by deploying a network management system. Subscriber mobility within the building, compatibility with user equipment of various manufacturers, and integration with external data and telephony networks are ensured.

Overall, option 3 is characterized by higher cost and deployment time compared to the solution under option 2. Similar to option 2, it is possible to provide the required values of controllability, stability, and security of LTE network. However, the failure of a single base station results in a denial of service for all subscribers of the network. Since the entire infrastructure is located inside the monitoring object, the solution is characterized by a higher level of security. Bandwidth depends on the number of sectors and is approximately equal to the capacity of the solution under option 2. Due to the distributed antenna system, a very high level of availability of communication services can be achieved.

The mobility of subscribers within the building is provided. The main advantage of the solution is high availability of services, and the main disadvantage is limited bandwidth and limited reliability due to centralization.

Option 4. The network of small indoor base stations

In contrast to option 3, this option involves placing small base stations (picocelles) in the monitored object. At the same time, a distributed antenna system is not required, so each picocell serves an area within a direct radio visibility. Together, all picocells provide the required accessibility level of communication services throughout the monitoring site. Picocells are connected to the enterprise LAN (Gigabit Ethernet or optical Ethernet). In this option, communication services are provided as needed. Since the system is completely controlled, it is possible to introduce additional traffic classes and prioritize subscribers (it requires additional scientific and technical support).

The solution reliability is provided by core and network backup and picocells. Unlike options 2 and 3, if one of the picocells fails, the others continue to serve the subscribers. It is possible to plan the system, in which, in case of any picocell failure, the required level of availability of communication services at the monitoring site will be provided. It is possible to implement the automatic adaptation of the neighboring picocells power when one of them is out of order.

Ensuring the solution stability is achieved by automatically adapting the system to the conditions of the interference situation. Different picocells can operate in the different frequency bands, providing redundant communication channels for user equipment. Ensuring the solution security is simplified by the localization of the signal inside the building and by the presence of trusted small LTE base stations made in Russia.

Since each picocell has the capacity equal to the macro-base station's sector capacity, the total system capacity is the maximum among all the given options, estimated at 1–10 Gbit/s, depending on the number of cells. The accessibility level of communication services can be arbitrarily high by placing additional picocells (with a rise in the solution cost).

System controllability is achieved by using a network management system. It provides mobility of subscribers within the monitoring site, compatibility with LTE subscriber equipment of the selected frequency range, and integration with local telephone, IP-networks of the object, and global networks, if necessary.

A feasible flowchart of the LTE communication system according to option 4 is shown in Figure 1.22.

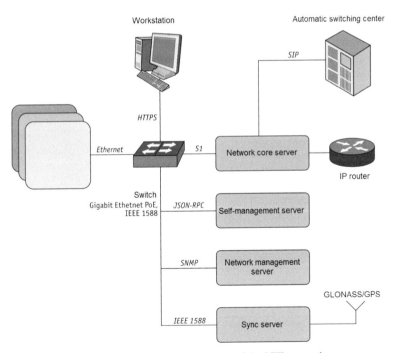

Figure 1.22 A feasible flowchart of the LTE network.

The solution is characterized by a higher cost compared to the solution under option 1. At the same time, the cost and term of the solution deployment is lower than that of the solution under option 3 (expensive installation of the distributed antenna system equipment is not required). The solution is characterized by a high reliability, stability, and capacity. There is trusted equipment developed by the Russian specialists.

Comparative characteristics of the above options for establishing a trusted LTE network are shown in Table 1.4.

Table 1.4 Comparative characteristics of the feasible options for establishing a trusted LTE network

	Option 1	Option 2	Option 3	Option 4
Services	Provided	Provided	Provided	Provided
Reliability	Low	Normal	Normal	High
Stability	Low	Low	Normal	High
Mobility	Provided under the operator network	Provided under the object and surrounding site	Provided at the site	Provided at the site
Capacity	Low	Normal	Normal	High
Security	Low	Normal	High	High
Availability	Normal	Normal	High	High
Controllability	Low	High	High	High
Compatibility	Provided	Provided	Provided	Provided
Integration	Provided with worldwide networks, not provided with local enterprise network	Provided	Provided	Provided
Permission to the radio-frequency spectrum use	Operator	Customer	Customer	Customer
Cost	Low	Normal	High	High
Setup time with operator	By agreement	Normal	High	Low

***

During February 24–27, 2014, the Russian delegation led by the Communications Minister of Russian Federation Nikolai Nikiforov took part in the World Mobile Congress (Mobile World Congress, MWC) in Barcelona. The main topic of the meetings with representatives of companies, including Ericsson, Qualcomm, Alcatel-Lucent, Nokia Solutions and Networks, and Vympel-Com, was the development of a broadband access within the Ministry project. As a reminder, Russian President Vladimir Putin signed on February 3, 2014 amendments to the Federal Law "On Communications", which launched a large-scale project to develop modern communication channels in the country settlements with a population of up to 250 inhabitants. The implementation of the project involves the construction of up to 200,000 km of fiber-optic communication lines. The plan was to cover high-speed connection to 97% of the country's population in 10 years. During the project process implementation and the ongoing program of television broadcasting digitalization in Russia, there may be a need for special client equipment that can receive digital signals of wireless communication networks, providing access to the Internet and digital television services.

At meetings with representatives of telecommunications companies, Nikolai Nikiforov discussed the readiness of their production and technical base for the development of such device. Communications Minister noted that Russia is making a big bet on the development of the mobile communications LTE that is the advanced technologies of next generation.

1.1.4 IIoT/IoT Technologies Development

From 19 to 20 September in the high-technology Russian city of Innopolis, an International Forum IoT World Summit Russia-2017 took place[21]. Scientists from 13 countries of the world took part in this forum, including Russia, the USA, the EU, Great Britain, Italy, Norway, South Korea, Singapore, etc, in particular, experts of Industry 4.0, leading technology engineers of IIoT/IoT solutions manufacturers, maintenance engineers, and cyber security specialists. During the lively discussions at more than 20 venues of the forum, the technical characteristics of the best IIoT/IoT solutions, advantages and disadvantages of communication and management technologies, and the features and recommendations for implementation and scientific and technical support were discussed in detail.

---

[21] http://iotworldsummit.ru/.

The significance of IoT development

International Forum IoT World Summit Russia (2017) was opened by the plenary session that was dedicated to the advent of the fourth industrial revolution. During the speeches, attention was drawn to the fact that a digitalization has already been successfully adopted by many large international production and/or distribution companies. For example, Rolls Royce Company uses cloud technologies such as Big Data and IIoT/IoT to optimize aircraft engines to reduce fuel consumption, which is the main expense of airlines, and the industrial automation manufacturer Rockwell Automation applies these technologies to the operational control of the supply chain, which allows identifying the emerging problems and respond them in a timely manner. Typically, the Internet of Things (IoT) or Industrial Internet of Things (IIoT) refers to any computer system or network that connects the surrounding real-world objects and virtual objects. In other words, it is a set of sensors and other IIoT/IoT devices connected by special communication channels. The typical components of some IIoT/IoT solution are described below (Figures 1.23 and 1.24):

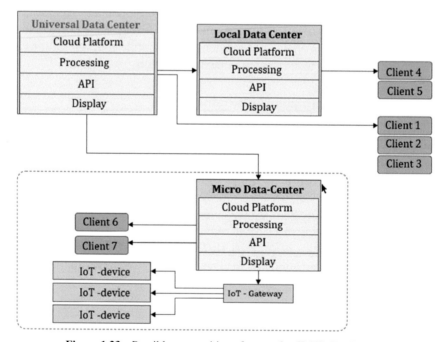

Figure 1.23   Possible composition of a complex IIoT/IoT solution.

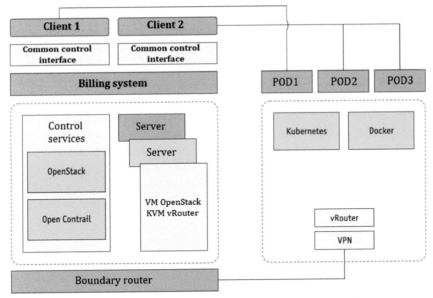

Figure 1.24 Block diagram of the IIoT/IoT-solutions Control Center.

- End IoT-devices located on the fog infrastructure;
- IoT–gateways for IoT–device data aggregation;
- Central node of IoT infrastructure;
- Communication channels and IoT devices communicating with the IoT gateway and the central node.

IoT devices are the end devices (sensors) located directly on the real-world automated objects. IoT-devices are able to capture the data about the real-world objects (e.g., gas sensor) and modify their activity (e.g., CNC machines) depending on the goals, objectives, and implementation. Data exchange between central node and IoT device is conducted by the communication channel (often wireless), transmitting the collected data and receiving control commands through IoT gateways or directly through the IoT application. According to Juniper Research, in 2017, the number of different IIoT/IoT devices (sensors, devices, field devices, etc.) was approximately 20 billion units, and by 2021, this rate is expected to increase to 46 billion units. In addition, a 25% annual increase in the number of machine-to-machine (M2M) connection is expected.

IoT gateways are located directly at the IoT device data collection point. The data collection points are situated in the communication nodes close to the sets of IoT devices, for example, in a manufacturing unit, in a bank, in an agricultural enterprise, in an apartment building, and in other places where a sufficient number of IoT devices and communication channels are in use. IoT gateways are cross-platform and may act like a host for IoT-devices (sensors) based on different hardware platforms. Besides, IoT gateways are the elements of IoT infrastructure and do not belong to the specific IoT application and can be used by multiple IoT applications (including those belonging to the different customers) simultaneously. This capability is provided by the use of micro-service segmentation, including the use of Docker-containers, and support of multitasking performance. The central node is a data center of one of three possible types. Data centers of the first type (universal (centralized) data centers) with a virtualization environment are based on the Open-Stack platform. Here, the virtual infrastructure is characterized by the following features:

- Typical components;
- Automatic deployment mechanism for virtual environment and its components (orchestration);
- Ability to monitor the state of the system;
- Flexible deployment of the various types of virtual server (including images from other virtualization environments);
- Advanced ways of providing fault tolerance (Live migration, VM evacuation, etc.);
- CDN and object storage support.

Data centers of the second (local data centers) and the third types (last mile data centers) are based on the same technological principles. Platforms KVM or OpenVZ are used as virtualization environments. Solutions similar to the data centers of the first type can be deployed (such as Kubernetes, alternative orchestrators, dockerize systems, etc.).

IoT control centers development

Data centers of traditional architecture not able to provide physical scaling and programmatically configure according to the performed tasks do not provide the required values of control, stability and productivity [31–35] because of the variety of devices and applications in IoT infrastructure and different types of data processing. Moreover, the construction of IoT/IIoT infrastructure control systems based on the known cloud platforms (Azure Microsoft, AWS Amazon, Google Cloud Platform, etc.) (Inset 1.5) leads to a conflict with the declared principle of independence from existing IIoT/IoT hardware–software solutions. A possible way to resolve this conflict is to use open-source solutions. Thus, the development of modern information technologies leads to the need to revise the known concepts of IIoT/IoT infrastructure management based on the appropriate data reduction centers [31–33, 36–38].

Inset 1.5

Famous cloud platforms AWS IoT for IoT control is a sort of the well-known cloud platform AWS from Amazon that nowadays dominates with 31% market share on the global market of the cloud service providers, outstripping the similar platforms of Google, Microsoft, and IBM[22]. This platform allows the integration of a typical IoT/IIoT devices into the cloud and it is a complete solution for developing, operating, and maintaining the related applications with the assistance of special service AWS Lambda, Amazon Kinesis, Amazon S3, Amazon Machine Learning, Amazon DynamoDB, Amazon CloudWatch, AWS Cloud-Trail, and Amazon Elasticsearch Service. Most of the known protocols are supported, including HTTP, WebSockets, and a simplified MQTT network protocol for working in low-bandwidth networks. However, belonging to the Amazon production and the need to meet certain requirements for cloud integration severely narrowed the potential range of available IIoT/IoT applications and devices. Currently, the number of users of the AWS IoT platform is just over 150 companies[23].

IBM Watson has advantages from competitive solutions with the ability of cognitive data processing with elements of forecasting and generating new knowledge in the subject area. This platform is based on modern cognomorphic and neural models, as well as methods of deep machine learning, predicting and recognizing images and patterns. Furthermore, Amazon and IBM offer a model with a closed license, but an open API for accessing services, which greatly limits the flexibility of deploying the platform. In 2016, IBM opened a global headquarters in Munich for the development of new technologies, Watson Internet of Things (IoT), based also on using blockchain and digital currency.

(Continued)

---

[22]https://www.canalys.com/newsroom/cloud-service-providers%E2%80%99-battle-drivesworldwide-cloud-infrastructure-market-42-q1-2017-0/.
[23]https://discovery.hgdata.com/product/aws-iot/.

Inset 1.5 Continued

Today, Watson Internet of Things (IoT) is a set of constantly developing services that allow organizing cognitive calculations for large-scale IoT-applications (more than 6,000 customers). For example, in the IBM project for the VISA payment system, personal assistants and POS terminals are connected to the platform as an IoT device. As a result, VISA users have access to payment services through the IBM cloud[24].

Ayla Networks, another well-known company, offered the industry first IoT platform based on Agile[25]. This multifunctional platform includes the following components: Ayla Embedded Agents, Ayla Cloud Services, and Ayla Application Libraries that allow connecting a standard set of IIoT/IoT devices to the cloud and deploy the appropriate applications for data collection, processing, and presentation.

According to DriveScale, currently the most common trend in the development of cloud computing is the tendency to aggregate, i.e., hyperconvergence. Dynamically developing market of IoT requires precisely hyperconvergent solutions. At the same time, Drive-Scale analysts note a new trend toward disaggregation. The infrastructure will continue to scale horizontally and the computing node will continue to be its base element, but the user will be able to add only the functionality that he needs, and this functionality will be simultaneously provided by a pool of nodes. This means that all applications will be able to access the memory pools, processing capacity, or capacity pools for data storage. Thus, the only object of virtualization is a network infrastructure, and at the next stage, it should be disaggregated so that the user can use only the amount of network infrastructure required by the applications from the shared pool. The answer to the described trends is the hybrid architecture of cloud solutions. Currently, hybrid clouds include solutions for hosting customer's own equipment at the data center facilities (colocation), solutions for shared hosting virtualization systems (shared hosting), solutions for dedicated virtual servers, mechanisms for providing virtual infrastructure to the customers, or off-the-shelf software as a service.

---

[24] https://iot.ru/riteyl/ibm-i-visa-dlya-transformiruyutiot-ustroystva-v-pos-terminaly-i-personalnye-pomoshchniki/.

[25] https://www.aylanetworks.com/newsroom/pressreleases/IoT-Platform-Provider-Ayla-Networks-Announces-39-Million-in-Series-C-Financing/.

It should also be noted that today the tendency to refuse the use of a proprietary software, in favor of free software, is becoming especially acute. Joe Hillerstein, a professor from Bentley University (USA) and co-founder of Trifacta Company, notes that companies increasingly adhere to the principle of openness and dynamically transfer applications and data from proprietary platforms to free ones. According to Forrester analyst Jeffrey Hammond, 76% of developers use some form of technology with open-source code. According to the data from Huawei, the most popular cloud service for today is IaaS (Infrastructure as a Service). And only the small part of companies consider PaaS (Platform as a Service) as a guide for the future. The development of the market shows that the open platform OpenStack becomes a leader on the market in the provision of virtual infrastructure services.

Business customers can choose the different solutions for infrastructure virtualization from a wide range of products that support OpenStack, to ensure independence from the software manufacturers and equipment. Thus, today, to support the operation of IoT infrastructures, it seems relevant to create a technologically integrated network of data centers built on a single virtualization platform. A key feature of that network is to present a special flexibility, scalability, and functionality for IoT applications located in data centers.

Thus, at present, all free types of data centers for IoT mentioned above are in high demand.

Type 1: Universal (centralized) data centers. These data centers will be core nodes of some unified distributed network of data centers. The provision of the typical virtual servers, including standalone and dedicated ones, provision of infrastructure services (IaaS), virtual servers with bundled software (SaaS), etc., will be the basic services of such data centers. It will be necessary to provide a proper scaling of computing capacity and a capacity for data storage, as well as scaling of communication channels.

Type 2: Local data centers. The main task of these data centers will be to bring the computational performance geographically closer to IoT providers located in the different regions of the Russian Federation. The main services of such data centers will be various SaaS-solutions (e.g., Billing solutions (also for communications providers) and CRM (customer relation management), EDMS (Electronic Document Management System), streaming media systems, ERP solutions, services to ensure the fault tolerance of the client infrastructure (including Hybrid/Private Cloud solutions), data backup service (Backup As A Service), etc.

Type 3: Last mile data center. Some small typed data centers for local data processing in various areas of the digital economy of the Russian Federation.

Development of the wireless standards for the Internet of Things

There are two main types of wireless technologies for the transmission of small scale data for a long distance (LPWAN): NB-IoT and EC-GSM-IoT developed by GSM Association and 3GPP implying the use of licensed spectrum of LTE-M and alternative technologies (LoRa, Sigfox, etc.) that work in unlicensed spectrum. With regard to technologies that use licensed spectrum and provide trusted environment for data acquisition and processing from IIoT/IoT devices, NB-IoT technologies have become very popular (by the end of July 2017, about 25 commercial networks NB-IoT (Inset 1.6) were launched in seven countries) and LTE-M (four commercial networks in two countries) (Figures 1.25 and 1.26, Table 1.6). The operators of KPN, KDDI, NTT DoCoMo, Orange, Telefonica, Telstra, TELUS, and others announced plans to deploy LTEM networks.

Inset 1.6
Narrow-Band IoT (NB-IoT) technology characteristics

In 2016, the international consortium 3GPP has launched a new Narrow-Band IoT (NB-IoT) communication standard in Release 13 (LTE Advanced Pro) operating in a bandwidth of 200 kHz width. NB-IoT is a wireless narrow-band version of the Low Power Wide Area (LPWA), intended for inter-machine interaction applications (M2M). The main manufacturers of equipment using NB-IoT technology include well-known foreign companies such as Huawei, Qualcomm, Intel Corporation, Nokia Networks, Verizon, Samsung Group, AT & T, and others.

The 3GPP consortium offers three Options for deploying the NB-IoT network: 1) NB-IoT Guard Band–a separate frequency spectrum is being detached for Narrowband IoT

2) In Band–the technology is placed in the "guard" frequency channel spacing of LTE networks

3) Stand Alone, NB-IoT, and LTE operate in the same bandwidth. Thus, NB-IoT can be deployed in the bandwidth GSM standard operates in after their refarming into LTE or in the "safeguard" between GSM and LTE.

Inset 1.6 Continued

Essentially, the maintenance and operation of solutions based on NB-IoT is simpler and cheaper than the advanced LTE and GSM networks have to date. Up to 100,000 NB-IoT devices can be connected to one cell of the base station, which is tens of times higher than existing mobile communication standards and opens up wide possibilities for the application of the Big Data technologies.

The NB-IoT standard allows increasing the coverage area significantly, while the NB-IoT data transfer rate reaches 200 Kbps, which is sufficient for devices transmitting the same type of small data.

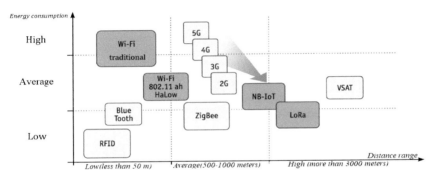

Figure 1.25 Evolution of communication standards for IIoT/IoT.

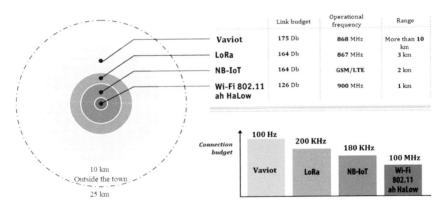

Figure 1.26 Comparison of known communication technologies for IIoT/IoT.

Table 1.5   Cost estimates of NB-Fi and NB-IoT ownership

Settings	NB	IoT
Availability of operators in the 4th quarter of 2018	Yes	
Forecast coverage in the 4th quarter of 2018	Less than 25%	
Monthly subscriber fee	20 Rub	
Extending the coverage of the blind area	Yes	No
Possibility of own network	Yes	No
Extra cost of the device	100 Rub	1,000 Rub
Cost of a cheap base station	5,000 Rub	500,000 Rub
Cost of a premium base station	50,000 Rub	2,000,000 Rub
Additional cost of the solution for 1,000 devices for the 6 years term of meters	150,000 Rub	2,440,000 Rub

In Russia, NB-IoT and LoRaWAN technologies are considered as more advanced ones (Figures 1.27–1.32), although "local" technologies such as LPWAN-NB-Fi (Narrow Band Fidelity), "Strizh" and others are also developing. Thus, in July 2017, the Association of the Internet of Things[26], which includes 34 companies and a number of leading research institutes and universities such as the Bauman MSTU, OJSC "MTT", JSC "Kaspersky Lab", etc., developed and proposed NB-FI, the draft of the Russian standard for communication for the Internet of Things to Rosstandart. It should be noticed that devices applying this standard will operate on the 868 MHz frequency, that is, in an idle band for either Russia or Europe (data transmission permits or licenses at this frequency are not required). NB-FI transmitters will communicate at a distance of 10 km and operate

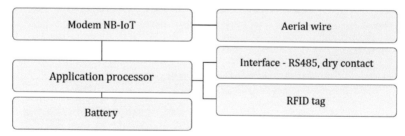

Figure 1.27   Telematic module scheme in operation in IoT network.

---

[26]https://iotas.ru/.

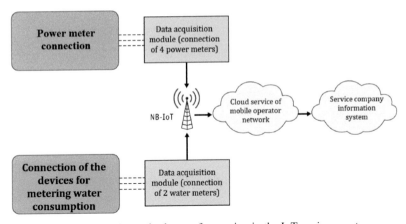

Figure 1.28 General scheme of operation in the IoT environment.

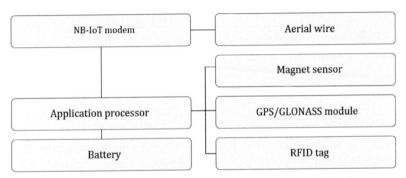

Figure 1.29 Telematic module scheme for the "smart parking".

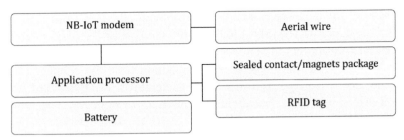

Figure 1.30 Telematic module scheme for the connection of sensors for opening doors and windows.

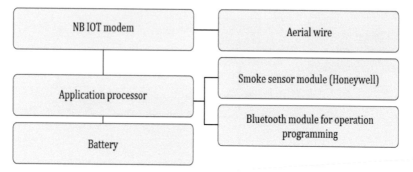

Figure 1.31 Telematic module scheme for smoke sensor.

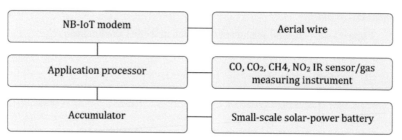

Figure 1.32 Telematic module scheme for CO, CO_2, CH_4, NO_2 gas concentration control.

on a single battery with a service life of 10 years. It is assumed that the radio sensors will be widely connected to a variety of devices: cars, household appliances, utility meters (water, electricity, and heat), security systems, home devices in a smart home, gaming devices, wearable device, etc. It is implied that all of these devices will "communicate" with each other and "coordinate" the actions. At the same time, approximate expenses for the support and development of the technology will be the following: the cost of the NB-FI standard base station is 100,000–150,000 Rub, the cost of the radio module for the device connection to the network is approximately 800 Rub, the cost of controllers for data transmission and acquisition from the counter is up to 200 Rub, and a battery–50–100 Rub. It is unknown what standard will become a leader in Russia, but there is something to fight for. The funds allocated for development are very large: in 2016, Russian customers spent 85 billion rubles (about 1.2 billion dollars) on products and services related to IoT, and this rate increased by 42% for the last year. At the end of 2016, more than 10.7 million IoT devices were connected to the Russian cellular networks (+ 60% per year).

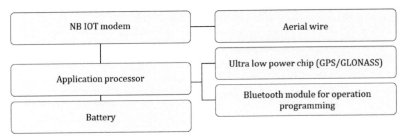

Figure 1.33 Telematic module scheme in operation in GPS/GLONASS systems.

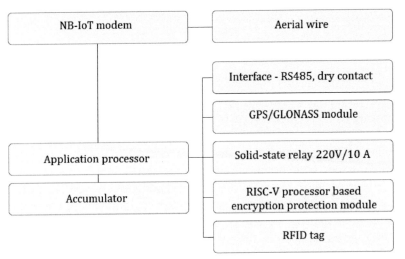

Figure 1.34 Telematic module scheme for encryption work defined in the standard GOST.

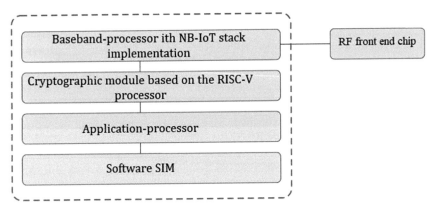

Figure 1.35 Chip scheme with encryption support defined in the standard GOST.

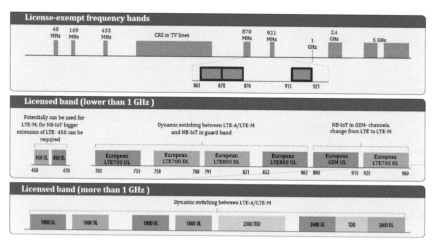

Figure 1.36 NB-IoT for nonstandard bandwidth.

According to IDTechEx forecasts, more than 40 billion devices will be connected to LPWAN networks in the nearest 10 years. At the same time, smart manufacturing, smart houses and cities, control over capital resources and tangible assets, agricultural efficiency monitoring, and many more will be the main application domains of these technologies. According to Riot Research, in 2017–2023, LPWAN market will grow in monetary terms with a CAGR of 45% and by the end of the forecast period, it will reach 19.1 billion USD. At the same time, the part of various LPWAN technologies will be changed significantly. Thus, the part of NB-IoT and LTE-M technologies using the licensed spectrum will be significantly increased, and as a result, they will dominate in the number of connections by the end of the forecast period. According to the evaluation of Lux Research's analysts, NB-IoT leadership will be due to a wider geographical coverage in comparison with other LPWAN technologies (Sigfox, LoRaWAN, etc.) and also because of the "size effect" – consolidated revenues of the operators considering NB-IoT as the priority of LPWAN technology amounted to 578 billion USD against 416 billion USD from operators LoRa and Sigfox in the aggregate. As a result, NB-IoT is projected to account for more than 90% of connections in the world by 2022. At the same time, LoRaWAN, most likely, will supplement NB-IoT in some applications.

Examples of the best domestic practices

It is well known that the implementation of innovations is easier if you have an example of how the others experience. Familiarization with the best practices of implementing the Big Data and IIoT/IoT technologies became the main goal of the IoT World Summit Russia-2017 forum. At the same time, as it turned out, domestic specialists already have something to share with their foreign colleagues. First, we are implementing a pilot project on NB-IoT together with MegaFon company and the city of Innopolis. The pilot zone has already been deployed. Second, the LoRaWAN technology pilot zone was deployed. Third, a joint protocol between Rostelecom, Tattelecom, and the Ministry of Informatization and Communications of the Republic of Tatarstan, concerning the test zone of 5G in Innopolis, was signed. Also the so-called smart city technologies (smart city) was discussed during the IoT World Summit Russia-2017 forum. According to Elena Dolgikh, CEO of SGM, the emergence of smart cities has become a response to the increase in the efficiency of urban services due to a very rapid urbanization.

Also LoRaWAN, a promising wireless technology, whose main advantages are low cost of device operation, long range of signal transmission (up to 15 km), low power consumption of radio modules, as well as developed mechanisms for ensuring information security, was discussed within the framework of the forum. The main areas of application of LoRaWAN include fire safety monitoring, traffic lights and street lighting control, gas and liquid pressure monitoring, traffic monitoring, etc. LoRaWAN data transmission technology allows the installation of transmitters in various devices, for example, in the utility meters (water, electricity, and heat meters). Also this technology can be implemented for the control of domestic manufacture in the agro-industrial sector, oil and gas production and processing, energy inputs production, etc. It is possible to design small transmitters and to embed them in the form of a sticker.

At present, the techno-economic paradigm is being replaced by Industry 4.0, a new technology revolution that is based on modern technologies of Big Data and Internet of Things (IoT). First, a large-scale "digitalization" of all spheres of human activity and significant development of the known methods of analysis and synthesis of systems with prescribed properties are expected, and then a qualitative and quantitative change of the world is predicted. At the same time, the volume of IP traffic will increase significantly, from 1.5 to 4.7 EB per year, as well as the actual data, from 175 to 600 EB per year[27]. It is interesting that most of the digital data will be generated not by people, but by the various hardware and software components of the IoT/IioT infrastructure, as well as by various computing devices, sensors, and smart sensors.

---

[27]http://www.cisco.com/c/dam/en/us/solutions/collateral/serviceprovider/global-cloud-indexgci/white-paper-c11-738085.pdf.

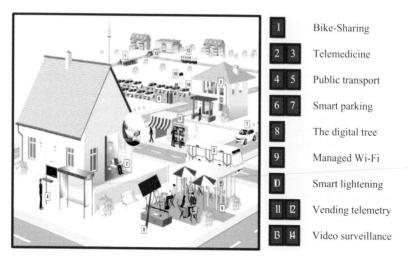

1	Bike-Sharing
2 3	Telemedicine
4 5	Public transport
6 7	Smart parking
8	The digital tree
9	Managed Wi-Fi
10	Smart lightening
11 12	Vending telemetry
13 14	Video surveillance

Figure 1.37   NB-IoT for nonstandard frequency bands.

According to Forrester analysts, the IIoT/IoT technology will have a significant impact on the digital economy of some technologically advanced countries of the world[28]. The tructural and functional complexity of modern IoT/IIoT systems will increase by several orders of magnitude. Cognomorphic models and neural networks [32–34, 36–38], as well as the other models and methods of artificial intelligence [31, 35] will play an increasing role here. Dockerizing and micro-service data processing technologies, implemented in a distributed computing environment, will be the basic technologies for building IoT infrastructures. Cloud (and fog) IoT control centers will be improved together with the development of gateways and IIoT/IoT end devices. In the face of growing threats to information security, the task of properly ensuring cyber security of the critical IIoT/IoT infrastructure of the Russian Federation will also require the solutions.

1.2 Typical Cyber Security Threats

1.2.1 Possible Scenarios of Cyber-Attack on the Information Infrastructure

The technique's toolkit that allows cybercriminals to hack the TCP networks is constantly growing and renewing. In this case, the main attack targets

---

[28]https://www.forrester.com/report/Predictions+2017+Security+And+Skills+Will+Temper+Growth+Of+IoT/-/E-RES136255/.

are interception (and, possibly, modification) of data transmitted through the network from one node to another, impersonation (an intruder node represents another node to take any privileges of impersonated node), unauthorized network connection, unauthorized data transmission (avoiding of the IP traffic filtering in networks protected by the firewalls), node forcing to transfer data at enhanced rate, and bringing the node into a state when it is not able to properly function, transmit, and receive the data (the so-called DoS–denial of service). To achieve their goals, intruders use sniffing, network scanning, and packet generation [14, 18, 23–25, 39, 40].

TCP network attack techniques
Sniffing
The Ethernet sniffing (the vast majority of local and federal networks apply this technology) is a trivial task: the only thing needed is just to translate the interface into a promiscuous mode. The programs that not only record all traffic in the Ethernet segment, but also filter it by the established criteria are easily accessible: for example, the freely distributed tcpdump or snoop programs.

Other network technologies are susceptible to FDDI and radio sniffing (e.g., Radio Ethernet). It is slightly more difficult to extract traffic from the dedicated and telephone lines, mainly due to a difficult physical access and a connection to such lines. However, it should be remembered that an intruder can attack and control an intermediate router and thus gain access to all transit traffic, regardless of the technologies used at the network access level.

It is possible to limit the Ethernet sniffing area by the network segmentation using the switches. In this case, without taking active actions, an intruder can intercept only the frames received or sent by the segment nodes to which he is connected. The only way to fight the Ethernet segment sniffing is data encryption. An intruder, who sniffs a network, can be detected using the AntiSniff system utility, which detects the network nodes whose interfaces were switched to sniffing mode. Note that the idea of sniffing as a safe activity that cannot be found does not correspond to a reality.

Network scanning
The network scanning aims to identify the computers connected to the network and determine the network services running on them (open TCP or UDP ports). The first task is performed by sending "Echo" ICMP messages by the *ping* program with a serial search of the network node addresses.

For greater security, an intruder can significantly stretch the scanning process in time ("slow scanning"); the same is true for TCP/UDP ports scanning. Also, an intruder can use "inverse mapping": in this case, not "Echo" ICMP messages are sent to the tested addresses, but other messages, for example, TCP RST-segments, answers to the nonexistent DNS queries, etc. If the tested node does not exist (turned off), an intruder will receive the message "Destination Unreachable: Host Unreachable" in response to the ICMP. Therefore, if this message was not received, then the corresponding node is connected to the network and works.

The *traceroute* program will help to determine the network topology and detect routers. Scanners, such as the *nmap* program, are used to define which UDP or TCP applications are running on the detected computers. Since the port numbers of all the main services are standardized, after determination, for example, that port 25/TCP is open, it is possible to conclude that this host is an e-mail server, etc. An intruder can use this information to attack at the application level.

Host TCP port scanning is performed in several ways. The simplest of them is to establish a TCP connection to the tested port using the *connect ()* function. If the connection was established, then the port is open and a server application is connected to it. The advantage of this method is the ability to perform scanning by any user and even without any special software: the standard *telnet* program allows specifying an arbitrary port number to establish a connection. A significant disadvantage is the ability to track and register such scanning: when analyzing the system log of a scanned host, numerous open and immediately interrupted connections are discovered, resulting in taking measures to increase the security level.

Half-open scanning does not have the described drawback, but it requires an intruder to create single TCP segments to bypass a standard TCP module (or, if already developed programs are used, at least superuser rights). In that mode, an intruder sends an SYN segment to the scanned port and waits for a response. Receiving the response segment with SYN and ACK bits means that the port is open; receiving a segment with the RST bit means that the port is closed. Having received SYN + ACK, an intruder immediately sends a segment with the RST bit to the detected port, thus eliminating the connection attempt. Since the connection was never opened (no ACK was received from an intruder), it is much more difficult to register such scanning. The third way is to scan with FIN segments. In this case, a segment with the FIN bit set is sent to the scanned port. The host must respond with the RST segment if the FIN segment is addressed to a closed port. FIN segments sent to a port in the

LISTEN mode are ignored by many TCP/IP implementations (the standard requires to send RST segments in LISTEN mode in response to segments that have an unacceptable ACK SN; nothing is said about the segments that have only the FIN flag). Therefore, nonresponse indicates that the port is open. The scanning method variants are to send segments with FIN, PSH, and URG flags (*Xmas scan*) or to send segments without any flags (*Null scan*). Naturally, the scanning with SYN segments provides more reliable results, but many firewalls may not allow SYN segments to be passed from the Internet to the internal network, skipping any segments without the SYN flag (e.g., the connections of the Internet hosts with internal hosts that are initiated from the Internet and forbidden, but the connections, which are initiated internally, are allowed).

The *tcplogd* program can register scanning attempts in different modes. To determine open UDP ports, an intruder can send a UDP message to the tested port. Receiving the "Port Unreachable" ICMP message (type 3, code 3) in response means that the port is closed. The scanner program can also determine the scanned node operating system according to the node reaction to specially designed nonstandard packets, for example, TCP segments with meaningless flag combinations or ICMP messages of some types.

An intruder, directly connected to a network segment, can use simple sniffing to determine the addresses of computers connected to the network and running UDP or TCP services on them. This network scanning form is more secure than test datagrams sending.

Packets generation

The generation of datagrams or free form and content frames is performed as simply as Ethernet sniffing. The *libnet* library will provide a programmer with everything necessary to solve the problem. The *libpcap* library provides tools for the reverse action of the packets retrieving from the network and their analysis. On many Internet sites, there are already developed programs that intentionally generate packets to perform any attack or scan a network (e.g., the *nmap* program mentioned above). Such program use does not require intruder to be either a qualified programmer or to understand the network operation principles, which makes many of the described attacks, especially denial of service attacks, widely available.

Data interception

The simplest interception form is sniffing. In this case, an intruder can obtain the useful information: user names and passwords (many applications transmit them openly), computer addresses in the network, including server addresses and applications running on them, the router address, the actual transmitted data that can be confidential (e.g., e-mail texts), etc.

However, if the network is segmented by switches, an intruder can only intercept frames received or sent by the segment nodes to which he is connected. Simple sniffing also does not allow him to modify the data transferred between two other nodes. To solve these problems, an intruder must take action to imbed him into the data transmission route as an intermediate node.

Fake ARP replies

To intercept traffic between the A and B nodes located on the same IP network, an intruder uses the ARP. It sends fake ARP messages so that each of the attacked hosts considers an intruder MAC address as its correspondent address (Figure 1.38).

Here is a possible plan of cyber-attack.

An intruder determines the MAC addresses of the A and B nodes: *ping IP (A); ping IP (B); arp-a*

An intruder configures additional logical IP interfaces on his computer by assigning them A and B addresses and disabling the ARP for these interfaces so that they are not "audible" in the network, for example:

ifconfig hme0: 1 inet IP (A) -arp
ifconfig hme0: 2 inet IP (B) -arp

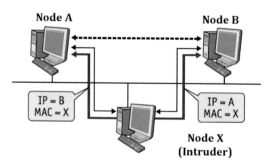

------▸   Traffic before fake ARP reports were sent
——▸   Fake ARP replies
——▸   Traffic after fake ARP replies were sent

Figure 1.38   ARP attack schema.

sets a static ARP table:

> *arp -s IP (A) MAC (A)*
> *arp -s IP (B) MAC (B)*

and allows the IP datagrams a retranslation (puts the computer into router mode).

An intruder sends a frame to the MAC address of the A node with a fake ARP reply, which informs that the B IP address corresponds to the X MAC address. In a similar way, the B node is reported that the same X MAC address corresponds to A IP address (Figure 1.23). The *libnet* library can be used to generate messages. Since the ARP is stateless, A and B nodes will receive ARP replies, even if they did not send requests. Then, an intruder periodically (it is enough to do so every 40 seconds) sends fake ARP replies to update the fake records in the ARP tables of the A and B nodes to keep these nodes unaware of the true addresses and prevent them from forming the corresponding ARP requests.

Fooled A and B nodes send their traffic through X node, assuming that they communicate directly with each other. An intruder is able to simply sniff traffic or change the transmitted data for his own benefit. The B node can be the network gateway in which the A node is located. In this case, an intruder intercepts all traffic between A node and the Internet. Also, an intruder may not transfer A node frames to B node at all, but impersonate B node, faking replies on its behalf and sending them to A node.

Network segmentation by switches is obviously not an obstacle to the described ARP attack.

In fake ARP replies, an intruder can specify a nonexistent Ethernet address instead of his MAC address. Later, he will be able to extract frames sent to this address from the network by sniffing or reprogramming his network card MAC address. It requires a little more effort, but it will help an intruder to completely cover up the tracks, since his own MAC address does not appear anywhere.

To detect ARP attacks, an administrator must maintain a MAC and IP address database of all hosts in the network and use the *arpwatch* program, which sniffs and notifies the administrator about the detected violations. If the network is segmented by switches, the administrator must configure them so that frames from all of the network segments are redirected to the segment where the administrator station is located, regardless of whom they are intended for.

A use of the static ARP tables, at least on the key nodes (servers, routers), will protect them from ARP attacks, however, at the expense of the overhead costs for maintaining these tables in an up-to-date state.

Fake router forcing

To intercept a traffic directed from some A node to another network, an intruder can force his address to the host as the router address, to which the data sent by the A node should be transferred. In this case, the A node will route traffic to an intruder's node, and the latter, after the data analysis and possible modification, will send it further to the real router.

Typically, a fake router forcing is performed by fake ICMP "Redirect" messages. In a fake message, an intruder announces his own address as the router address (Figure 1.39).

To eliminate the given attack possibility, it is necessary to disable the processing of "Redirect" messages on the host. Despite the fact that this action is contrary to the requirements for hosts, it looks quite reasonable, especially for the hosts in networks with a single gateway, but not all operating systems can support such a switch-off.

If an intruder wants to intercept the traffic between P network nodes and Q network nodes, and he is located not in any of the P or Q networks, but on the route between them, then he can try to fool the routers. Routers do not respond to ICMP "Redirect" messages, so for a successful attack, it is

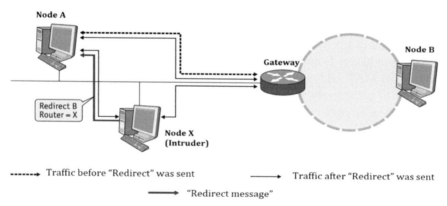

Figure 1.39   Fake router forcing by ICMP "Redirect".

necessary that they use some routing protocol. In this case, an intruder can generate fake routing protocol messages to switch the required routes to himself. For example, X node accepts broadcast RIP messages sent by nodes A (vector P = 3) and B (vector Q = 2) and sends a message with vector Q = 1 to the individual A router address, and a message P = 2 to individual B address (Figure 1.40).

A situation is possible when the value of a vector declared, for example, by the B router, Q = 1. In this case, X cannot immediately suggest the best route, but is able to apply the following technique. First, by selecting a pause in sending of RIP messages by B router, X on behalf of B sends to A Q = 16 vector, which will make A router remove the route to the Q network from its table, since before that A sent datagrams to Q via B. Immediately after that, X sends the Q = 1 vector on its own behalf, and A sets the route to the Q network via X. The subsequent Q = 1 vectors from B will be ignored because they do not offer a better route. In the same way, the EIGRP and OSPF protocols can be attacked.

For the performance of this attack, the X node must be on the same network with A and B routers, so placing hosts in the transit networks should be avoided. However, the X node can be a hacked router, so the host's absence in the transit network does not mean that it is impossible to perform the described attack.

A protection against such attacks is the authentication of routing protocol messages, for example, by a national algorithm similar to MD5. A BGP feature in the context of the discussed attack is the TCP use and the BGP

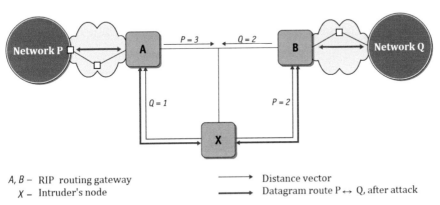

A, B – RIP routing gateway
X – Intruder's node

——→ Distance vector
——→ Datagram route P ↔ Q, after attack

Figure 1.40 Forcing a fake X RIP router to intercept traffic between P and Q networks.

neighbors IP locations. Thus, to introduce a fake routing information into the BGP database of A router, an intruder cannot act on his behalf or on behalf of a nonexistent node: he must impersonate B router (subsequently, the latter must be somehow eliminated from the process). Two possible attack schemes are noted on BGP.

BGP neighbors from the different autonomous systems should be available to each other directly, which means, to be in the same IP network. Therefore, to perform an attack, an intruder must also be in the same network. Technically, such situation is possible, but practically it is unlikely. Since it is more probable that the links between autonomous systems are dedicated channels that are physically inaccessible to anyone except for the network administrators. The "dedicated channels" here describe not the network technology that is used (it can be an Ethernet network, as in many traffic exchange points), but the network physical isolation and the impossibility of the unauthorized connection to it.

In contrast to external BGP connections, the BGP neighbors in one autonomous system are not required to be in the same network. It opens much greater opportunities for an intruder to act. Depending on where he is located according to BGP routers, an intruder can apply the various impersonation methods. For example, taking into account that in order to send datagrams to each other IBGP, neighbors can use the results of the internal routing protocol, an intruder can pre-attack the internal routing protocol by closing traffic between networks where BGP routers are located (e.g., P and Q networks in Figure 1.40) and modifying the BGP connection data for his own purposes.

The malicious attack on the BGP seems complicated; nevertheless, such attacks are possible. TCP segments authentication will help to avoid difficulties.

Impersonation

Suppose that the A node exchanges IP datagrams with the B node, and the nodes identify each other by the IP addresses specified in the datagrams. Assume further that B node has special privileges when interacting with A, that is, A provides B some service that is not available to other Internet hosts.

An intruder on the X node, who wants to receive such service, must impersonate the B node. Such actions are called impersonation of B node by X node.

Speaking about the services, we refer to UDP or TCP; in other words, this is about the impersonation of UDP messages or TCP connections. Often, together with impersonation, an intruder makes denial-of-service attacks against the B node to exclude the latter from the network communication process.

The A, B, and X hosts can be arranged in a different way, according to each other, that affects which impersonation methods an intruder will use.

If an intruder can intercept traffic from A node to B, then the impersonation task is almost done. Receiving from the A node the UDP messages or TCP-segments addressed to the B node, an intruder does not transmit them to their destination, but instead he forms packets on behalf of B. At the same time, B node is unaware of the event, so there is no sense in denial-of-service attack. As a reminder, the traffic interception is possible when:

- A, B, and X are in the same IP network (ARP attack);
- A and X are in the same network, and B is in another (fake router forcing);
- A and B are in different networks, and X is on the route between them (or includes itself in the route by attacking the routing protocol).

In other cases, an intruder cannot intercept data transmitted from A to B. Still, nothing prevents him from sending to the A address fake datagrams on behalf of B, but A node will send the reply packets to B node, bypassing an intruder. An important circumstance in these conditions is whether the X node has the ability to sniff these reply packets or whether an intruder is forced to work blindly.

Let us consider the most difficult case: intercepting and sniffing of data sent from A to B are impossible (Figure 1.41). This case is the most common: X node is on a network that has nothing to do with A and B nodes and is not located between them (A and B can be located in one or in different networks).

It is emphasized that impersonation without feedback is meaningful only if it is enough for an intruder only to transfer data to the A node on behalf of the B node, and the following A node reply does not matter anymore. A classic example of such attack is when an intruder sends TCP segment to the login program port containing some command to the A node operating system. The A node executes this command, assuming that it came from the B node.

TCP connection impersonation without feedback

If UDP messages impersonation without feedback remains trivial (an intruder should only fake a datagram addressed from the B node to the A node and

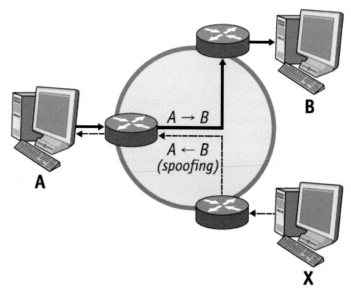

Figure 1.41   Impersonation without feedback.

send it to its destination), then in the case of TCP, everything is not so simple. Before sending a data segment to the A node, the X node must establish a connection with it on behalf of the B node. Recall how the connection is established: the X node sends the segment with the SYN bit to A, where the start ISN(B) number is defined. The A node responds to the B node with the SYN segment, by which it acknowledges the receipt of the previous segment and sets its initial ISN(A) number. An intruder will never receive this segment.

There are two problems here: first, the B node received from A the replies to the SYN segment that it has never sent, will send to the A node the segment with the RST bit, thereby cancelling an intruder effort. Second, the X node will still not be able to send the next segment to A (this should just be the data segment), because in this segment, the X node must acknowledge the receipt of the SYN segment from A, which means to put ISN (A) + 1 value to the "ACK SN" title field of its segment. But an intruder does not know the ISN (A) number because the corresponding segment has gone to the B node.

The first problem is solved relatively simply: an intruder performs a denial-of-service attack against the B node on the assumption that the B node was not able to process segments coming from A. For example, the B node can be attacked by a large number of SYN segments from the nonexistent nodes. However, to an intruder's relief, it may turn out that the B node is simply turned off.

To solve the second problem, an intruder must be able to predict ISN(A) values. If the A node operating system uses some function to generate ISN values (e.g., linear dependence on the system clock), then by opening several test connections in a row to the A node and analyzing the ISN values sent in the SYN segments from A, an intruder may try to define this function empirically.

A good TCP implementation should use random numbers for ISN values (a more detailed consideration of this issue can be found in RFC-1948). Despite the requirement deceptive simplicity, the problem of guessing ISN numbers remains relevant to this day.

So, below is the attack scheme of a TCP connection impersonation without feedback (Figure 1.42).

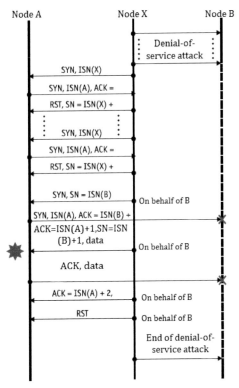

Figure 1.42 The attack scheme of a TCP connection impersonation without feedback.

1. An intruder disables the B node.
2. An intruder makes several trial attempts to establish connections with the A node to obtain ISN (A) values sequence from it. Immediately after the receipt of the SYN segment from A, an intruder breaks the half-established connection by sending a segment with the RST flag and, having analyzed the obtained ISN (A) values, determines the formation law of these values.
3. An intruder sends an SYN segment to A on behalf of B.
4. The A node responds to the B node with its SYN segment, confirming the SYN receipt from B, and indicates the ISN (A) value for this connection. An intruder does not see this segment.
5. On the basis of the previously obtained data, an intruder predicts the ISN (A) value and sends to A segment on behalf of B containing an ISN (A) + 1 acknowledgment and data for the application process. Having received this segment, the A node considers the connection with B installed and transfers the received data to the application process. The attack goal is achieved. The data can be, for example, the command that A executes, because it came from the trusted B node.
6. The A node sends an acknowledgment of the data receipt and, possibly, its data to the B node. An intruder will not receive these segments, but they do not interest him (under the task conditions). To close the connection correctly, an intruder can blindly send to the A node on behalf of the B node confirming the receipt of one octet (ACK SN = ISN (A) + 2) and then send the segment with the FIN flag. Thus, the data transfer channel from the X node (the same B) to the A node is correctly closed. In order to completely close the connection, an intruder must confirm the FIN segment receipt from A (as well as all the data that preceded this segment). Clearly, he cannot do this, because generally he knows neither the data amount nor the time of FIN segment sending from A. However, since the data transmitted from A to B do not have a value for an intruder, he simply sends to A a segment with the RST flag, thus completely closing the connection.
7. An intruder completes the "denial of service" attack against the B node.

TCP connection desynchronization

The X intruder located on the same network segment with the A and B nodes or on the route between A and B can desynchronize the TCP connection between A and B to establish full control over the connection. In other words, an intruder will be able to act on behalf of both A and B.

A TCP connection desynchronization will be defined there. When the connection is established, each A and B node knows which octets can be sent with which numbers to the correspondent at the moment: if the last acknowledgment sent by the A node was $ACK_{A \to B}$ and at the same time the A node announced the $W_{A \to B}$ window, then A expects from B the octets with $SN_{A \to B}$ numbers, falling into the defined window, that is:

$$ACK_{A \to B} \leq SN_{B \to A} \leq ACK_{A \to B} + W_{A \to B}$$

Similarly, in the B node, it is expected from A:

$$ACK_{B \to A} \leq SN_{A \to B} \leq ACK_{B \to A} + W_{B \to A}$$

If, for example, for any reasons the A node receives from B a segment with a number $SN_{B \to A}$, that does not fall into the window, then this segment is destroyed, and in reply A sends to B the segment with $SN_{A \to B} \leq ACK_{B \to A}, W_{A \to B}$, which would indicate to the B node, which octets A expects to receive. Note that, most likely this segment does not contain data, but nevertheless the number should be indicated, where $SN_{A \to B}$ is a number of the next data octet that A will ever send to B.

Suppose that an intruder somehow managed to bring down the counters of the A and (or) B nodes so that the above inequalities are no longer satisfied (how that can be done will be discussed below). Further, the notation of the form $SN_{A \to B(B)}$ will be used, which means "acceptable $SN_{A \to B}$ from the B point of view".

Now, if B sends a segment to A with a certain $SN_{B \to A(B)}$ number, adequate from the B point of view, but not already falling in the window in the A node, then A returns to the B node an acknowledgment with its value $ACK_{A \to B} = SN_{B \to A(A)}$. However, in this segment, there is $SN_{A \to B(A)}$ number, which now B considers as not falling in its window and sends to A $SN_{B \to A(B)}$, $ACK_{B \to A} = SN_{A \to B(A)}$ acknowledgment. As before, $SN_{A \to B(B)}$ number is unacceptable for A, and the A node again sends B an acknowledgment, and this cycle, called ACK storm, theoretically extends to infinity, but in fact it lasts until one of the ACK segments is not lost in the network. The stronger the storm, the greater the network load, the higher the percentage of losses and therefore, the faster the storm stops. So, in the desynchronized state, any attempt to exchange data causes only the ACK storm, and the segments themselves with the given connection members are destroyed.

At the time, an intruder, who knows the "correct" numbers from the both nodes, undertakes the intermediary functions (Figure 1.43). He sniffs

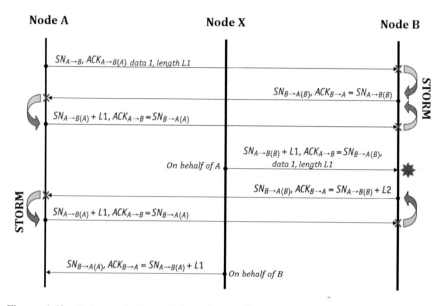

Figure 1.43 Actions of the malicious intermediary in the desynchronized connection between the A and B nodes (L1, L2–the amount of data in the forwarded segments).

the network, detects a segment with data of octets length L, directed, for example, from A to B, changes the number in it to the B node expected $SN_{A\to B(B)}$ number, and, after recalculating the checksum, sends a segment to B on behalf of A.

After that, an intruder sends an acknowledgment to the A node on behalf of B to this segment containing the correct $SN_{B\to A(A)}$ number from the A point of view. During this exchange, as a side effect, two ACK storms occur: the B node, which receives from A the original segment with $SN_{A\to B(A)} \neq SN_{A\to B(B)}$, initiates the first one and the second occurs when the A node receives from B a segment confirming the data receipt from X, and in this segment, $SN_{B\to A(B)} \neq SN_{B\to A(A)}$.

Clearly, an intruder does not launch an attack to simply retranslate the segments, which he can sniff. Nothing prevents him from changing the data contained in the segments or adding his own (Figure 1.43), which is displayed as "data 2" having a L2 length of octets, while the original data is denoted "data 1" with an L1 length of octets. For example, if there is a session of the telnet program and A sends a command to B, then an intruder can insert another command into this segment. He will receive his task result by sniffing the reply $SN_{B \to A(B)}$ segment, directed from B to A, which A does not perceive due to the mismatch of ordinal numbers, since $SN_{B \to A(B)} \neq SN_{B \to A(A)}$. But an intruder, removing the result of his task execution from this segment, will send what is left (i.e., the original task result) to A on behalf of B already with an acceptable serial $SN_{B \to A(A)}$ number.

An intruder can generally ignore the segments sent by the A node and send only his data to B, receiving replies by sniffing and reacting to them, but in this case, the A node will notice that B does not respond to its commands, and may be concerned.

Considering the way how an intruder can desynchronize a TCP connection.

Early desynchronization (Figure 1.44): an intruder, while sniffing, detects the moment when the connection between A and B is established, on behalf of A, disconnects by the RST segment, and immediately opens it again, but with new ISNs. The procedure will be examined in detail.

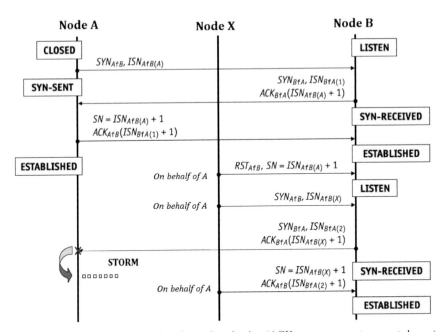

Figure 1.44 Early TCP connection desynchronization (ACK storm segments are not shown).

1. First, A sends to B $SYN_{A \to B}$, $ISN_{A \to B(A)}$ segment, then B corresponds to the $SYN_{B \to A}$, $ISN_{B \to A(1)}$, $ACK_{B \to A}(ISN_{A \to B(A)} + 1)$. Upon the receipt of segment, A passes to the ESTABLISHED status and sends B an $(ACK_{A \to B}(ISN_{B \to A(A)} + 1)$ acknowledgment segment.
2. At this point, an intruder sends A to B $RST_{A \to B}$ segment and then $SYN_{A \to B}$, $ISN_{A \to B(X)}$, segment, containing the same port numbers, but another $ISN_{A \to B(A)} = ISN_{A \to B(X)}$ number unacceptable to A.
3. When these segments are received, the B node closes the established connection with A, and then immediately opens it again and sends to A $SYN_{B \to A}$, $ISN_{B \to A(2)}$, $ACK_{B \to A} \times (ISN_{A \to B(X)} + + 1)$ segment, where $ISN_{B \to A(2)}$ is a new starting sequence number $ISN_{B \to A(1)} \neq ISN_{B \to A(2)}$.
4. The A node does not perceive this segment because the serial numbers mismatch, but on the behalf of A, an intruder sends to B the $SN_{A \to B} = ISN_{A \to B(X)} + 1$, $ACK_{A \to B} \times (ISN_{B \to A(2)} + 1$ segment.

After this, both A and B nodes are in the ESTABLISHED state, but the connection is desynchronized (Inset 1.7).

Inset 1.7		
Point of view	Sequent $SN_{A \to B}$	Sequent $SN_{B \to A}$
A node	$ISN_{A \to B(A)} + 1$ (A will send)	$ISN_{B \to A(1)} + 1$ (A is waiting to receive)
B node	$ISN_{A \to B(X)} + 1$ (B is waiting to receive)	$ISN_{B \to A(2)} + 1$ (B will send)

Let us note that, in violation of standards, some TCP implementations send the RST segment in response to receiving the RST segment. In this case, desynchronization by the described method is impossible.

Desynchronization with zero data (Figure 1.45): an intruder, waiting for the moment when the connection is in an inactive state (the data is not transmitted), sends to the A node on behalf of B and to the B node on behalf of A fake segments with data, thus causing desynchronization. It is necessary that the sent data is "zero", which means that the application-receiver should just ignore it and does not send any data in response. This desynchronization method is suitable for Telnet connections. First, they are often in an inactive state, and second, the Telnet protocol has a "no operation" command. The segment containing a random number of such commands will be accepted by the application and completely ignored.

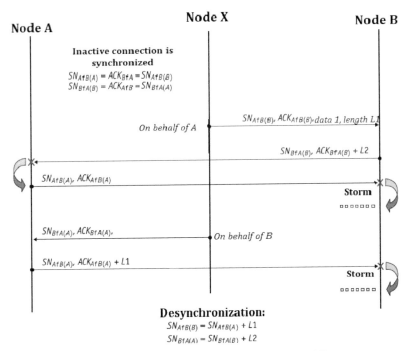

Figure 1.45 TCP connection desynchronization with zero data (ACK storm segments are not shown).

	Inset 1.8	
Point of view	Sequent $SN_{A\to B}$	Sequent $SN_{B\to A}$
A node	$ISN_{A\to B(A)}$ (A will send)	$ISN_{B\to A(B)} + L2$ (A is waiting to receive)
B node	$ISN_{A\to B(X)} + L1$ (B is waiting to receive)	$ISN_{B\to A(B)}$ (B will send)

After the described procedure, the following situation occurs (Inset 1.8).

Impersonation by desynchronization is a relatively simple and a very effective attack. It allows an intruder to establish the full control over a TCP connection without using fake ARP, ICMP, or routing protocols, without denial of service attacks, which can be detected by the network administrator or the hacked node. Such an attack detection is possible by sniffing for ACK storms.

To protect against the described attacks, the router (gateway, firewall) connecting the network to the external network must be configured to prevent skipping of the packets that:

- Arrive to the external interface, but have the sender address from the internal network;
- Arrive to the internal interface, but have the sender address from the external network.

The first case corresponds to the situation when the A and B nodes are located on the internal network, and an intruder is situated on the outside and tries to send a datagram to the A node supposedly from the B node. The second case corresponds to the situation when an intruder is in the internal network, and the A and B nodes are outside. It is emphasized that the proposed measures will not protect against all varieties of impersonation: for example, when the X node is in the same network as the A or B node, or all three nodes are located in the same network.

The good algorithm of random ISN generation protects from attacks in the absence of feedback, but is useless if an intruder can see segments transmitted from A to B.

In general, only data encryption or segment authentication can guarantee protection against impersonation.

Unauthorized network connection

For an unauthorized network connection, an intruder must have a physical capability for such connection. An intruder's next step is to configure the TCP/IP stack parameters of his computer.

Sniffing (of the network segment) will give an intruder a lot of useful information. In particular, he can determine which IP addresses network nodes have, and use ICMP "Echo" requests (ping) to determine which addresses are not used (or the computers are turned off). After this, an intruder can assign himself an unused address.

Finding the default router IP address is possible by interception of the frames with datagrams directed to IP addresses that do not belong to the network. These frames are directed to the router MAC address. Obviously, from time to time, the network nodes generate ARP requests about the router MAC address. The replies to these requests, sent by router, contain both its MAC address and its IP address. If an intruder knows the router MAC address and intercepts such responses, an intruder will determine the desired IP address.

An intruder can also use the ICMP "Router Advertisement/Solicitation" messages to detect a router.

An intruder can send an ICMP "Address Mask Request" message to determine the network mask. In response, a router must send the network mask in the "Address Mask Reply" message.

If the router does not support "Address Mask Request/Reply" messages, an intruder can apply the following simple method. By arithmetic calculations, he defines a minimal network, including its own address and the resulting router address, and then assigns himself this network mask. For example, let the address assigned by an intruder be X = 10.0.0.57, and the router address be G = 10.0.0.1, that is, writing the last octet in binary form:

$$X = 10.0.0.00111001$$
$$G = 10.0.0.00000001.$$

The maximum common part of both addresses (which means that the required minimum network, including both addresses) is:

$$N = 10.0.0.00XXXXXX.$$

Hence, N = 10.0.0.0/26, and the mask is 255.255.255.192.

All datagrams sent outside this minimal network will be transferred to the router. If the mask is not properly defined and, in fact, an intruder is in the network, for example, 10.0.0.0/16, and he sends the datagram to the node 10.0.1.1, the router will accept this datagram from an intruder and simply transfer it to the destination node on the same network.

Certainly, it is likely that an intruder will incorrectly determine the IP address to assign and he will be outside the network. In addition, the possible configuration of several IP networks in one Ethernet segment complicates an intruder's task. However, after a trial-and-error period, an intruder has all the chances to determine the necessary parameters for configuring his host.

If the network has a DHCP server that provides IP addresses to everyone, then it fully configures an intruder node without any effort on the part of the latter. This event will be registered in the server log.

To prevent an unauthorized network connection, the administrator must use a static ARP table on the router (and key hosts) and the arpwatch program. The static ARP table will not allow an intruder to get any datagram from the node that uses it, since an intruder's MAC address certainly is not in the table. The arpwatch program notifies the administrator about the node appearance with the unknown MAC address.

However, an intruder, having determined the computer IP and MAC addresses in his network, will wait for his shutdown (or he will launch a denial-of-service attack against him, which leads to the attacking host failure to operate), and all his possible actions will be attributed to the attacked host.

Unauthorized data exchange

In order to ensure internal (corporate) network security at the gateway, it is possible to use filters that prevent the passage of certain datagram types. The datagrams can be filtered by the IP addresses of the sender or receiver, by protocol (the IP datagram "Protocol" field), by the TCP or UDP port number, by other parameters, and by a combination of these parameters.

Let us regard two techniques that an intruder can apply to penetrate some filters.

Tunneling

Suppose that an intruder wants to send data from the X node to the A node outside his network, but filtering rules on the router prohibit sending datagrams to the A node (Figure 1.46). At the same time, it is allowed to send datagrams to the B node, which is also outside the protected network.

The B node is used by an intruder as a retranslator of the datagrams directed to A. To complete this action, it creates a datagram sent from X to B, in the "Protocol" field of which the value 4 ("IP") is placed, and as the data, this datagram carries another IP datagram, directed from X to A. The filtering router skips the generated datagram, since it is addressed to the allowed B node, and the IP module of the B node extracts from it the attached

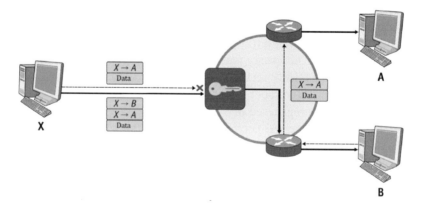

Figure 1.46   Tunneling through the filtering router.

datagram. Noting that the attached datagram is not addressed to B, the B node sends it to its destination, to the A node (Figure 1.46).

The described operation is called tunneling. For its implementation, an intruder may (or may not) have conspired with the B node, depending on what the node is and how it is configured.

The same result can be achieved by applying the IP "Loose/Strict Source Routing" option.

The datagram sender address cannot be hidden, so if the router does not also pass datagrams coming from A (i.e., it filters at the sender address), it is impossible to deceive it in the above way. In this case, an intruder has only one-way communication with the A node.

It is obvious that tunneling can be used in the opposite direction to penetrate to the protected network from the Internet. At the same time, the X node is on the Internet and the A and B nodes are in the protected network, and the B node is allowed to receive datagrams from external networks.

To protect against the tunneling, the router should be prohibited to transmit to the external network a datagram with a "Protocol" field = 4 and datagrams with options. Recall that the tunneling is used to connect networks that support multicasting to the MBONE network through the networks that do not support multicast routing. In this case, all routers of the protected (internal) network system must support multicast routing, and encapsulation and extraction of the group datagrams should be performed directly on the filtering router.

Tiny fragment attack

In the event that a fragmented datagram arrives at the filtering router input, the router examines only the first datagram fragment (the first fragment is determined by the value of the IP header field "Fragment Offset" = 0). If the first fragment does not satisfy the skip conditions, it is destroyed. The remaining fragments can be easily skipped without spending on them the filter computational resources, since without the first fragment, the datagram cannot still be collected at the destination node.

When configuring a filter, a network administrator often has a task to allow connections to TCP services initiated by the internal network computers, but to prohibit the connection establishment of the internal computers with external ones on the latter initiative. To solve the task, the filter is configured to block TCP segments that enter from the external network and have the SYN bit set. Segments without this bit are easily passed to the protected network, since they can refer to a connection already established earlier on the initiative of the internal computer.

Consider how an intruder can use fragmentation to bypass this restriction, that is, transfer a segment with an SYN bit from an external network to an internal one.

An intruder forms an artificially fragmented datagram with a TCP segment, with the first datagram fragment having a minimum data field size of eight octets (recall that the fragment sizes are indicated in eight-octet blocks). The datagram data field contains a TCP segment starting with the TCP header. The first eight octets of the TCP header contain the source and destination port numbers and "Sequence Number" field, but the flag values do not fall into the first fragment. Consequently, the filter will skip the first datagram fragment, and it will not check the remaining fragments. Thus, the datagram with the SYN segment will be successfully delivered to the destination node and it is transferred to the TCP module after the assembly.

A datagram example of two fragments

Fragment 1 (IP header is grayed out, there are eight octets of TCP header in the data field) (Inset 1.9, Fragment 1).

Fragment 2 (there is the rest of the TCP header with the SYN flag set in the data field) (Inset 1.9, Fragment 2).

Inset 1.9

Fragment 1

IP header
MF = 1, Fragment Offset = 0

Source Port	Destination Port
Sequence Number (SN)	

Fragment 2

IP header
MF = 1, Fragment Offset = 0

Acknowledgment Sequence Number (ACK SN) = 0							
Offset	Reserved	-	-	-	-	SYN	-	Window
Checksum							Urgent
Options							Padding

The penetration through the filter, described above, is called "Tiny Fragment Attack" (RFC-1858). Its use in other cases (to bypass other filtering conditions) does not make sense, since all other "interesting" fields in the TCP header and other protocols are in the first eight octets of the header and, therefore, cannot be moved to the second fragment.

To protect against this attack, the filtering router certainly should not inspect the content of all datagram fragments except the first one. This would be equivalent to assembling datagrams on the intermediate node that will quickly absorb all the router computing resources. It is enough to implement one of the two following approaches:

- Do not skip datagrams with "Fragment Offset" = 0 and "Protocol" = 6 (TCP) whose data field size is less than a certain value sufficient to accommodate all "interesting fields" (e.g., 20);
- Do not skip datagrams with "Fragment Offset" = 1 and "Protocol" = 6 (TCP): the presence of such datagram means that the TCP segment was fragmented in order to hide the certain header fields and that there is the first fragment with eight data octets somewhere. Despite the fact that in this case the first fragment will be skipped, the destination node will not be able to compile the datagram, since the filter has destroyed the second fragment.

The second fragmentation aspect, interesting from the security point of view, is the overlapping fragments. Consider an example of a datagram carrying a TCP segment and consisting of two fragments. The first fragment (the IP header is highlighted in gray, there is the full TCP header, without options, supplemented with zeros to a size that is a multiple of eight octets in the data field) (Inset 1.10, Fragment 1).

Fragment 2 (here is a part of another TCP header, starting with the ninth in the order of the octet, in which the SYN flag is set in the data field) (Inset 1.10, Fragment 2).

It is visible that the second fragment is superimposed on the first fragment (the first fragment contains octets 0-23 of the original datagram, and the second fragment starts with octet 8, because the "Fragment Offset" = 1). The behavior of the destination node that received such datagram depends on the IP module implementation. Often, when assembling a datagram, the data of the second overlapping fragment is written over the previous one. Thus, when

Inset 1.10							
Fragment 1							
IP header MF=1, Fragment Offset=0							
Source Port						Destination Port	
Sequence Number (SN)							
Acknowledgment Sequence Number (ACK SN)							
Data Offset	Reserved	-	ACK	-	-	-	-	Window
Checksum						Urgent Pointer=0	
0							

Fragment 2							
IP header MF=0, Fragment Offset=1							
Acknowledgment Sequence Number (ACK SN)							
Data Offset	Reserved	-	-	-	-	SYN	-	Window
Checksum						Urgent Pointer=0	
Options							Padding

assembling the datagram in the example, the TCP header overwrites the fields starting with "ACK SN", according to the values from the second fragment, and the result is the SYN segment.

If the first of the approaches, described above, is used to protect against "Tiny Fragment Attack" (the first datagram fragment inspection), then by applying the overlapping fragments, an intruder can bypass this protection.

A router using the second approach will successfully resist the "Tiny Fragment Attack" with the overlapping fragments.

Forced enhanced data rate

The congestion control mechanism implemented by the TCP allows the malicious data receiver to force a sender to transfer a data with a multiply enhanced rate. As a result, for his own needs, an intruder selects the sending server resources and the computer network, slowing or blocking the connections of other network interaction participants.

Attacks are carried out by the means of a specially organized transmission of data reception acknowledgments (ACK segments) by an intruder. All attacks described in this paragraph exploit the following tacit assumption embedded in the TCP: one TCP connection participant fully trusts the other party that it operates in strict accordance with the same protocol specifications as the first one.

In the following subparagraphs, we will assume that the A node sends data to the B node, and the latter tries to force the A node to transmit data at an enhanced rate.

Acknowledgments splitting

Let the A node, after the connection is established, be in a slow start mode. The *cwnd* overload window is equal to one *full-size* segment, which is sent to B (Figure 1.47). However, B instead of responding with one receipt acknowledgment of the whole segment (ACK SN = 1001) sends several acknowledgments with increasing ACK SN numbers (e.g., 300, 600, and finally 1001), as if confirming the receipt of the segment in parts.

The TCP standard is not violated, because TCP confirms the reception of the octets, not the segments. However, the slow start algorithm does not explicitly imply that the recipient sends no more than one acknowledgment on receipt of each next segment, and as a result, the *cwnd* window is increased

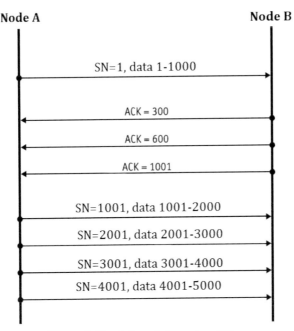

Figure 1.47   Acknowledgments splitting.

on one *full-size* segment with the arrival of each acknowledgment, regardless of how many octets it receives.

As a result, in the normal B node behavior in the next step, the sender would have increased *cwnd* window only on one segment and would have sent B the next two *full-size* segments (SN = 1001 and 2001). But the B node, having sent three acknowledgments instead of one, would cause the A node to increase the *cwnd* value to four and send four segments instead of two.

In general, if a *full-size* segment contains N octets, then the B node can send $M \leq N$ acknowledgments. Simple arithmetic calculations show that at the *i*-th step (i.e., the $RTT \times i$ time, where RTT is the access time), the A node will not send $N \times 2i$ octets, as it would be with the normal recipient behavior, but about $N \times M^i$ octets, if the B node takes care of declaring the proper recipient window size. The typical segment data field size is 1,460 octets. The most aggressive recipient behavior (1,460 acknowledgments for each segment) will theoretically result in the fact that after the third step, the B node can get 2.9 GB of data. Certainly, speed cannot grow infinitely–it will be limited by the capabilities of the network and the sender node.

Fake acknowledgment duplicates

Let us regard again the situation when the A node is in a slow start after establishing a connection. The A node sends one segment to B (SN = 1–1000), but the B node, instead of responding with ACKSN = 1001, responds with a series of fake ACKSN = 1 acknowledgments (Figure 1.48).

Acknowledgment duplicates preparation at the A node includes the fast retransmission and fast recovery algorithms. The latter is particularly opportune. Recalling that the fast recovery algorithm is based on the assumption that each received duplicate of the previous acknowledgment, except for the one that speaks about the segment loss, also implies that some other *full-size* segment has left the network. As a result, the *cwnd* window, measured in *full-size* segments, is temporarily increased by the number of duplicates received until the acknowledgment, different from the previous ones, is received.

Since acknowledgment duplicate itself does not carry any information about the receipt of which segment caused the sending of this duplicate, nothing can save the sender from the fake acknowledgment duplicates generated

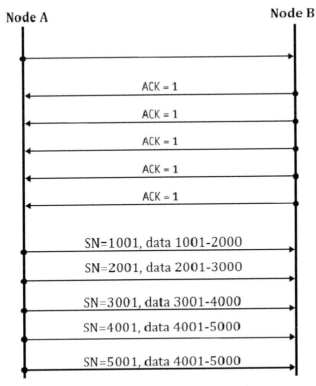

Figure 1.48   False acknowledgment duplicates.

by the recipient. Thus, the receiver causes the sender to unreasonably increase the *cwnd* window and enhance data sending.

In fact, the A node will send data at the rate at which B generates acknowledgment duplicate. At some point in the A node, the retransmission timer for SN = 1–1000 will trigger since it has not been confirmed. By that time, the B node will respond to this by sending a cumulative acknowledgment to all the data received and switch to generating fake duplicates of this last acknowledgment, again forcing the sender to unreasonably increase *cwnd*.

Premature acknowledgments

Another kind of attack is based on the fact that the recipient can send in advance the acknowledgment of the segments that are not yet accepted by them, making the sender to believe that the data has already been delivered, which will result in an increase in *cwnd* and early new data transfer.

Note that although the receiver can predict the sender segment limits (when sending a large data amount, the sender, as a rule, places the data in *full-size* segments that have the same volume) and accordingly form premature acknowledgments, high accuracy is not required. Acknowledgment of any data frame, as already noted above, leads to an increase in *cwnd* window and new *full-size* segment transfer. Moreover, if the recipient "overachieves" and confirms what has not yet been sent, the acknowledgment will simply be ignored by the sender and will not lead to any unpleasant consequences.

Unlike the two previous ones, the attack with premature acknowledgments destroys the mechanism of the data transfer reliability: if any of the data segments sent from A to B is lost on the route, this segment will not be retransmitted, since it was already previously confirmed by the receiver. However, the HTTP and FTP application protocols, by which most of the data is transferred in TCP/IP networks, provide the ability to request from the server not the entire file, but its specific parts (most servers support this feature). Therefore, using the described attack and having received the data body from the HTTP or FTP server at an enhanced rate, later an intruder can "patch holes" formed due to segments loss by partial transfer requests.

The described attacks are especially effective when transferring relatively small data amounts (files), when the entire file can be transmitted explosively at one time. Experiments have shown that the download file speed is increased by several times. The operation of competing TCP connections (available in the same communication channel) is practically blocked, because due to the sharply increased traffic intensity, other connections diagnose the congestion state and take appropriate measures to reduce the data transfer rate, effectively freeing the channel for an intruder.

To implement the described attacks, a relatively small (several dozen lines) modification of the TCP module on the B computer is required. To protect against the acknowledgment splitting, it is enough to refine the TCP module implementation: allow increasing *cwnd* only after receiving the acknowledgment covering the whole segment. Adequate protection against two other attacks does not exist. To solve this problem, an extra *cumulative nonce* field to the TCP header can be added, which will act as a segment identifier. Using this field, legitimate acknowledgments should refer to the segment (segments), the receipt of which caused the sending of this acknowledgment. It should prevent the sending of fake duplicates and acknowledgments of the segments that have not yet been received. For details, please read the source, but it is unlikely that the TCP design has been changed.

Denial of service

Denial of service (DoS) attacks seem to be the most common and easy to perform. The attack purpose is to bring the hacked node or network to a state when the data transmission to another node (or data transfer in general) becomes impossible or extremely difficult. As a result, users of the network applications running on the attacked node cannot be served, hence the attack name. DoS attacks are used both together with other attacks (impersonation) and on their own. DoS attacks can be divided into three groups:

- Attacks by a large number of formally correct, but possibly fake packets aimed at depleting the node or network resources;
- Attacks by specially designed packets, causing a general system failure due to program errors;
- Attacks with fake packets that cause changes in the system configuration or state, which leads to the data transfer impossibility, connection dropping, or dramatic decrease of its efficiency.

Host or network resources depletion
Smurf

The attack is a generation of "Echo" ICMP response storm directed at the attacked node. To create it, an intruder sends several fake "Echo" requests on behalf of the victim to the several networks broadcast addresses that act as amplifiers. A potentially large number of nodes located in the amplifier networks and supporting the processing of the broadcast "Echo" requests simultaneously send responses to the attacked node. As an attack result, the network in which the victim is located, the attacking node itself, as well as the amplifier networks, can be temporarily blocked by a response messages storm. Moreover, if the attacked organization pays for the ISP's services according to the traffic received, its costs can significantly increase.

For the hacked node and its network, there are no adequate ways to protect against this attack. Obviously, blocking ICMP messages by a router at the input to the attacked network is not a sufficient solution to the problem, since the channel that connects the organization to the ISP remains is vulnerable to attacks, and more specifically, it is usually the bottleneck in the organization's work with the Internet. As ICMP messages were delivered by the provider to the organization router, they are payable.

The *smurf* attack can be detected by analyzing traffic on the router or in the network. The attack sign is also a full load of the external channel and failures in the host work within the network. When the attack is detected, it is necessary to determine the "Echo Reply" message sender addresses (these are amplifier networks), to establish their administrative affiliation in the Internet registry and to ask administrators to take protective measures for the amplifiers that will be described below. The attacked network administrator must also contact his ISP with the attack notification. The provider can block the "Echo Reply" message transmission to the channel of the attacked organization.

To eliminate *smurf* attacks, the following protective measures can be taken by both potential amplifiers and network administrators, in which an intruder is located:

- Disable the routing of the datagrams with a broadcast destination address between the organization's network and the Internet;
- Disable the node processing of the "Echo" requests aimed at the broadcast address;
- Disable the routing of the datagrams sent from the internal network to the Internet, but having an external sender address.

With that in mind, note that each network can be in any of three roles: an intruder, an amplifier, or a victim, therefore taking measures to protect other networks. We can hope that their administrators are sufficiently qualified and take similar measures that will help to protect us.

SYN flood (Neptune) and Naptha

A common SYN flood attack is to send SYN TCP segments to the attacked node in an amount larger than the server can process at the same time (this number is not large, usually it is several dozens).

Upon receipt of each SYN segment, the TCP module creates a TCB block, which allocates certain resources to the future connection service and sends its SYN segment. He will never receive a reply to it. (In order to cover up the tracks and not to make themselves troubles by ignoring the response SYN segments, an intruder will send his SYN segments on behalf of a nonexistent sender or several randomly selected nonexistent senders.) A few minutes later, the TCP module eliminates the connection that was never open. However, if at the same time an intruder generates a large number of SYN segments, he will fill all the resources allocated for servicing opening connections, and the TCP module will not be able to process the new SYN segments until it is freed from the malicious user requests. By sending new requests constantly, an intruder is able to keep the victim in a blocked state for a long time period. To stop the attack impact, an intruder sends a segment series with the RST flag, which eliminate half-open connections and free up the attacked node resources.

The attack purpose is to bring the node (server) to a state where it cannot accept requests to open connections. However, some bad-designed systems as a result of the attack do not only stop opening new connections, but they cannot support already established ones, and in the worst case, they freeze up.

An SYN flood attack can be defined as holding a large number of connections on the attacked node in the SYN-RECEIVED state. Recently, it has been shown that a large number of connections in the ESTABLISHED and FIN-WAIT-1 states also causes server failures. In different systems, these failures are represented in different ways: process table attack and file descriptor overflow, impossibility to open new connections and break already established, freezing of the server processes or the entire system.

Similar vulnerabilities of TCP/IP stacks were collectively called Naptha. It is emphasized that the Naptha attack execution does not require from an intruder the same costs for maintaining TCP connections as from the attacked node. An intruder does not create TCB blocks, track connection states, and start application processes. When a TCP segment is received from the attacked node, an intruder simply generates an acceptable response based on the flags and values of the received segment header fields to switch or hold the TCP connection on the target node in the desired state. Certainly, an intruder, in order to hide himself, will send segments on behalf of a nonexistent (disabled) node and receive the response segments from the attacked node by sniffing.

Full protection from the described attacks does not exist. To reduce node exposure to the attack, the software that allows setting the maximum connection number that can be opened, and the list of allowed clients (if applicable), must be used. Only the required ports can be open (in the LISTEN state), and the rest of the services should be disabled. The operating system must be resistant to Naptha attacks. When performing the attack, there should be no denial of service to users and services not related to the attacked.

It is also necessary to analyze the traffic in the network to detect the attack start, the sign of which is a large number of the same segment type and the blocking of the malicious segments by a filter on the router. Cisco routers offer a TCP Intercept mechanism that mediates an external TCP client and a TCP server located in the protected network. When the SYN segment is received from the Internet, the router does not retranslate it to the internal network, but responds itself to this segment on behalf of the server.

If the connection with the client is established, the router also establishes a connection with the server on the client's behalf and, in the further segments transfer in this connection, acts as a transparent intermediary, which neither the client nor the server suspects. If the response from the client for a certain time has not arrived, the original SYN segment will not be sent to the sender.

If the SYN segments begin to arrive in large numbers and at an enhanced rate, the router goes into the "aggressive" mode: the response time from the client is drastically reduced, and each newly arriving SYN segment results in the excluding one of the previously obtained SYN segments from the line. When the connection request flow intensity decreases, the router returns to normal, "standby" mode.

The TCP intercept use actually transfers the load of the SYN attack fighting from the attacked host to the router, which is better prepared for this.

UDP flood

The attack is a flooding that attacked network with the UDP messages storm. To generate a storm, an intruder uses UDP services to send a message in response to any other. Examples of such services are *echo* (port 7) and *chargen* (port 19). On behalf of the A node (the source port is 7), an intruder sends a message to the B node (the destination port is 19). The B node responds with a message to port 7 of the A node, which returns a message to port 19 of B node, etc., to infinity (in fact, obviously, until the message is lost in the network). Intensive UDP traffic makes the A and B nodes operation difficult and can create the network congestion.

The echo UDP service can also be used to perform a *fraggle* attack. This attack is completely similar to *smurf*, but less popular within an intruders due to little efficiency.

To protect against attacks such as UDP *flood*, all unused UDP services on the network nodes should be disabled (note that we will hardly ever need *echo* and *chargen* services at all). The filter on the router gateway must block all UDP messages except those that are sent to the authorized ports (e.g., port 53-DNS).

Note that the use of intermediate systems to implement attacks related to the directed packets storm generation (such as *smurf*) is proved to be a very fruitful idea for intruders. Such attacks are called distributed (distributed DoS, DDoS). For their implementation, intruders create daemons that capture intermediate systems and subsequently coordinate the creation of the packets directed to the attacked node (ICMP "Echo", UDP, TCP SYN). The systems are captured by errors in the network service programs. The examples of such distributed systems are *TFN* and *trin00*.

Fake DHCP clients

The attack is to create the large number of fake requests from various nonexistent DHCP clients by an intruder. If the DHCP server allocates addresses dynamically from a certain pool, then all address resources can be spent on fake clients, which means that legitimate hosts cannot be configured and will lose access to the network.

To protect against this attack, the administrator should maintain a table on the DHCP server for mapping MAC and IP addresses (if possible). The server must issue IP addresses in accordance with this table.

System failure

DoS attacks of this group are not associated with any TCP/IP stack protocol problems, but use errors in their software implementation that are relatively easy to fix. A brief overview of the most well-known attacks will be given, bearing in mind that in the near future they are likely to be only a historical interest. On the other hand, there is no guarantee that new errors will not be detected. To protect against them, we must use the latest versions of operating systems and keep track of their updates.

Ping of death

The attack is to send a fragmented datagram to the attacked node, the size of which after the assembly will exceed 65,535 octets. Recall that the length of the IP datagram header "Fragment Offset" field is 13 bits (i.e., the maximum value is 8,192) and the fragment offset is measured in eight octets. If the last datagram fragment designed by an intruder has, for example, the "Fragment Offset" = 8,190, and the length is 100, its last octet is in the collected datagram at position 8,190 К 8 + 100 = 65,620 (plus at least 20 octets of the IP header), which exceeds the maximum allowed size of the datagram.

Teardrop

The attack uses the error that occurs when counting the fragment length during the datagram assembly. Suppose that the A fragment has offset 40 ("Fragment Offset" = 5) and length 120 and the B fragment has offset 80 and length 160, that is, the fragments are superimposed, which, in general, is allowed. The IP module calculates the length of the B part that does not overlap with A: (80 + 160) - (40 + 120) = 80 and copies the last 80 octets of the B fragment to the assembly buffer. Suppose that an intruder constructed the B fragment in such a way that it has a displacement of 80 and length of 72. Calculating the overlap length gives a negative result: $(80 + 72) - (40 + 120) = -8$, which, given the representation of negative numbers in machine arithmetic, is interpreted as 65,528, and the program starts to write 65,528 bytes into the assembly buffer, overflowing it, and wiping it adjacent memory area.

Land

The attack is to send to the attacked TCP SYN segment, whose sender IP address and port match with the recipient address and port.

Nuke

The attack is to send the urgent data to the attacked TCP port (the "Urgent Pointer" field is used). The attack hit Windows systems through port 139, where the application process did not provide the capability of receiving urgent data, which led to the system collapse.

The node configuration or state modification
Traffic redirection to a nonexistent node

Earlier, various traffic interception methods by an intruder were considered. Any of them can also be used to redirect the data sent by the attacked node (or the whole network) "to nowhere", resulting in victim connectivity loss with selected nodes or networks.

In addition, let us note that by applying the routing protocol in the network, an intruder can generate fake protocol messages containing incorrect or limiting values for some fields (sequence numbers, recording age, metrics), which will lead to violations in the routing system.

Long network mask forcing

If the host receives a mask for its network via the ICMP "Address Mask Reply" message, then by creating a fake message with a long mask (e.g., 255.255.255.252), an intruder will significantly restrict the attacked host connectivity. If the network "narrowed" by an intruder does not have a router by default, the victim will be able to send data only to the nodes that fall under the imposed mask.

Currently, hosts are dynamically configured using DHCP (which, however, opens more opportunities for an intruder: by pretending to be a DHCP server, he can completely distort the TCP/IP stack settings of the host being attacked). However, it should be checked how the system will respond to the ICMP "Address Mask Reply".

TCP connection desynchronization

The attack was discussed earlier. Obviously, if after the desynchronization implementation an intruder does not function as an intermediary, data transfer via the attacked TCP connection will not be possible.

TCP connection release

If the A node established a connection to the B node, an intruder can force the A node to close the connection early. To do this, it is sufficient to send to the A node a fake RST segment or an ICMP "Destination Unreachable" message on behalf of the B node.

In order to form an acceptable RST segment for the A node, an intruder must sniff a connection to put in the RST segment the SN number that is located within the accessible window of the B node, otherwise the A node will ignore the fake segment.

The reaction of the A node TCP module to receiving ICMP "Destination Unreachable" messages is not clearly defined by the standards, although RFC-1122 states that the IP layer should transmit the corresponding message to the TCP module, and in turn, the latter is required to react to it somehow. Also, RFC-1122 says that when receiving such a message with Code 2 ("Protocol Unreachable") or 3 ("Port Unreachable"), the TCP module should eliminate the connection to which this message relates.

Data transfer forcing at a reduced rate

An intruder can force the TCP module of the A node to reduce the data transfer rate to the B node in the TCP connection in the following ways.

- Send to the A node on behalf of the intermediate router a fake ICMP message "Source Quench". RFC-1122 claims that the IP layer should inform the TCP module of the message receipt, and the latter should respond by reducing the data transfer rate, and it is recommended that the TCP module switches to the slow start mode like when retransmission timer triggers.
- Send to the A node on behalf of the intermediate ICMP router fake "Destination Unreachable: Datagram Too Big" messages. When using the Path MTU Discovery algorithm, the TCP module, receiving such messages, will reduce the size of the sent segments to the number specified in the ICMP message, or until it reaches the set minimum. If we strictly follow the standard, then the minimum datagram size, which any IP module must transmit without fragmentation, is 68 octets. In this

case, if no options are used in the IP and TCP headers, the payload will be 28 octets per segment. In addition, when the "Datagram Too Big" message is received, the slow start mode is enabled, although without decreasing the cwnd window value.

- Send to the A node on behalf of the B node fake TCP acknowledgments duplicates. Thus, an intruder forces the A node, first, to include the fast retransmission algorithm and send the segments already received by the B node, and second, to go into the congestion-avoidance mode, thereby reducing the data sending rate.

***

In conclusion, the possible security measures that can help in the fight against the attacks described here are given. The security administrator, based on the network security policy, is able to determine what measures are necessary and acceptable in each particular case.

Filtering on the router

Filters on the router are used to disable the datagram skipping that can be used for malicious attack.

- Disable skipping of the datagrams with a broadcast destination address between the organization's network and the Internet.
- Disable skipping of datagrams sent from the internal network (network of the organization) to the Internet but with an external sender address.
- Disable skipping of datagrams arriving from the Internet but with an internal sender address.
- Disable skipping of the datagrams with the "Source Route" option and encapsulated datagrams (IP datagram inside the IP datagram).
- Disable skipping of the datagrams with ICMP messages between the organization's network and the Internet, except for the required ("Destination Unreachable: Datagram Too Big" for the Path MTU Discovery algorithm, as well as "Echo", "Echo Reply", "Destination Unreachable: Network Unreachable", "Destination Unreachable: Host Unreachable", "TTL exceeded").
- On the client access server, allow skipping of the datagrams directed only to the IP address assigned to the client (or from this address).
- Disable skipping of the datagrams with UDP messages sent to or from echo and chargen ports, or to all ports other than used (often only port 53 for DNS service).

- Use TCP Intercept to protect against SYN flood attacks.
- Filter TCP segments in accordance with the security policy: all services are allowed, except for forbidden ones, or all services are prohibited, except those allowed (for the application service description, see Section 1.5). If there are no hosts on the internal network that are supposed to be accessed from the Internet, but internal hosts are allowed to access the Internet, we should prevent TCP SYN segments from going online from the Internet to the internal network, as well as skipping datagrams with "Fragment Offset" = 1 and "Protocol" = 6 (TCP).

Router protection

Measures to protect the router are designed to prevent attacks aimed at violating the scheme for routing datagrams or for a router capture by an intruder.

- Use the authentication of routing protocol messages using the MD5 algorithm.
- Filter the routes stated by client networks, the provider or other autonomous systems. Filtering is performed in accordance with the organization's routing policy. The routes that do not comply with this policy are ignored.
- Use a static ARP table of the network nodes on the router, as well as on the switches.
- Disable all unnecessary services on the router (especially, the so-called "diagnostic" or "small" TCP services: echo, chargen, daytime, discard, and UDP: echo, chargen, discard).
- Limit access to the router console or the dedicated administrator work-station, use password protection; do not use telnet to access the router on the network that can be sniffed.
- Use the latest versions and software updates, follow the security bulletins produced by the manufacturer.

Host protection

Host protection measures are used to prevent attacks that attempt to intercept data, deny service, or intruder penetration into the operating system.

- Disable processing of "Echo" ICMP requests directed to the broadcast address.
- Disable processing of ICMP messages "Redirect", "Address Mask Reply", "Router Advertisement", "Source Quench".

- If the hosts on the local network are configured dynamically by the DHCP server, use the DHCP server to map the MAC and IP addresses and give the hosts predefined IP addresses.
- Disable all unnecessary TCP and UDP services (transfer ports from the LISTEN state to CLOSED).
- To serve incoming connections, use a supermodel such as tcpserver or xinetd, which allows setting the maximum number of simultaneous connections, a list of allowed client addresses perform DNS address validation and register connections in the log file.
- Use a program such as tcplogd, which allows us to track hidden scanning attempts (e.g., half-open connections).
- Use a static ARP table of local network nodes.
- Use the security features of application services used on the host.
- Use the latest versions and software updates, follow the security bulletins produced by the manufacturer.

Preventive scanning

The security administrator must know and use an intruder method and tools and conduct a preventive scan of his object network in order to detect security weaknesses before an intruder does.

1.2.2 Threats and Consequences of Wireless LAN IEEE 802.1x Implementation

The wide spreading of wireless networks in the local enterprise networking concerned the information security issues. Let us see what typical threats and implications of Wireless LAN IEEE 802.1x deployment are encountered by Russian enterprises today [6, 7, 14, 18, 23–25].

State of the art

Initially, the developers of the Wireless LAN standards and, in particular, the IEEE 802.11 family were in a preferred position, since by the time the work on the above standards began, the IT community had already formed the common approaches to information security. Moreover, "cyber underground" has already started to form and there was and not a lot of time before the occurrence of a black market of services in the information system hacking. Thus, the developers were informed of potential threats and vulnerabilities of wired LAN protocols. It was also assumed that the task of transferring wired technologies to radio air involves the use of an open accessible communication environment (with minimal equipment investments). For this reason, some of the security risks arising from the wireless networks use in Russian enterprises have been timely reduced. However, the uncontrolled growth of security interest (both by apparent intruders and by researchers, including cryptographers, who unintentionally opened the way to breaking) was still underestimated by Wireless LAN developers. In support of this, we can cite the increasing incidence of deciphering encryption schemes in wireless networks from 2009 to 2016.

Capabilities of protection technologies

The interaction of the wireless traffic protection subsystems (Figure 1.49). The existing wireless network security standards and protocols are aggregated in Table 1.6.

The master key source used to generate/regenerate the whole keys set (for data encryption, integrity control checksums, encryption of the key exchange process) can be:

Static password that is identically stored on the client PC and access point (Pre-Shared Key scheme–PSK). In this case, the master key (PMK) is generated based on multiple repetition of the irreversible SHA-1 hash conversion over the passphrase and network identifier (SSID) by the algorithm PBKDF2;

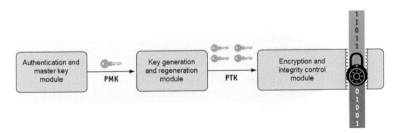

Figure 1.49 Interaction of wireless traffic protection subsystems.

Table 1.6 Wireless network security standards Master key source subsystem

Protocols	Master Key Source Subsystem	Key Generation and Regeneration Subsystem	Encryption and Integrity Control Subsystem
WEP	PSK	<KEY+IV>	RC4/CRC32
WPA	PSK EAP	TKIP	RC4/MICHAEL
WPA2	PSK EAP	CCMP	AES/CCMP

One of dozens of Extensible Authentication Protocols (EAP) schemes, for example, EAP-TLS, EAP-TTLS/MSCHAPv2, PEAPv0/EAP-MSCHAPv2, PEAPv1/EAP-GTC, PEAP-TLS, EAP-SIM, etc. In this case, the master key (PMK) is generated as a "side effect" of the authentication process and changed from session to session.

Key generation and regeneration subsystem

After the master key determining, the key generation subsystem (and regeneration in TKIP case) performs causes generating secondary key information.

- WEP standard does not perform any additional transformations over the master key and just adds it with a pseudo-random or incrementally increasing 24-bit initialization vector (IV) for each packet (within the limits of the unrepeatability IV). For encryption algorithms with US export restrictions for cryptographic algorithms, the total key length was 64 (40 + 24) bits. If encryption algorithms did not have restrictions, then its total key length was 128 (104 + 24) bits.
- WPA (TKIP) and WPA2 (802.11i) standards generate 512-bit algorithm session key file (PTK) based on the master key (PMK) and two unique pseudo-random numbers (one is transferred to the client PC by the access point, and the other, back to front, is transferred to the access point by client PC) by HMAC-SHA1 algorithm:
  - for data encryption algorithms;
  - for integrity control algorithms;
  - for message of key rollover safeguarding, etc.

WPA and WPA2 standards use a 48-bit initialization vector (IV) and support automatic key regeneration, which the system starts upon the certain events occurrence (with the initialization vector reset to 0).

Encryption and integrity control subsystem

WEP standard established the RC4 stream cipher as the cryptographic algorithm used. It made the numerous implementations to victims of this solution. It is no coincidence that in the intermediate WPA standard on the way to achieve the IEEE 802.11i/WPA2 security level, almost all aspects of key management of the future standard were implemented, but it was not affected by the crypto algorithm itself. The reason for this is many implementations that could not provide the block cipher AES execution efficiently (because of the hardware base limitations) and with the required speed.

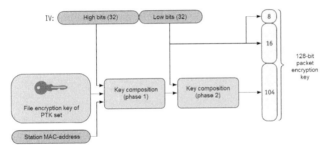

Figure 1.50 The keying scheme.

As an integrity control measure, the WEP standard used a linear cyclic CRC-32 checksum, added to the packet before encryption. Unfortunately, the characteristics of this algorithm, designed to search and fix random errors in communication channels, were completely unacceptable for campaign against obfuscations.

WPA/TKIP standard introduced a whole range of dramatic changes aimed at maximizing system stability without changing the encryption algorithm (RC4) and the key size. The use of the session key material (PTK) for the packet keying itself is shown in Figure 1.50. The generative phase 1 is building more computationally time-consuming than phase 2, since it is required to be executed only once for every 216 packets transferred.

To control the message integrity based on one of the PTK keys using a simple but nonlinear function, a 64-bit MIC block is calculated for each packet, added to the package with the CRC-32 block (inherited from WEP standard) before the encryption procedure.

WPA2 standard implemented a fundamentally different scheme of simultaneous confidentiality and integrity based on the CCMP (Counter-Modewith CBC-MAC) and block crypto algorithm AES with a 128-bit key generated from the PTK. However, the need for a separate calculation of the MIC block has disappeared.

The most dangerous attacks

The main methods of information and technical impacts on the wireless local area network (Figures 1.51–1.54) from the hackers include:

- Passive traffic wiretapping;
- Traffic wiretapping with the generation of service packets (ARP, deassociation, etc.);
- Traffic wiretapping with the radio-frequency range in order to increase the amount of legitimate traffic;

- Falsification of some packages on behalf of a legitimate station;
- Denial-of-service attack against the station;
- Complete impersonation from the names of the legitimate station;
- "Hacker-agent" attack (false access point) to the address of the selected station;
- Denial-of-service attack against the access point;
- Complete impersonation on behalf of the access point.

We will analyze the attacks with the highest degree of risk in the context of the classification of the object of attack (the attacking protection subsystem).

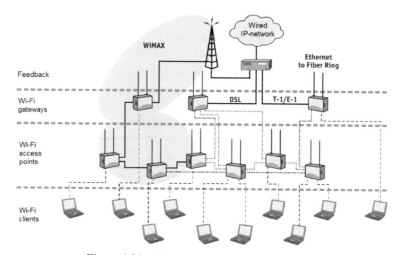

Figure 1.51   Wireless technology classification.

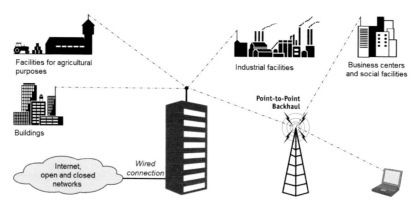

Figure 1.52   Example of an enterprise wireless network.

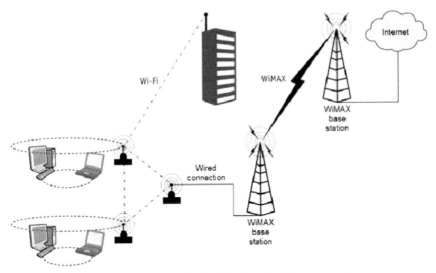

Figure 1.53   Example of a distributed wireless network.

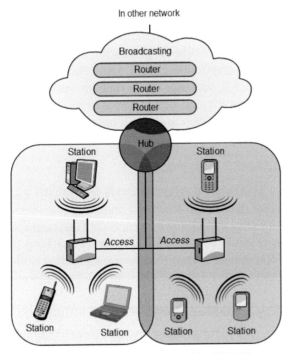

Figure 1.54   Enterprise networking WLAN.

Attacks on the authentication subsystem and the master key source

Attacks based on an access to a common key (PSK)

Systems with a common pre-shared key (*PSK*) a priori contain a vulnerability to the actions of an internal intruder who has access to his instance of this key with respect to other users. The WEP standard supports only the above scheme (a single key for the entire network), WPA and WPA2 standards specify the support of unique predefined keys with their binding to the MAC address in the access point base, however, and the use of a single key for the entire network is widely used for these standards. The master key breaking is sufficient to generate PTK transport keys of any network station, and consequently, to directly implement threats to breach the confidentiality and integrity of the information transferred by it.

In the case of programmatic availability of the pre-shared key from other computing processes, there is an additional threat of compromising the entire network in PSK scheme in case of hacking by other attack vectors (e.g., using the Trojan program, etc.) of one of the stations sharing a common PSK.

Attack on LEAP cryptographic vulnerability

One of the widely used authentication schemes–LEAP from Cisco Systems–contains the algorithm vulnerability for generating session keys. Despite the choice of a robust query-response authentication scheme without passing a clear-text password, LEAP does not use a random vector ("salt") and, in addition, has a very limited space of input values in one of the encrypted parts. All this allows us to efficiently select a password based on previously computed arrays of the typical processed data.

Password attack

In addition to those listed above, a general threat for systems with user input of a password phrase (PSK in the form of a passphrase or user password in EAP schemes) is a dictionary attack. The speed of password scanning for all three standards is extremely low, but if there is any kind of additional information about the password or if the persistence policy is violated when choosing a password phrase, it can become a significant risk to the system.

Attacks on the key generation/regeneration subsystem

WEP keys mining (for RC4 stream cipher) using vector IV

Cryptographers have developed a whole family of statistical attacks on the WEP encryption subsystem, based on the fact of an unsuccessful selection of the RC4 stream cipher and especially the principles of initialization of its original vector IV. The first attack option (called FMS, which gave the name to the whole family) was proposed by S. Fluhrer, I. Mantin, and A. Shamir in 2001, which required the interception of about 1 million packages. Modern implementations (e.g., PTW attack (E. Tews, R. Weinmann, and A. Pyshkin) require the interception of only 40,000–60,000 packets. In some software for hacking Wi-Fi networks, these implementations are supplemented with algorithms provoking legitimate hosts to generate additional service traffic, reaching the required volume of intercepted packets in just a single minute.

Attacks on the encryption and integrity control subsystem

Byte-mining source text in WEP standard

Using the cryptographically unstable integrity control CRC-32 function in WEP standard resulted in a very trivial way of byte-by-byte finding of the source text of the intercepted packet in those cases where the access point's reaction somehow distinguishes a valid or incorrectly selected checksum. The attack, called Chopchop, begins with the original correct packet by cutting the last byte and trying to guess the value of the cutoff byte of the source text by selecting the correct checksum. After the correct last byte mining, the second-to-last byte is cut off, and so on. Each byte mining takes 128 attempts (at the average) to send a cutoff packet to the access point.

Byte-mining source text in WPA standard

In 2008, German researchers M. Beck and E. Tews showed that the Chopchop attack can be performed for WPA if the device supports QoS service provided by IEEE 802.11e standard. QoS service allows us to route data streams through various channels; each channel has IV counter independent of the others. Normally, one channel is used to transfer packets at a time, leaving the remaining channels to allow rapid transfer of expedited data. By intercepting an encrypted packet that is likely to have a high IV value on the channel, we can go to a less loaded channel with a lower IV value. Thus, when switching to another channel, it becomes possible to resend a packet of a repetitive cipher sequence, while the channel counter value will not change. Thanks to QoS services, it is possible to retransmit fake packets of one cipher sequence several times. Moreover, if the MIC value is incorrect, the counter value does not change, so the next packet can be sent along the same channel. Thus, the Chopchop attack on WPA checks the success of the assumption of the overlaid sequence by sending MIC messages about the incorrect checksum value.

Due to the fact that the recovery of long packets takes a long time, for which either vector IV or session keys can change, short messages such as ARP and DNS packets were used as the basis for the attack. For example, for the ARP, the number of bytes of the unknown text is 14: two bytes of the ARP packet, eight bytes of MIC, and four bytes of the checksum. The attacker restores the MIC and the checksum by performing the attack 12 times. This procedure takes at least 11 minutes, because each time we need to observe a 60-second interval, in order to not start the mechanism of MIC-drop and the formation of new session keys.

This attack allows to fake encrypted ARP packets, as well as to cipher the integrity control key from the key packet (PTK) because the MICHAEL algorithm is a reversible function.

Byte-mining source text in WPA standard using MitM

Japanese researchers T. Ohigashi and M. Morii have further strengthened the attack described above. Due to the introduction into the network between the attacked subscriber and the access point of the "hacker-agent" station in the modified version, the attacker was able to mine the byte of the source text (based on the package with the higher IV) without using the QoS service functions. The remaining attack parameters correspond to the original Chopchop variant for WPA.

Denial of service attacks

Denial of service attacks can be carried out by attackers at both the first (physical) and second (channel) levels by generating malicious traffic in order to disrupt the normal mode of the wireless network. To date, the most known methods are:

- Noisy radio-frequency bandwidth;
- Filling the network with false beacons;
- Filling the network with false requests for EAP authentication to the access point;
- Filling the network with false EAP deauthentication commands to the subscriber address (since this command is transferred in an unencrypted form).

The result of such impacts can be a significant decrease in the bandwidth of the wireless network or its complete inoperability.

Feasible attack implications
Violation of the transferred information confidentiality

The implication of the successful attack implementation in order to disclose the source text of the transferred network message may be:

- Confidential internal information (documents, files, e-mails, etc.) disclosure to an intruder;
- Disclosure of the service network information on the network structure and topology, the network protocols used, and the software versions used, which may lead to the appearance of new additional attack vectors on its part;
- Disclosure of the password information of the application protocols transferred directly (FTP, POP3, telnet, SNMP, etc.) or reversibly recoverable (AUTH-BASIC HTTP); this threat is very serious, given that according to the statistics more than half of users have the same password for different services, and therefore there is a possibility of indirect disclosure of passwords transmitted in a crypto-secure form (e.g., authentication in the enterprise domain).

Violation of the integrity

An attack that results in a violation of the transferred information integrity may be aimed at both direct modifications of the critical data (the attack object) and malicious modification of service packets for the further development of attack vectors. Here, the most interest for the attacker can be provided by:

Preparation for the attack of "hacker-agent" is redirection of the victim's data streams through the equipment controlled by the attackers;

Interruption in connection initiation process to reduce the authentication/encryption scheme established between the victim subscriber and the access point to the level of any vulnerable protocol.

Session interception

Attacking the class of intercepting a session is very serious in its feasible implications, although it is quite difficult to implement a threat. Substitution of a legitimate subscriber station by an attacker station in the middle of a steady session, for example, of any application protocol, gives the attacker all the powers and privileges of the disconnected user without the need for an authentication step. The further actions of an attacker in the application system can have catastrophic implications.

Phishing

The substitution of a regular way of passing data from a subscriber (on a network or application layer) allows an attacker to gain control over all data processed by user on the network. In a result, password information can be disclosed, the subscriber workstation can be infected with malicious code, unauthorized access to information systems, and modification of information in them can be performed.

Illegal use of services

A hacker can use services available from the wireless network attacked by him that it does not fall under the previous kinds of implications and thus does not cause direct damage to the enterprise information security. More often, it causes only economic harm, since sessions service access will pay by an attacked enterprise. Implications scales usually depend on the cost of the unauthorized services.

Network infrastructure unavailability

Denial of service attacks can cause a significant damage to the enterprise if its business process model is heavily dependent on the information transfer network availability, and the planned wireless accessibility/continuity planning has not been undertaken with respect to the attacked wireless network. A long successful attack aimed at disabling the network infrastructure can lead to the most valuable implications for the enterprise.

***

In general, the situation in the wireless network security is constantly changing (and usually not for the better), and there are periodic publications about new attacks on the wireless LAN subsystems of the IEEE 802.1x. Currently, only a combination of WPA2 + TLS, WPA2 + EAP-TTLS, and WPA2 + PEAP security mechanisms (with the compulsory banning of weaker security schemes) has not been published in open sources of any types of attacks (except vocabulary). Unfortunately, it cannot ensure Wireless LAN security for long life cycle of the mentioned systems. We have to admit that there is a possibility of new attacks against these mechanisms, and, most likely, not directly to cryptographic algorithms that have already proven themselves well, but to serving protocols, as it happened, for example, with the WPA scheme. These attacks aimed at weakening known crypto protection schemes.

An additional negative aspect of the whole situation in information security is that at present, practically any theoretical attack is realized in a generally available software implementation in a matter of days and hours. As a result, advanced (including cryptographic) attacks become available to a wide range of people, even those with minimal knowledge of wireless networking.

In order to avoid the undesirable implications listed above, one should heed the following recommendations:

- Use wireless LANs only in information systems with low requirements for accessibility and survivability;
- In systems with high demands on confidentiality and integrity of transferred information, when using wireless local systems, use additional cryptographic facilities at higher layers of the OSI model: for example, IPSEC (at level 3) or SSL/TLS (at the layer 4 of the OSI model);
- If necessary, use software and hardware to detect, prevent, and suppress unauthorized effects on wireless networks, for example, based on immune technologies.

1.2.3 Cyber Security Threats of Corporate Digital and IP-ATX (Private Automatic Telephone Exchanges)

The successes of microprocessor technology led to a mass transition of medium- and large enterprise ATXs (PATXs) to digital call processing in the 1990s. Moreover, at present, there is one more quantum leap in enterprise telephony–from "traditional" digital ATX to IP telephony. At the same time, the proportions of all three technologies (analog, digital, IP telephony) among working PATXs have almost equaled. With the undisputedly greater range of features offered to a subscriber by a digital telephone line, it is necessary to take into account a whole range of issues related to the protection of voice information, generated by digital processing of calls and voice.

The digital scheme of messaging (both control and voice) in practice does not eliminate the threats, typical for traditional schemes, but generates whole classes of new threats of confidentiality compromise. Perhaps the only advantage of digital (including IP) voice processing in this aspect is the potential readiness of the scheme for the transparent software tools implementation for voice information cryptographic protection. However, this process with regard to PATX general (nonspecial) appointment has just started to develop. This section analyzes the most probable threats to the confidentiality of the voice information in digital and IP PATXs [7, 6, 18, 23–25].

The connection threat within the switching matrix

Digital signal processing is fraught with the unauthorized use ("branching") of voice traffic within the switching matrix without any unmasking options. The fact of copying cannot be traced; it does not cause any changes in the transmitted signal amplitude, or distortions associated with the delay in transmission. This is a qualitative change between digital telephony systems and the previous-generation systems.

Almost all major equipment developers for the PATX implemented in the software some or other copying capabilities of voice traffic, if the wiretapping party has the appropriate authority defined by the telephone exchange administrator. In some cases, this is a true three-party conference call with the incoming voice channel disconnected from the wiretapping party; in other cases, this is a branch of the stream according to a special scheme when dialing a certain number. Some information security researchers break out the so-called "police regime", which allows outsourcing the same operations, when dialing from a city telephone network a certain local number and the access code. Let us consider the implementation of these technologies in some ATX models.

AVAYA definite model of digital enterprise ATX realizes the opportunity of hidden copying the voice information within the scope of service observing, which is positioned as a means to control the progress of telephone operator work, especially in call centers. This function can be activated by sending a warning signal to the voice channel every 12 seconds about the fact of wiretapping by the third party, and without it. Setting up the wiretapping privileges is performed from the administrator's console using the group principle: each subscriber line corresponds to the priority class COR, and the permission or prohibition of wiretapping is determined for each pair of classes in the matrix form. The wiretapping is activated by dialing the service access code and then the subscriber number and can be assigned to one of the function keys of the wiretap. In addition, with some setting, it is possible to access the function from external lines, for example, from the urban telephone network.

The IP telephony server CallManager by Cisco Systems Inc. also provides the opportunity to include in the conversation the third subscriber who has proper authority (both with and without a warning signal). The function is called Barge In, and it has the following two different schemes of technical implementation.

1. A scheme based on software and hardware, built-in all enterprise IP telephones with two lines. When a conference call (including one-party,

i.e., wiretapping) is received, the IP wiretap independently performs the branching and mixing of two voice streams (the primary one is toward the subscriber and the secondary is in the direction of the wiretap) by the hardware on the second line. At the same time, with the consistent setting, no warning signals are added to the primary voice stream; moreover, no information about the connection is displayed on the monitored IP device. This scheme is limited to only one wiretapping connection and only to the broadband (64 kbit/s) G.711 codec, but does not introduce any unmasking distortions into the voice stream.

2. The scheme is based on the dedicated firmware of the IP telephony server conference call.

When a request is received, the IP telephony server closes the voice traffic in both directions (which has passed directly up to this point between the IP devices) to the conference call device and, with its help, mixes and branches the data (in this case, already for an unlimited number of wiretaps and beyond, depending on the codec, used by the subscribers). The scheme disadvantage in comparison with the first option is the audible distortion ("voice failure") at the moment of switching flows.

The setting of wiretapping privileges performed separately for each wiretapping line (a set of lines that have the right to connect, including the quiet one, to the conversation) is directly indicated. Thus, an intruder receives the administrator privileges of the digital PATX (e.g., through a successful attack on his personal computer) and provides insight into almost unlimited opportunities for quietly wiretapping to telephone conversations.

The threat of indoor conversation wiretapping, applying auto-reply

Digital and IP devices as complex computer devices introduced another class of threats of voice information leakage associated with the opportunity of remote (and under some conditions–unauthorized) microphone switching and transfer of conversations conducted indoors on a digital channel. As a first one, let us consider an option that is not related to the undocumented capabilities of the devices themselves, i.e., the widespread option "Auto Answer". In case of its activation, the called device delivers one (often shortened) call signal when a call comes in, and then automatically turns on the microphone and the loudspeaker so that the subscribers can communicate with each other via a speakerphone or using a headset.

When setting the "Auto Answer" option, depending on the calling line (intercom), it starts to present a real threat of indoor wiretapping conversations. An intruder who has received PATX administration privileges can create an intercom group by including the attacked line and his number on it, change the call signal from his line to record silence, and thus get the opportunity to wiretap the conversations indoor by making a call to that line.

The scheme has the following minor demasking features:

- Depending on the device model, the fact that the microphone is switched on may be reflected by indicators;
- Line will be busy when we try to call from outside;
- There is a risk that the wiretapped subscriber will pick up the handset to make a call.

However, all these do not exclude the opportunity of performing a successful and quietly wiretapping, especially in situations where there is an active discussion of a particular issue in the room, and the telephone set is installed in such a way that its indicators are not visible.

The threat of undocumented device control features

The undocumented device features (especially IP-devices) are another threat to the confidentiality of a speech information in the secure places. The IP telephone software is a complex software package that implements, among other things, the TCP/IP stack and can contain the following:

- Undocumented features made by developers for testing or at certain development stages of new device functionality;
- Errors in the implementation, which lead, for example, to the vulnerabilities of the "buffer overflow" class and allow us to gain complete control over the software of the device before it is rebooted.

An example of the first group threat is the capability in one of the software versions to send the managing XML message CiscoIPPhoneExecute to the most popular IP phone models 7940 and 7960 of Cisco Systems Inc. Among other features (dialing, keystroke emulation, etc.), it could include the microphone of the device and transfer all voice traffic to the IP address specified in the XML message.

The threat of IP traffic interception at the network transmission time

Different options of the implementation of traffic interception threats are traditional for computer networks that use broadcast segments (Ethernet, including switched, radio, Ethernet, etc.) in their structure and create another level of feasible attacks on IP telephony systems. If there is no traffic encryption at the network or higher layers of the OSI model, then there are several options for violating the transmitted message confidentiality.

If the intruder does not have the administrative privileges to active network equipment, the most effective among the switched Ethernet networks is the ARP spoofing attack, which performs the modification of the routing table on the channel (MAC) level using specially formed ARP packets. In addition, change over the switch to hub mode using a large number of false packets (MAC storm) may cause the disclosure of a certain part of the transmitted information. Note that this method has significant demasking properties, which are expressed in a sufficiently sharp decrease in the network quality.

When an attacker obtains the administrative privileges on commuting or routing equipment (e.g., as a result of an attack on the administrator's computer or when his password is search-attacked during his password transmitting in clear form), he has much more powerful means of IP traffic interception as follows:

- Ability to activate on the switches of mirrored (SPAN) ports, receiving an exact copy of traffic transmitted on certain ports;
- Use of other technologies for "branching" traffic from network equipment manufacturers:
- ERSPAN (Encapsulated Remote SPAN) protocol, which encapsulates each intercept packet into a packet
- GRE protocol, which allows transferring it over IP networks without any limitations on the range;
- IP Traffic Export option, which provides a "branching" of traffic when it is routed at the third level of the OSI model.

ERSPAN and IP Traffic Export support the ability to fine-tune the filtering of intercepted packets, which allows copying the traffic only from certain groups of IP devices.

Wireless networks in the absence of the strong encryption algorithms are also a potential source of a disclosure of the voice traffic transmitted through them.

The threat of data spoofing in the IP telephony supervisory channel

The method of centralized management of IP telephone calls (implemented in the PATX) contains another possible way for subscribers to intercept their conversations. When the IP connection is established, the initial exchange of information containing subscriber numbers, their names, technical capabilities of the devices, etc., including the IP addresses of the end devices, goes between the IP telephony servers (Figure 1.55).

At this stage, the data spoofing is possible (by tools, the network layer attacks) about one or both IP addresses to introducing the intruder's computer into the chain of voice traffic transmission on the principle of a transparent proxy server.

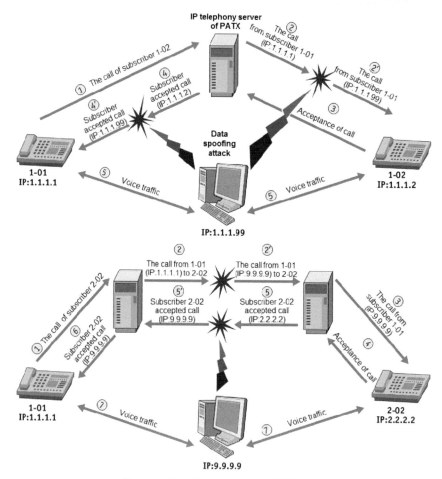

Figure 1.55   IP connection establishment.

This class of attacks remains completely invisible at the application level, since the user usually does not see the network coordinates of the remote subscriber, and the protocol stack is not able to detect the fact of spoofing and can be detected only through a special network traffic monitoring.

In general, the prerequisite for the appearance of such attacks is the following: in modern IP telephony protocols (H.323, SCCP, etc.), the terminal equipment when receiving and transmitting the voice stream is slave with respect to the server of the PATX and relies entirely on the information communicated to it in the manager channel (e.g., it does not check the correspondence between the sender's IP address and the voice stream recipient within the same conversation). The problem of providing protection against the introduction of a proxy server into the voice stream raises the issue of ensuring the integrity of data transmitted in the supervising channel by persistent cryptographic methods.

***

The change in the telephony technologies for the information objects from analogue to digital and then to IP devices has generated many new threats to the privacy of voice information transmitted by the PATXs and circulating indoors with fixed telephone equipment. This problem requires the development and implementation of new tools and techniques for monitoring the operation mode of the PATX and their distributed components, as well as techniques for monitoring unauthorized impacts and anomalies in the computer networks transmitting IP telephony traffic. Some of the above methods were developed and tested by the authors. The analysis of the obtained results clearly demonstrates the relevance of the direction of voice information protection discussed in this section in digital and IP-based PATXs.

1.2.4 Threats and Security Profile of the Mobile Operating System OS Sailfish and Tizen

This section analyzes the requirements to create a trusted mobile platform running Sailfish and Tizen–open-source operating system based on the Linux kernel. Up-to-datedness of developing the mobile OS Sailfish and Tizen security profile is explained as follows. At present, the efficiency of state organizations and companies significantly increases with the use of an appropriate mobile software solution. At the same time, the features of the known information security tools (MDM, VPN, antivirus software, etc.) are not enough to build a secure mobile service platform due to the lack of the trusted mobile OS [19, 41].

A comparative analysis of modern mobile OS functionality, such as iOS, Android, Windows, Blackberry, and Tizen, has shown that the required security profile for creating trusted mobile platforms can be developed based on the Tizen OS core, as it inherits the best system architectural solutions of MeeGo, LiMo, and Bada and supports hardware platforms on ARM and x86 processor architectures. At present, Tizen is an open operating system based on the Linux kernel, which already manages a wide class of enterprise mobile devices, including specialized solutions for a distance education, traditional smartphones, and Internet tablets, as well as numerous devices of Industrial Internet of Things (IIoT) and of the Internet of Things (IoT) (on-board computers, smart TVs, digital cameras, sensors, etc.). It is essential that hardware solutions and platforms are currently supported. These platforms are made by Intel, Samsung, and other manufacturers, which are part of a Technical Steering Group (TSG) and are active participants in the Linux Foundation and Tizen Association.

At present, mobile OS Tizen OS and Sailfish OS have received the greatest popularity in the Russian information security. It is significant that the mentioned operating systems inherited the best structural and functional properties of the perspective MeeGo OS by Finnish Nokia (2010–2012) (Figure 1.56). Commercial prototype MeeGo OS appeared in 2010 and quickly gained popularity among mobile application developers. Unfortunately, in 2012, the support of this progressive OS was suspended. After buying the Nokia division, Microsoft refused to continue working on the MeeGo OS to develop the own mobile platform Windows OS.

At present, the Tizen OS development is supported by the international association of professional developers Tizen (with the participation of the Linux Foundation). The association includes a number of well-known companies, the leading of which are Samsung and Intel. The Tizen OS development continues based on open source (Figure 1.57).

At the same time, some of the Tizen OS components are closed (e.g., the Tizen SDK development kit). JQuery and JQuery Mobile JavaScript libraries are widely used to develop Tizen applications. Since version 2.0, development of applications in C++ based on the Open Services platform from Bada has been supported. Also, the Tizen SDK allows the use of HTML5 and related Web technologies. In October 2015, the association "Tizen.Ru"[29] was created to develop and extend the Russian local version of the trusted

---

[29]http://tizen.ru/#/.

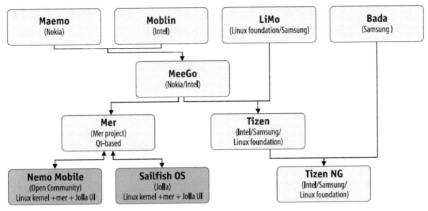

Figure 1.56 History of the Tizen OS and Sailfish OS development.

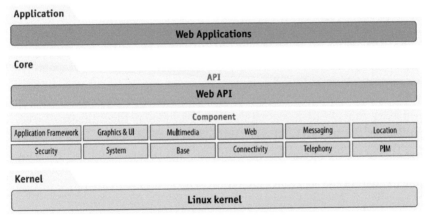

Figure 1.57 Original architecture of the Tizen OS kernel.

mobile OS (the authors of this article are among its experts). The association included a number of well-known foreign and domestic companies: Samsung Electronics Rus Company LLC, GNU/Linux Center LLC, research institute of safety-related system integrated safety and security support system LLC, Infotecs JSC, Kaspersky Lab JSC, Institute for system programming of RAS, etc. In the same year, the Samsung Tizen 2.x OS obtained the first FSTEC of Russia certificate No. 3441, which certified the compliance of this system with the requirements for the fourth level of the undocumented feature absence control, as well as the availability of built-in information security against unauthorized access.

It is significant that this allowed the development of mobile devices running Samsung Tizen 2.x OS to secure restricted access information that does not contain national security information. At present, creating trusted solutions and digital services continues mainly in the field of advanced technologies of the Internet of Things (IIoT/IoT), as well as Industry 4.0 as a whole.

The development of the second OS–Sailfish OS–was initiated to create mobile devices with a closed user interface by the independent Finnish company Jolla. The rest of the operating system was supplied with an open source. Jolla was created in 2012 by the leading developers of Nokia, who have the skills and experience to create a progressive MeeGo OS within the creation of the advanced smartphone Nokia N9, which has not lost its popularity to the present day. In late 2013, Jolla and Yandex announced a scientific and technical cooperation in the field of mobile technology. In 2016, another Russian company Open Mobile Platform joined Jolla.

In November 2013, Jolla introduced its first smartphone, and in November 2014, the first tablet on Sailfish OS 2.0. In February 2017, the version of the operating system for the smartphone Sony Xperia X was presented in Barcelona at the Mobile World Congress.

In 2015, Open Mobile Platform LLC introduced the first Russian version of Sailfish Mobile OS RUS, which was later included in the Register of Russian Software. In early 2017, it was certified by the Russian Federal Security Service (FSS) in the AK1/KS1 class to meet the requirements for securing confidential information from unauthorized access in automated information systems located in the Russian Federation. The Certificate of Conformity of the Federal Security Service of Russia SF/119-3068 dated February 20, 2017 certified that Sailfish Mobile OS RUS can be used to protect information that does not contain national security information. In addition, the cryptographic information protection facilities (CIPF) for this OS were developed, which is also certificated by the Federal Security Service of Russia in the KS1 class (Certificate of Conformity SF/114-3069 of February 20, 2017).

As a result, now Sailfish Mobile OS RUS represents one of the best Russian trusted mobile operating systems, the merits of which are the following capabilities:

- Reliably monitor the file system and boot loader integrity;
- Fully lock the device (without the unlock ability), if the file system or boot loader integrity is violated;
- Assign and monitor user rights in accordance with enterprise security policies (both at the OS level and with MDM management systems);

- Prevent unauthorized software installation (including by the user itself) from untrusted sources, etc.

A comparison of the main characteristics of the Tizen OS and Sailfish OS is given in Inset 1.11.

Inset 1.11

At present, the Russian versions of the trusted mobile Tizen OS and Sailfish OS are the open cross-platform applications based on the Linux kernel and are distinguished by the developed multi-level architecture and a sufficient functionality of built-in information security tools.

Tizen 2.4 runs on Linux 3.14.25, and Sailfish OS and the corresponding SDK are based on the Linux kernel and Mer. In particular, Sailfish 2.0 runs on the Linux kernel 3.6.11 (for comparison, Android 6.x runs on the Linux kernel 3.18.10 and higher). Tizen OS officially used on two models of smartphones: Samsung Z1 and Samsung Z3, operating system Sailfish OS used on smartphones Jolla. There is business firmware Sailfish OS for the following devices: Jolla1, JollaC (*Intex AquaFish*), Turing Phone, Fairphone2 (for comparison: Android is installed on more than 24,000 unique models of devices, iOS–on 30 unique models). Sailfish OS is more convenient in terms of compatibility with Android, for example, it is possible to launch an arbitrary .apk package for Android. Sailfish OS can be ported to an Android device. Tizen OS and Sailfish OS are characterized by a closed server infrastructure and similar methods of service integrations (location, push notifications, etc.). The OS includes a number of closed components, for example, user interface elements, hardware drivers, basic applications, and application compatibility libraries with Android. By default, the basic applications of these systems are closed. For this reason, the complexity of the adapting applications, such as "Contacts", "Email", and "Calendar", is high enough.

However, in general, the Sailfish OS source code compared to Tizen's one is more open.

To develop applications running Tizen OS, a closed development kit Tizen SDK in the programming language C ++ (basic logic) and HTML5 (UI) was developed. There are two sets of APIs Tizen: WebAPI based on the Web Framework (a match of Javascript and HTML5), which is positioned as the main development environment for applications.

(Continued)

Inset 1.11 Continued

Sailfish OS applications can be developed using Samsung IDE based on Eclipse, and also Jolla SDK based on library Qt in C++ (basic logic) and QML (UI). Python 3.x is also supported. The user interface is based on the closed source code of Silica (from the Qt library). We note that Sailfish OS includes Mer projects (as well as Nemo and Qt Creator as an IDE) with open source, using various licenses (mainly LGPL). Mer is a Linux distribution for mobile systems, optimized to support development using Qt and HTML5. Tizen OS is somewhat behind the Sailfish OS in terms of integration with the server side infrastructure (there are a number of technical problems with the use of VoIP connectors, "Calendar" and "Contact"), as well as the fact that Tizen does not have the ability to build the installed image from the public repository.

In turn, Sailfish OS does not have a centralized source code repository. From the point of patent purity view, the Tizen OS and Sailfish OS almost do not differ from each other (e.g., Sailfish OS is protected by the corresponding Jolla patents). Currently, Tizen OS leaves behind Sailfish OS in the readiness to meet Russian requirements in information security. Both OSes use the digital distribution of applications as an application store like the App Store for the iOS system and the Play Market for Android. At the same time, Tizen OS supports several repositories for application distribution, which is similar to the Android infrastructure. Applications are signed with the developer and a distributor key (application store). The application security is carried out by the owner of the store resource before deployment. Sailfish OS has an official Jolla store (*Harbour*), similar to the infrastructure for iOS.

Analysis of security requirements

The development of the security profile was carried out based on the original Tizen OS version 3.0 (2013), which supports multi-user mode for 64-bit architecture of Intel and ARM processors.

At the same time, the requirements of All-Union state standard "Information Technologies, security of mobile devices, and methods and means" (Figure 1.58).

While developing the Tizen OS Security Profile, were took into account the requirements for the OS in terms of information security (minimum security class is the sixth, maximum is the first) [42]. For example, for

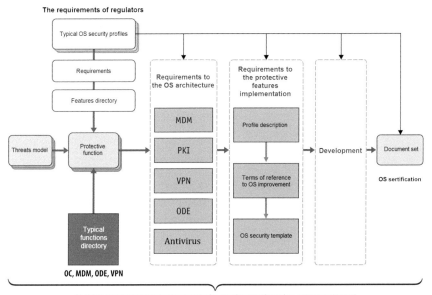

Figure 1.58 Requirements for the Russian Tizen OS Profile.

the OS of the sixth protection class, was considered that they are used in information systems that are not the state information systems, the personal data information systems, the public information systems, and are not purposed for processing restricted access information contained national security information.

For the OS of the fifth protection class, it was taken into account that their use is limited to the state information systems of the third security class* in the absence of interaction of these systems with information and telecommunication networks of international information exchange, as well as the state information systems of the fourth security classes, the personal data information systems, the level of personal data protection** in the case of the urgency of threats of the third type**, and the lack of systems interaction with information and telecommunications networks international exchange of information, as well as the need for the fourth level of personal data protection.

For the OS of the fourth class of protection, it was taken into account that their use is limited to state information systems of the third protection class in the case of their interaction with information and telecommunications networks of international information exchange, as well as state information systems of the first and second security classes*, personal data information systems, if necessary ensuring the third level of personal data security** in case of the urgency of the second type threats** or the interaction of these systems with info-telecommunication networks of international information exchange, as well as if necessary to ensure the first and second levels of personal data security**, in general use information systems of the second class***. In addition, the recommendations and requirements for the safety functions of a number of other methodological documents of the FSTEC of Russia were taken into account.

Analysis of security threats

The conducted analysis of the security threats for the mobile Tizen OS required standoff of the following typical information security threats to [41]:

- Access to information, processed on the mobile device from the external violators by overcoming or disabling mechanisms for identifying and/or authenticating the operating system;
- Intentional unauthorized access to the file system objects and the mobile device interfaces from the internal intruder, for which the requested access is not allowed;
- Receiving by the violator of the access to the information processed on the mobile device during the period when the authorized user left the mobile device without completing the access session to the OS (without blocking the mobile device);
- Restriction of the user access to the resources of the mobile device on which the OS is installed or to the resources of the corporate information system due to the long-term retention of the computing resource of the operating system in the loaded state by the violator making multiple requests requiring a large amount of time for processing them;
- Inaccessibility of the computing resources (CPU time, RAM) for critical services of the operating system and application software (applications) functioning in the mobile device due to inefficient distribution of resources between service and application flows (without regard to the degree of their criticality);
- Unauthorized or wrong information deletion from the memory of a mobile device running under OS control;
- Leakage or an unauthorized modification of information in the RAM of the mobile device used by the various processes and the data streams they generate;

- Unauthorized entry of changes to the operating system's registry (or other configuration data directory) that affect the operation of individual services, applications, or the operating system of the mobile device as a whole by the intruder;
- Recovery (mining) of an authentication information of administrators and users of the mobile device OS;
- Use of identification and initial authentication information corresponding to the user account of the mobile device OS by an intruder;
- Unauthorized modification of security event logs of the mobile device OS by the violator;
- Unauthorized access to information due to the use of nonstandard software by users of the mobile device;
- Violation of the integrity of the mobile device OS software components;
- Disabling and/or bypass the OS components that implement information security features replacing it with a bootable operating system by intruder;
- OS data integrity violation (including settings of information security mechanisms);
- Unauthorized access of the intruder to the authentication information of administrators and/or mobile device OS users;
- Unauthorized entry by the violator of changes to the OS security event logs due to access to log files in the OS operating environment using special software tools that provide the ability to process the OS security event log files;
- Unauthorized copying of information from the mobile device memory to removable computer storage media (or outside the information system) by the mobile device administrator or user for the purpose of further misuse;
- Interception of control commands from the server OS to the network client OS with their subsequent unauthorized modification or deletion;
- Performance decrease of the mobile device OS due to the introduction of redundant software and its components into it.

To neutralize these information security threats, the corresponding security mechanisms and functions that were part of the Tizen OS Security Profile [41], were developed, the feasible architecture of which is shown in Figure 1.59.

A comparative analysis of the security mechanisms of the Tizen OS and Sailfish OS is given in Inset 1.12.

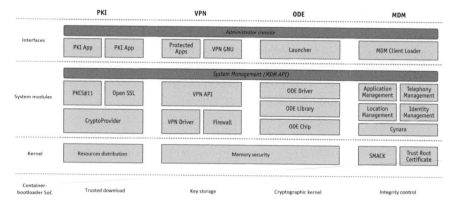

Figure 1.59   Feasible Tizen Security Profile architecture.

Inset 1.12

Comparative analysis of the original security mechanisms for mobile Tizen OS, Sailfish OS, Andriod, and iOS

Operating system security mechanisms

In the Tizen OS, both standard (UNIX user identifier, file system permissions, discretionary access control (DAC)) and additional security mechanisms (SMACK, MAC access control, etc.) are implemented. In Sailfish OS, only standard security mechanisms are implemented.

On Android and iOS, both standard and additional information security mechanisms (MAC, additional functions of the Trusted BSD kernel for iOS, the policy of mandatory SELinux access control (starting with version 5.0), etc.) are implemented.

Application security mechanisms

Tizen OS applications run on behalf of an unprivileged user. Each Web application runs as a separate process inside the Web application runtime. SMACK rules are used to further enforce separation of applications (sandbox).

Sailfish OS runs its own applications on behalf of a user who has the intermediate privileges and nonoriginal applications on behalf of a separate, unprivileged user identifier, and also supports the installation of Android applications in the emulation environment (the entire Android security model is not implemented).

Inset 1.12 Continued

Android OS and iOS. Android runs all applications on behalf of an unprivileged user (sandbox), iOS runs its own and nonoriginal applications in two isolated MAC sandboxes, but with the same unprivileged user ID.

Access control architecture

Tizen OS uses the SMACK access control system along with a hierarchical set of privileges (OS vendor, mobile platform operator, developer, etc.). This is the most reliable and secure access control architecture compared to other operating systems.

Sailfish OS uses the standard mechanism for sharing Unix user accounts.

Android OS and iOS. Android uses a sandbox (separate user ID with additional MAC policies to secure sensitive system services is assigned to each application), and iOS uses the functionality of the Trusted BSD kernel MAC.

Multi-user mode

Tizen OS supports multi-user mode. The access control system reliably monitors the management and data connections in each process.

At present, Sailfish OS does not support multi-user mode. Applications run under the same common user ID.

Android OS and iOS. Android supports multi-user mode, starting with version 5.0, and iOS is not designed to work in multi-user mode.

Network security tools

The Tizen OS includes a standard IPsec VPN client by Samsung with a closed source code. It also supports TSL connections using the standard OpenSSL library.

Sailfish OS offers TSL connections using OpenSSL. In addition, there is an unoriginal VPN solution based on OpenVPN.

Android OS and iOS. They support standard TSL-connections; in addition, the Android OS supports PPTP VPN, iOS-L2TP, IPsec, and OpenVPN.

Local data protection

Tizen OS offers an API for cryptographic protection of local media data, while the standard operating system configuration does not support transparent disk encryption modes, as well as full cryptographic protection of local media data.

(Continued)

Inset 1.12 Continued

Sailfish OS does not provide any special tools for memory protection (integrity control, cryptographic protection, etc.).

Android OS and iOS. Support the established mechanisms of cryptographic protection of local media data.

Cloud storage protection

For the Tizen OS, third-party applications are designed to provide protection for cloud storage.

For Sailfish OS, third-party applications have also been developed to help protect cloud storage.
Android OS and iOS. Do not support the cloud storage protection.

Digital rights managment (DRM)

Tizen OS supports DRM only for partners using a special API.

Sailfish OS does not support DRM.

Android OS and iOS. Android supports DRM through Media DRM interface, and iOS has Fair Play for DRM support.

Trusted download module

Tizen OS supports trusted platform module (TPM) with a key generation and memory blocks along with the secure boot functions.

Typically, Sailfish OS does not support trusted platform module (TPM).

Android OS and iOS. They support trusted platform modules (TPMs).

Infrastructure security

Tizen OS supports a large number of repositories for various applications; the latter are signed by developers and suppliers (in the application store).

Sailfish OS supports a similar application implementation scheme. The official application store Jolla (Harbour) is the only official provider of its own applications.

Android OS and iOS. Android OS supports a large number of repositories for applications. At the same time, Google regularly conducts application security audits in the Google Play store. iOS only supports Apple's official app store. Applications are signed centrally after the individual quality control of each application and the appropriate security audit procedure.

Inset 1.12 Continued

Organizational aspects of security

The Tizen Association established a special team to respond to cyber security incidents.

Manufacturers of Sailfish OS have created a special contact center for rapid response to computer security incidents.

For Android and iOS, special teams have been created to respond to computer security incidents, and related 24-hour contact centers.

As a result, the Russian safe OS Tizen profile was created. It is significant that this profile supplemented the basic open code of the Tizen OS kernel with special functions and security procedures that meet the requirements of the Russian security systems.

This allowed us to meet the corresponding requirements set by the regulators to trusted mobile solutions. In particular, the following requirements for the networked client operating systems were implemented: OS installation includes only modules directly used in the information system:

- Services for receiving application software from resources in external information and telecommunication networks are excluded or disabled;
- Information storage services on resources in external information and telecommunication networks are excluded or disabled;
- Information protection from unauthorized access (UA), including verification of the integrity of application software modules in accordance with the requirements of FSTEC of Russia;
- Encryption of information processed on the mobile device using cryptographic methods of information protection in accordance with the requirements of the FSS of Russia, etc.

In the course of the work, a threat model for mobile OS was developed. The typical security functionalities of the mentioned OS, as well as the known information security tools (MDM, ISF from UA, CIPF) were singled out and significantly supplemented, which allowed to take into account the minimum requirements of the FSTEC security systems of Russia and the FSS of Russia.

Then, the Tizen Security Profile architecture, which meets the maximum security requirements, was developed and proposed. The relevant terms of reference for the development of the safety profile and its "default" values have been prepared. A package of documents was prepared for the certification of the domestic security profile of the Tizen OS with a view to its further use within the framework of distance educational technologies in accordance with the information security requirements of the FSTEC of Russia and the FSS of Russia.

1.3 Cyber Security Threats Monitoring Necessity

This section states that the number of vulnerabilities in the information systems of the Russian subway annually increases. The reasons for the security problems of the information resources of the industry and the facts of the successful cyber-attacks on various subway systems are given. The problems of the cyber security of the subway are outlined, the security systems used in it are described, and the conclusion is made that the level of security of the Russian metro is not inferior to many urban transport systems in other countries. Measures are proposed to further improve the security of the critical information infrastructure of the industry in question.

1.3.1 Cyber Security Incidents Factual Account

On October 24, 2017, there was a powerful and already fifth cyber-attack of Bad Rabbit that year in several CIS countries. As a result, a number of the most important strategic objects of transport infrastructure suffered. The cyber security scenario was typical for this malware type: the virus encryptor had a code similar to the code NotPetya, which was sent to the workstations by the attackers at the same time. Further, users of infected workstations were offered to update the Abode Flash Player. If this action was confirmed, the malicious software was downloaded. At the same time, access to the critical infrastructure was blocked, and for its unlocking, a requirement (via screen instructions) was to transfer a certain amount to a specified account using an anonymous Tor network.

This was an extremely unpleasant situation for any information resource owner, but when the target of attackers becomes critical infrastructure objects, the situation turns to become dire. This raises the question: to what extent the Russian objects are classified in this category protected from this kind of cyber-attack, considering this issue on the example of transport infrastructure in the whole and of Russian subway, in particular.

According to the Information Security Center of Innopolis University[30], there is a rapid increase in the number of new vulnerabilities found in the

---

[30]https://university.innopolis.ru/research/tib/csirt-iu/.

Russian subway information systems. Therefore, from 2005 to early 2010, only 10 vulnerabilities were found, and after the Stuxnet worm's appearance in 2011–2012, there were more than 150 vulnerabilities. In 2016–2017, there were reports on the detection of 1,000 previously unknown vulnerabilities, with about 75% of them being of high and critical risk. This value is much higher than the same parameter for commercial information systems, which indicates a low level of CPCS information security in general [12]. At the same time, SCADA and HMI systems (Figure 1.60), programmable logic controllers and field devices, various applications, and technological protocols are of the greatest interest for attackers [3, 6–9, 12, 43, 44].

The solutions from Siemens Company[31] are among the commercially successful ones in the Russian market [3, 6–9, 21]. The most common technologies for data transfer are Modbus (RTU and TCP/IP), Profibus/Profinet, which have a market share of approximately 33% each, followed by OPC with a 25% share. In the programmable logic controllers segment, Siemens solutions are leading (31%), followed by Schneider Electric (11%), ABB (9%), Allen-Bradley (7%), and Emerson (5%). MS Windows is still leading among the operating systems used with the CPCS.

Let us consider that the USA and European metropolitan railways are leading in terms of the number of CPCS accessible from the Internet, while 50–75% of such SCADA systems are vulnerable and can be hacked [3, 7, 8].

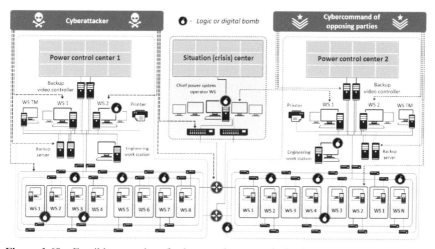

Figure 1.60 Feasible scenarios of cyber-attacks to a typical subway management system.

---

[31]http://www.tadviser.ru/index.php/*Статья*:IIoT_-_Industrial_Internet_of_Things_

The third place is occupied by the metropolitan railway of Asia (more than a third of the total number of CPCS are accessible from the Internet and more than 60% of them are vulnerable). In Russia, the availability of subway information systems from the Internet is even lower. The vulnerable are about 50% of CPCS accessible from the Internet (less accessibility from the outside and a more compact controlled area). At the same time, most of the security problems of the Russian subway information systems are due to the presence of hardware and software bugs (including the so-called "police modes") in Western equipment, configuration errors (use of expired factory passwords), lack of continuous updates, insufficient means of monitoring hidden channels management and ensuring the sustainability in general, noncompliance with security policies, etc. [3, 7–9, 17, 21, 43, 44].

Known examples of cyber-attacks

At the beginning of 2015, the "Maxima Telecom" company implemented a unique and advanced project in the Moscow subway. It was a mobile Wi-Fi network based on Cisco advanced solutions (access points Cisco Aironet 2600 series, controllers Cisco 2504/2960 series and routers Cisco series 809F, and switches Cisco Catalyst series 6500 as a core network). Moscow became one of the first cities where free Wi-Fi became available to passengers inside wagons in trains on all 12 city lines. Points of wireless Internet access were equipped with more than 5,400 wagons. At the same time, each of the trains is connected to the network at a speed of up to 100 Mbps. The network bandwidth is more than 20 Gbit/s (Figure 1.61). More than 1.2 million people connect to Wi-Fi network of the Moscow subway every day and pass up to 70 Tbyte [3, 7–9, 17, 12].

May 28, 2015, Wi-Fi access points in the Moscow subway were hacked[32]. As a result, users connected to the network displayed pornographic content in the browser, and there was no Internet connection (the press office of the Internet service provider "MaximaTelecom" did not recognize the fact of the cyber-attack).

On March 23, 2016, members of the terrorist group DAISH (ISIL) (outlawed in Russia) attacked the Wi-Fi network of the Moscow subway. (The MaximaTelekom press office also did not recognize the fact of cyber-attack). When trying to connect to a Wi-Fi network on the Kaluzhsko–Rizhskaya line between the Rizhskaya and Alekseevskaya stations, the DAISH flag and a

---

[32]http://www.rbc.ru/rbcfreenews/556700709a7947a0a5193a2d.

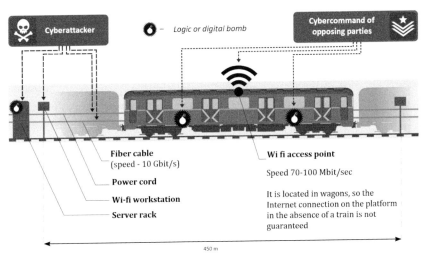

Figure 1.61 Feasible scenarios of cyber-attacks on the Moscow subway Wi-Fi.

threat message ("Yesterday was Brussels, today is Moscow"[33]) appeared on the screen of mobile devices. In November 2015, after the terrorist attacks in Paris, a similar case occurred. Then, the passengers of the subway, trying to connect to the network, also saw a similar message.

Topical issues of cyber security

According to the main performance indicators and the security level, the Russian subway looks good against the backdrop of global urban transport systems, especially European and American ones. The matter is that the Russian subway (covers seven cities of Russia) goes and is operated autonomously, and in most large foreign cities a multimodal system is adopted, i.e., the metropolitan is organizationally and financially integrated into the general structure of urban transport: ground and underground subway, railways, buses and trolleybuses, tunnels, and bridges. Other organizational structure does not disrupt the Russian subway in terms of protection and information security to be a strong medium among foreign transport conglomerates [3, 8, 9, 12, 17, 19, 20, 23, 24, 44, 45].

It should be noted that over the past 3 years, one of the most advanced systems for ensuring the antiterrorist security of passengers has been created in the Russian subway, which is constantly developing and improving.

---

[33]https://www.securitylab.ru/news/480345.php.

At the entrance to the subway station, inspection zones appeared with modern equipment. On the platforms are installed explosion-proof containers and detectors of explosive vapors, as well as equipment of high radiation detection. The subway security inspectors with special skills to identify suspicious persons in the crowd are staffed. In addition, projects are implemented to ensure cyber security and the subway protection from terrorist threats, including a comprehensive automated information support system for the antiterrorist protection of the underground (KASIP AZM), which includes advanced intelligent systems of video surveillance, access control, secure radio communications, standard information protection from NTD and cryptographic protection, antivirus protection, detection and prevention of cyber-attacks, etc.

In addition, there are many other problem areas for security, which should be paid the closest attention precisely in terms of their cyber security.

The subsystem of video monitoring, which provides:

- Operational picture of the facility state with high timeliness and informative value;
- Recognition and automatic control of video images;
- Archiving and processing of video information; the historical data release on requests;
- Operational management of video capture devices from the operator desk of a single situation center (SSC).

The subsystem of digital operational communication, which provides:

- Speech communication and digital information exchange between the situation center and services, arrays, entities, etc.;
- Reception, processing, verification, and registration of emergency reports (emergency situations) received from citizens through emergency loud-speaking communication channels, wireless access channels, via short text messages (SMS) or via the Internet on SSC workstation;
- Automatic and controllable broadcast to all subway staff members about emergency situations, generation of the warning scenarios, text, voice, and video messages, interfacing with the All-Russian Integrated Public Notification and Warning System of Russian Emergency Ministry (OKSION), broadcast to the responsible employees of the Emergency Ministry and security ministries, dispatchers of emergency services through city, cellular, and other networks of dial telephony, text messages, and e-mails.

The subsystem of automated and integrated control, which provides:

- Centralized (using the SSC), and, if necessary, distributed, integrated management of forces and assets;
- Operational situation analysis, issue the orders and directives based on management object location;
- Disseminating the setting and orders to the action officer, performance control.

The subsystem of documentation, which provides:

- Automatic documentation of messages received via emergency messages, communication channels;
- Automatic documentation of SSC operator actions;
- Storage of audio-recordings of negotiations and video recordings of the video monitoring system with time reference;
- Ability to create both automated and manual reports, statements and emergency certificates, the single service actions, entities, etc.;
- Automatic documentation of the monitoring system data for managing objects and related information systems;
- Ability to exchange information with other information systems.

The subsystem of technological control and management (Figure 1.62), which provides:

- Component performance monitoring;
- Remote control and configuration of the end devices;
- Device self-test and results output to the system operator workstation;
- Operational indication of equipment failures and the output of diagnostic messages;
- Restriction of various system user groups' rights.

The subsystem of information security from unauthorized access and cryptographic protection itself, which provides:

- Integrity control and authorized access to video information, databases, and archive;
- Secure data transfer in open telecommunications networks;
- User authorization based on group security policies;
- Detection, prevention, and recovery due cyber-attacks;
- Antivirus protection, etc.

The cyber security of the subway and transport infrastructure in general is a relatively new activity for Russian information security specialists.

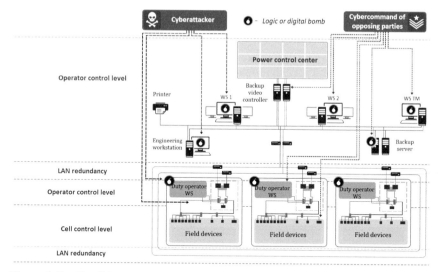

Figure 1.62   Feasible scenarios of cyber-attack on the typical subway supervisory system.

First of all, the issues of ensuring the availability and security of open networks when they are used in alarm systems, as well as the knowledge consolidation and the accumulation of practical experience in responding to cyber security threats, including automation and telemechanic systems protection from group and mass cyber-attacks, are topical.

According to the report of the Information Security Center of Innopolis University, during the summer and September 2018, the average flow of security events was up to 57 million events per day. About 15,000 of them are associated with the identification of malicious content that did not cause damage, and about 1,000 with massive infection of a large number of critical infrastructure devices, and the share of critical incidents was more than 18.7%, which means that almost every fifth incident was critical. This dynamic correlates with the overall increase in the intensity of group and mass cyber-attacks on Russian critical infrastructure. In our opinion, the attacks in 2017 (Wanna Cry, Petya, and Bad Rabbit) are just trial experiments to destabilize Russian critical infrastructure.

At the same time, the number of unknown and, accordingly, undetectable cyber-attacks is still between 40 and 60% of all feasible. In order to discover them, all-new methods are required, the so-called invariant approaches to the detection and prevention of cyber-attacks. Thus, at present, projects and works on security audit are especially relevant for the Russian subway,

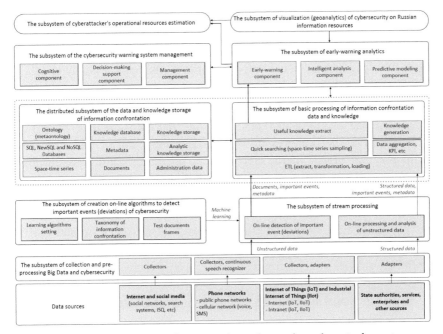

Figure 1.63 An example of a prospective early warning cyber-attacks center.

and also on the creation and development of special operational centers for responding to cyber-attacks (Figure 1.63).

We note that in the USA and the European Union, there are about 50 large centers for detection and prevention of and recovery from cyber-attacks only in the transport infrastructure, while the whole of the Russian Federation has 5 corporate and 20 state centers belonging to the state CADPR system (system for cyber-attacks detection, prevention, and recovery), and only one industrial center (CPCS of Kaspersky Lab[34]) for responding to information security threats.

Moreover, most Western and Russian systems have one common problem: they run only upon cyber-attack.

This problem may lead to the fact that after a sufficiently powerful (previously unknown) cyber-attack, there will be nothing to secure. In addition, malware constantly mutates and develops so rapidly that developers, who create antiviruses and associated systems for detecting and preventing intrusions and anomalies, often do not have time to issue appropriate signatures.

---

[34]https://university.innopolis.ru/news/cyber-security/.

Moreover, the classic signature method of the most known antiviruses and IDS/IPS systems, which involves scanning and capturing the corresponding images or patterns of data traffic and then comparing against the signatures of malicious software and the launching of security measures, is already insufficiently effective. At present, all-new methods of combating cyber-attacks are in demand:

1. The so-called profiling methods, implying the early creation of some standard of the normal system operation, comparison of this standard with the real system parameters in real operating conditions and, if necessary, the immediate launch of appropriate cleaning and protection procedures;
2. Invariant methods that combine multi-model approaches and techniques for recognizing both known and previously unknown group and mass cyber-attacks based on deep semantic invariants of the CPCS routine behavior.

1.3.2 Need for Joint Initiatives of Society and States

This was the motto of the fourth Conference "Industrial Control Systems" (ICS) organized by Kaspersky Lab at Innopolis University (Innopolis, Tatarstan). The key topic of the event was modern trends in ICS cyber security.

In the recent years, the number of cyber-attacks on industrial facilities has significantly increased. Therefore, the information security companies endeavor to offer their industrial solutions to ensure cyber security.

The conference was held with the support of the Ministry of Communications and Mass Media and the Ministry of Industry and Trade of the Russian Federation, the Ministry of Communications and Information Technology of the Republic of Tatarstan, the leading enterprises of the military industrial sector, the leaders of the high technologies city of Innopolis, and Innopolis University.

The forum near Kazan brought together more than 200 leading Russian and foreign experts and experts of industrial security systems, as well as a number of the heads of large enterprises of the electric power, nuclear, oil and gas, transport, engineering, and other branches of the modern economy.

As it was pointed out by the high-level speakers, attacks on industrial systems have become a reality and constitute a global threat not only for the economics, but also for the lives of people. In this situation, joint efforts of the state, the expert community, and specific specialists from different economic sectors are required.

During the two conference days, the participants learned what threats to industrial systems are coming to the fore today and will be relevant in the near future, and received recommendations on what should be done to successfully protect them. In addition, Kaspersky Lab announced the opening of the first in Russia, the center for responding to computer incidents on industrial and critical objects–Kaspersky Lab ICS-CERT.

In touch with law enforcers

According to Andrey Dukhvalov, the head of the Kaspersky Lab's Advanced Technologies Department, the main goal of Kaspersky Lab ICS-CERT is to coordinate the actions of automation systems manufacturers, owners and operators of industrial facilities, as well as researchers in information security. CERT will collect information on the vulnerabilities found, incidents, and actual threats, and based on these data provide recommendations for the protection of industrial and critical infrastructure facilities.

In addition, ICS-CERT is going to hold consultations on the requirements of the state and industry regulators in the field of information security of industrial facilities. Also, the center specialists will be able to assess the level of industrial automation system protection and conduct an investigation of information security incidents. The ICS-CERT data and services will be available free of charge to interested enterprises around the world.

It is expected that the main center customers will be the producers of ACS TP components, national CERTs, and industrial enterprises operating in a wide range of sectors: energy, engineering, oil and gas, metallurgy, construction materials, transport, etc. Also, Kaspersky Lab ISC-CERT is ready to collaborate with third-party researchers of information security, government agencies, and international law enforcement organizations.

The cyber fighters gap

We can say that the world has been shaken by the massive hacker attacks on enterprises in recent years. As an example, we will call to memory some of those cyber-attacks.

In 2012, Saudi Arabian company Saudi Aramco, one of the world's largest oil producers, acknowledged that it was subjected to cyber-attack. The infected message came as an e-mail, as a result of which the work of 35,000 computers was paralyzed. Oil shipments inside the country were suspended.

In December 2014, the German steelmaking plant was attacked. The attacker actions led to numerous failures of the CSCP components and to a shutdown of the entire system.

A year later, a group of hackers broke into the Ukrainian energy company. The intruders carefully planned and coordinated the multi-level cyber-attack. As a result, there was a 6-hour power cutoff in five regions and disabling the remote control system.

However, despite the fact that the damage from the cyber criminals activities is growing every year, the conference participants found disappointing conclusions that not only in Russia, but also in the whole world today, there are very few training centers for the basics of industrial cyber security.

What is the reason for the unpopularity of the cyber-speciality institutions, not only in Russia, but also in the whole world, and with their obvious necessity?

According to experts, the society is only ripening a request for quality solutions in cyber security of industrial facilities. Therefore, at the time when the standards of information security office have already been formed and successfully applied, while Russian universities prepare highly qualified specialists for them, it is not yet possible to speak about the similar achievements in industrial cyber security. Of course, this specialty will eventually form, but experts say, it will not be a one-time process. Although, according to forum participants, already now individual training centers have begun to show interest in industrial cyber security and began to appeal to companies working in this direction with requests to help in the development of the appropriate courses.

The conference participants spoke about the Russian solutions gaps to protect enterprises from cyber-threats. Moreover, it turns out that not only Russia suffers from a lack of comprehensive solutions in the sphere of industrial cyber security. In May 2016, 40 companies around the world declared that industrial cyber security is also in their sphere of interest.

Although, from the point of view of experts from Kaspersky Lab, there are only five or six companies on the planet whose developments in the CSCP protection are worthy of close attention.

Sometimes the lack of professionalism in industrial cyber security goes to ridiculous, because often to prevent hacker-attacks on industrial facilities, security specialists use conventional antiviruses. This, in the expert opinions, is not entirely correct, since there are differences between information systems and CSCP. The decision that proved itself to be a reliable defender of information systems, unfortunately, can be completely useless in disrupting hacker-attacks on enterprises.

There are attacks one by one

In general, the current situation with the enterprise cyber security in Russia is far from prosperity. In 2017, Kaspersky Lab recorded 2,202,506 antivirus software actuations in enterprises PC, which include 104,663 malicious software. By the way, the Republic of Tatarstan, where the conference was held, is in the top 5 regions of the Russian Federation in terms of the number of attacked computers in enterprises PC (4% of all cyber-attacks). In Tatarstan, in comparison to many other subjects of Russia, modern and effective solutions are actively used to protect industrial process control systems. However, the trend is the more potentially attacked systems there are, the more incidents there will be.

Here, of course, the legitimate point: why do the nuclear cycle enterprises, defense complexes, and oil and gas and energy companies often still become victims to hackers?

The possible reasons:

- Typical "office" cyber security facilities, which security experts prefer to use for the CSCP protection, are ineffective;
- Low awareness level of security specialists and business leaders about the industrial cyber security problems, based on a mixture of rumors, conjectures and facts, as well as the lack of reliable information;
- Shortage of specialists, few good practices;
- There are no persons responsible for CPCS information security at the enterprises;
- Illusion of sufficient security, a wrong opinion about the inaccessibility of the facility due to the presence of a guard post, video cameras, and barbed wire around the perimeter.

The potential of the Russian cyber market

The conference participants agreed that the market of industrial cyber security has considerable features. It should be noted that interest in the very topic of industrial cyber security is constantly growing.

For example, the annual conference "CPCS Cyber Security 2016: The Time to Play Together" was conducted by Kaspersky Lab for the fourth time. If initially there were delegates from 50 enterprises, this year 100 companies took part in the event. And the organizers of the forum express their confidence that their number will only increase.

In addition to the mandatory program, the conference also hosted the final stage of the All-Russian contest for the detection of vulnerabilities in models with real industrial equipment–Capture the Flag (CTF).

1.3.3 Capture the Flag Competition on Vulnerability Detection

Within the framework of the 4th Industrial Control Systems (ICS) Cyber Security Conference 2016 organized by Kaspersky Lab, Capture the Flag (CTF), a final stage of competition in discovering the vulnerabilities of models based on real industrial equipment, was held.

At Innopolis University (Republic of Tatarstan), four teams selected following the results of preliminary qualifying stages, from Dolgoprudny (Moscow region), Yekaterinburg, Novosibirsk, and Saratov, fought for the main prize.

The CTF tournament resulted in a learning break-in of a simulated power system built on the basis of a microgrid architecture–its work has been stopped in less than a day. This time won the team Filthy Thr33 from Ekaterinburg. The battle CTF is a team game, the main purpose of which is to capture the opponent's flag. The CTF format can be used in paintball, computer games, and computer security. In cyber security, CTF competitions could be held remotely (on the Internet) and face to face. In the first case, teams compete in the Internet, from the different parts of the world. Each team member (including the captain) must be over 18 years of age. Such competitions almost always last more than 24 hours in the nonstop format. At face-to-face competitions, teams gather in one place, each at their own table. Such competitions last for 7–8 hours. Often the practice of combining formats is practiced: for example, the qualifying stages for some CTF competitions take place on the Internet, and the final is in the city of the competition organizers. The victory in the qualifying online round gives teams the right to participate in the full-time competition.

There are several formats of CTF competitions in cyber security. First, it is a task-based (or jeopardy) format–when players are given a set of tasks (tasks) to which they want to find the answer and send it. The aim is a "flag": it could be a set of characters or any phrase. For a correctly executed task, the team receives a certain number of points. The harder the task, the more points it takes to get the correct answer. All tasks in the CTF competitions in the task-based format can be divided into several categories: for example, administration, cryptography, and steganography tasks; assignments to find Web vulnerabilities and favorite tasks from joy category–entertaining tasks of various subjects. One of today's favorite activities from the joy category is to make a collective selfie. And this activity usually performs with great success in many CTF competitions!

The second format of the CTF competition is the classic format. In the classical scheme, each team receives a dedicated server or a small network to maintain its functionality and security. During the game, teams receive points for the correct operation of their server's services and for the stolen information (also "flags") from the servers of rivals.

Currently, team competitions on information security of the CTF format have become widespread not only in Russia, but throughout the world. However, CTF competitions on industrial safety of SCADA differ in their features. This is a new kind of competition that is rapidly developing and becomes popular.

Hackers from all countries

The current finals were preceded by qualifying competitions for 154 teams, mostly from Russia. Participants from Belarus, Ukraine, Kazakhstan, Turkey, Sweden, Romania, India, and China aspired to try at breaking the power system on the basis of a microgrid architecture.

The participants of the competition had to get inside the power system, to find out the object schema presented in the form of the mount that imitates the model of the digital substation distributed in Russia, built in accordance with the IEC 61850 standard. Further, it was necessary to gain control over the management system, to turn off or disable the components responsible for the safety and continuity of electricity transmission, to achieve misinformation of the dispatch center, and finally to damage the physical equipment, for example, to make a short circuit on the ETL model.

For each task, the participants have awarded a certain amount of points that are different depending on the complexity of the task.

The game is over

As a result, those teams that carried out their tasks, in breaking the energy system prepared by CTF's organizers of the competition, better and faster than others, got a set of industrial devices that were integrated into a micro network. The conditions of the final tournament were as close as possible to the real ones, for example, at the solar station included in this micro network, the controller was used, which is widely used in such systems. Finally, separate components of the power system, as well as its architecture, in general, were checked. The winning team–Filthy Thr33 from Ekaterinburg – was the first to succeed, which means, it violated the power system, arranged a short circuit, modeled the local damage, and thus deprived consumers of the energy source (in the contest was modeled a small town and factory). As a prize, each team member received a program-defined radio HackRF One, which is capable of transmitting or receiving radio signals with a set of aerial wires.

The completed CTF tournament "Kaspersky Lab" was held for the second year in a row, and, according to experts, it becomes a good tradition of the industry conference on industrial cyber security.

According to Vladimir Dashchenko, a senior researcher of threats in the critical infrastructure of Kaspersky Lab, the conditions of the CTF tournaments organized by the laboratory are always as close to the reality as possible, because their purpose, first of all, is to understand the potential vulnerabilities and shortcomings in the critically important infrastructures and industrial systems, which the whole world uses today.

1.3.4 Security Operations Center (SOC) Key Role

Typically, a system solution such as the security operations center (SOC) means a comprehensive information security management (IS) system that allows to implement in practice the following operations:

- Monitoring the cyber security events;
- User activity audit;
- Vulnerability management/configuration control;
- Management of information security incidents;
- Control of the compliance with the requirements of legislation, international and industry standards, internal corporate policies, etc.

Therefore, with the assistance of the Cyber Security Threat Monitoring Center, it is possible to assess the current state of cyber security (the analysis task) [9, 45–50] of the metropolis critical information infrastructure, as well as to promptly eliminate the identified deviations to properly provide the required level of IS (synthesis problem) [47, 48, 51–54].

Nowadays, the considerable experience has been accumulated in using SAP, IBM, HP, and other foreign and domestic SOC and SIEM vendors to provide information security to critical information objects [9, 45, 47, 48]. Let us regard the difficulties that arise here and the way how they can be resolved in practice by the example of localization and adaptation of SAP solutions (Figure 1.64).

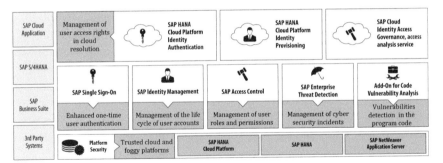

Figure 1.64 Example of localization and adaptation of SAP solutions.

These standard solutions include the following task solutions:
- *Access control:*

    1) SAP GRC Access Control;
    2) SAP Identity Management;
    3) SAP Single Sign-On;
    4) SAP Cloud Identity Access Governance.

- *Ensuring information security:*

    1) SAP UI Masking;
    2) SAP UI Logging;
    3) SAP Add-on for Code Vulnerability Analysys;
    4) SAP Solution Manager Configuration Validation and Early Watch Alert;
    5) SAP Security Optimization Service;
    6) SAP Enterprise Threat Detection.

- *Control of compliance with the requirements of regulators:*

    1) SAP Global Trade Services;
    2) SAP Nota Fiscal Eletronica.

- *Counteraction to fraud:*

    1) SAP Fraud Management;
    2) SAP Business Partner Screening;
    3) SAP Tax Compliance.

- *Risk management, organization of internal control and internal audit (GRC–Governance, Risk, and Compliance):*

    1) SAP GRC Risk Management;
    2) SAP GRC Process Control;
    3) SAP GRC Audit Management.

Additionally, the third-party solutions can be applied, including Greenlight (SAP Regulation Management and SAP Access Violation Management) and NextLabs (SAP Technical Data Export Compliance, SAP Enterprise Digital Rights Management, SAP Dynamic Authorization Management), and others.

Here, the listed solutions have to combine the following functionalities.

SAP GRC Access Control. The solution is designed to manage the roles and authorities of users in SAP-based systems at the level of business and technical roles, allows implementing the access risk control in accordance with the segregation of duties principle.

SAP Identity Management, Cloud Identity Access Governance. The solutions are designed to manage the life cycle of accounts and control access rights.

SAP Single Sign-On. The solution is intended for enhanced single user authentication in SAP systems.

SAP UI Masking. The solution is designed for confidential data in SAP-based systems viewing protection. It technically prevents users from seeing critical information.

SAP UI Logging. The solution is designed to obtain data concerning person, time, which resources in SAP-based systems, and from what location one had access. Provides registering the events of access to confidential information.

SAP Add-on for Code Vulnerability Analysys. The solution is designed for static analysis of source code written for SAP-based systems.

SAP Solution Manager Configuration Validation and Early Watch Alert. Solutions are part of the SAP Solution Manager and have, along with other functionality, the ability to check the security settings of SAP systems.

SAP Security Optimization Service. The service allows evaluating the current state of information security of SAP systems in terms of manufacturer's recommendations, etc.

It is significant that among the listed solutions, there is a solution that can be used to solve the problems of creating a system of intelligent cyber security monitoring centers for the critical information infrastructure of the megapolis (Security Operations Center–SOC) and SAP Enterprise Threat Detection (hereinafter, SAP ETD). However, before starting to elaborate in detail this solution, let us consider the main basic components of SOC (Figure 1.65).

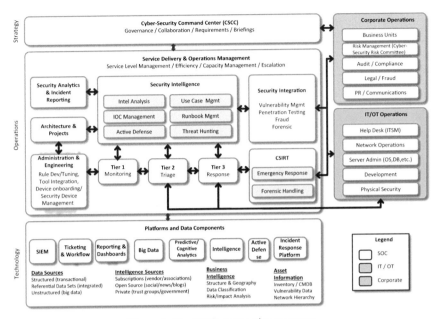

Figure 1.65 SOC perspective structure.

The main components of the SOS

Modern SOC is a complex organizational and technical system, which includes employees, processes, and relevant technical solutions. Here, among the technical solutions, the Security Information and Event Management (SIEM) system is identified and correlated (Figure 1.65).

The main sources of cyber security events gathering within the organization and their subsequent analysis for modern SIEM systems are the following:

- Network traffic (NetFlow, deep packet inspection, full packet capture);
- Application logs, security systems, network infrastructure, operating systems, database management systems (Log Management);
- User activity (user behavior analytics/user and entity behavior analytics UBA/UEBA).

For better understanding of the current status of SAP solutions in comparison with traditional solutions for SOC, let us consider the solutions of the three leaders of the quadrant Gartner: IBM, HP, and Splunk. Comparing Gartner's reports over a 11-year period (2008–2018), it is clear that HP and IBM retained their positions at the top of the quadrant, while Splunk was among the leaders (Figures 1.66 and 1.67).

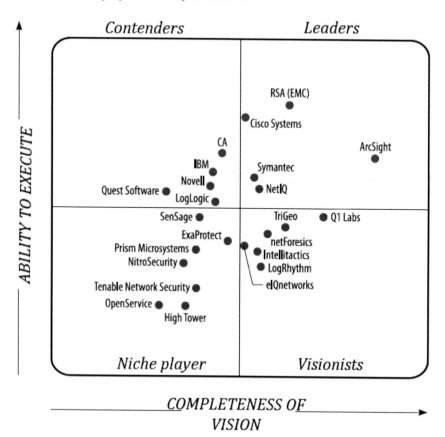

Figure 1.66   The "magic" SIEM Gartner quadrant, 2008.

What has stimulated these companies to become the leaders?

- *IBM.* The basis of IBM QRadar is the competence of the company in the field of network traffic analysis (NetFlow). Further development–the implementation of Log Management analysis and User Activity Analysis (UBA/UEBA)–allowed IBM QRadar to capture technological leadership in the SIEM marketplace. The emergence of IBM Watson for Cyber Security makes the IBM Qradar solution unattainable for competitors in the next 3–5 years.
- *Hewlett-Packard.* The basis of HP ArcSight is the company competence in the Log Management. The gap in the analysis of user activity is currently being addressed by HP through its partnership with Securonix.

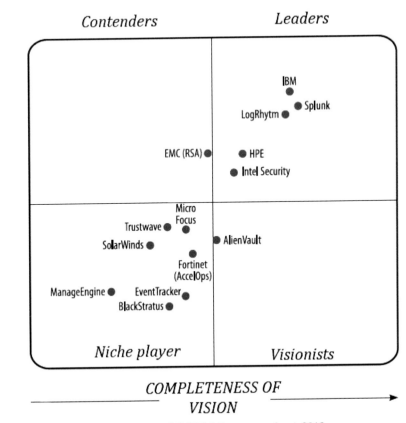

Figure 1.67 The "magic" SIEM Gartner quadrant, 2018.

- *Splunk*. The basis of the rapid take-off of the Splunk App for Enterprise Security solution is the unique competence of the company in the Log Management. Gap in the analysis of user activity Splunk solved through the acquisition of the Caspida company in 2015.

Let us consider that the previously mentioned technologies were adopted for the development of the Open Segment of the National Early Warning System on the Cyber-Attack on the Critical Information Infrastructure of the Russian Federation [45–47, 51, 55, 49] and are being developed in the following directions:

- Improvement of the UBA/UEBA functional;
- Possibility of event correlation development in real-time mode;
- Development of models of predictive analytics (prediction/warning);

- Improvement of methods and algorithms for detecting and preventing anomalies;
- Increasing the sources of events;
- Storing massive data;
- Convergence of various methods of information processing;
- Development of opportunities for cognitive analytics, and models and methods of artificial intelligence in general.

In Figures 1.68 and 1.69, Big Data on cyber security processing typical scheme shows possible examples of such improvement (e.g., SAP and IBM solutions).

Improvement of SAP ETD

Now back to the solution of SAP ETD, which was originally designed to identify incidents of information security in the SAP system landscape. SAP ETD collects information about cyber security events at the application level, which is not available to most manufacturers of traditional SIEM solutions (Figure 1.70). The main goal of the product is to protect the SAP application suite, the SAP S/4HANA system landscape and the SAP HANA database.

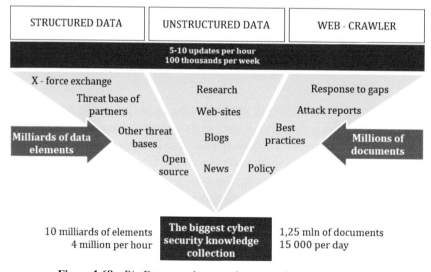

Figure 1.68   Big Data on cyber security processing typical scheme.

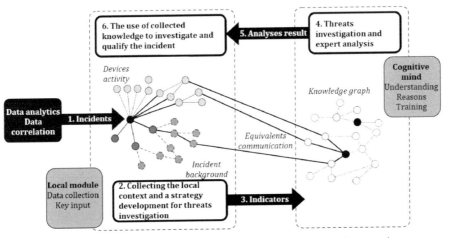

Figure 1.69 Analysis cognitive technologies of Big Data on cyber security.

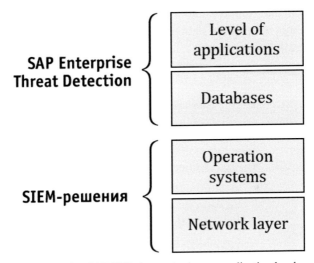

Figure 1.70 SAP ETD data gathering on application level.

The main purposes of this system are listed as follows:

• Cyber-attacks detection;
• Identification of confidential information leakage;
• Detection of unauthorized actions of internal users;
• Monitoring the state of cyber security;
• Incident investigation.

Typical functional capabilities of the solution include:

- Detection of attacks at the application level;
- Creating your own attack detection templates (in addition to available ones);
- Development of templates without programming;
- Correlation of historical data in real time;
- Ability to download data from other sources (in addition to SAP), for example, FireEye, Trend Micro, etc.;
- Innovative technology for analyzing logs.

Figure 1.71 Integration scheme of SAP ETD with HP ArcSight shows an example of integrating SAP ETD with HP ArcSight (Figure 1.72).

Currently, the following typical scenarios for the work of SAP ETD for monitoring the cyber security threats of the critical information infrastructure of the metropolis have already been tested:

- Identification and suppression of information pilfering (detection of uploading large amounts of data or frequent uploads);
- Logging into the system under a fake account (identifying accounts with a short lifetime);

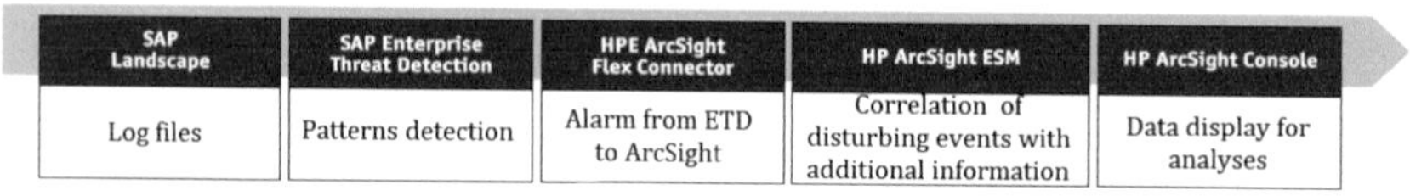

SAP Landscape	SAP Enterprise Threat Detection	HPE ArcSight Flex Connector	HP ArcSight ESM	HP ArcSight Console
Log files	Patterns detection	Alarm from ETD to ArcSight	Correlation of disturbing events with additional information	Data display for analyses

Figure 1.71   Integration scheme of SAP ETD with HP ArcSight.

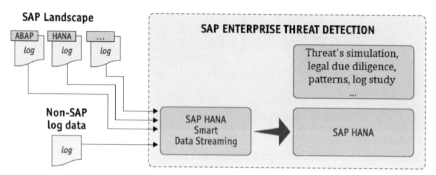

Figure 1.72   SAP ETD Architecture basis.

- Identifying the facts of hiding the actions of users and administrators in the system (changing security settings, for example, disabling audit settings);
- Detection and neutralization of the cyber-attack consequences, type "Denial of service";
- Detection of the facts of "rough" password mining;
- Identification and suppression of attempts to penetrate the productive system (attempts to connect from test systems to productive systems), etc.

As a result, the authors of the book have brought the number of templates in version 1.0 of SAP ETD to 40 and are continuing to develop them further [45–47, 49, 51–53, 55] from only 12 preset templates intially. The possible architecture of this system is shown in Figure 1.72, and its interface is shown in Figure 1.73.

It is essential that this forced development of templates for detection, prevention, and neutralization of both known and previously unknown group and mass computer attacks identifies destructive program impacts and bookmarks (NDV) and investigates incidents of cyber security in real (quasi-real) time. It became possible (Figure 1.74) to use it in practice:

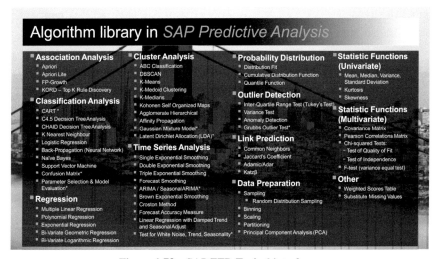

Figure 1.73 SAP ETD Typical interfaces.

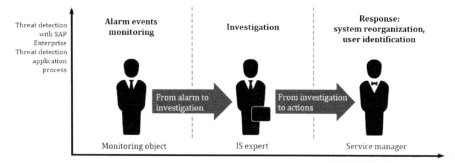

Figure 1.74   Threat detection with SAP.

- Filters bringing to the normalized view of event logs (a series of filters is called path);
- Visual forms of representation of the filtered values for the choice of the outstanding value;
- Various forms of representation of threshold values, etc. [45–47, 51, 55, 49].

The authors of the book have successfully adapted and developed the best foreign ones (IBM, HP, SAP, etc.) and domestic solutions (ASO NGO ECH-ELON (Comrad and Rubicon), Positive Technologies (SIEM), Solar Security (JSOC), Informzashchita (SOC+), Kaspersky Lab ICS-CERT, etc.) to create a system of distributed intelligent monitoring centers for cyber security threats of the critical information infrastructure of Russian megacities in the framework of the State Program "Digital Economy" of the Russian Federation. In particular, the SAP solution, SAP ETD, is characterized by the joint use of the advantages of existing SAP and third-party SIEM technologies, as well as the use of innovative technologies for detecting threats of cyber security (Enterprise Threat Detection), a high-performance real-time database HANA 2.0, predictive analytics, storage, and processing of large amounts of SAP S/4HANA data, etc.

In the future, to successfully use the technological advantage in the high-performance processing of Big Data and Big Data Analytics in networks and Internet/Inranet and IIoT/IoT systems and the creation of truly integrated SIEM systems based on SAP ETD, it is proposed to significantly develop the import capabilities of data from various third-party applications and systems. We note that, at the current time, it is already possible to import data into SAP ETD from systems of such manufacturers as FireEye and Trend Micro.

Further, it is proposed to realize the possibility of importing data on identified security events, at least from the following typical information security tools:

- Firewalling (ME);
- Detection, prevention, and neutralization of the consequences of cyber-attacks and anomalies (IDS/IPS);
- Protection of AWP and AC from unauthorized access (NID);
- Cryptographic protection of information (VPN, PKI, HSM, etc.);
- Antivirus protection and antispam protection;
- Internet traffic control;
- Analysis of security and conformity control;
- Control of confidential information leaks (DLP);
- Access control to information systems and data external drive;
- Users and administrators activity control;
- Antifraud;
- Technological protection of information in the process control system and IIoT/IoT systems etc.

2

MSSP/MDR National Operator Development

This chapter reveals the limiting possibilities of national MSSP/MDR operators for solving the problems of the Digital Economy and considers the possible ways of providing cyber security services for digital enterprises. The necessity of creating some universal monitoring platform MSSP/MDR for working with Big Data is substantiated. It is expected that such a platform will form the basis of the next generation MSSP/MDR technology core in the interests of federal and regional executive authorities, as well as large business. The relevance of the development of a "cloud" and/or "foggy" system for monitoring cyber security threats based on collection, processing, storage, and analysis of Big Data is shown. It is significant that such systems will allow to work at high data transfer rates and to guarantee processing of very large inbound information flows in real time with minimal delays. New approaches to knowledge management of cyber security and the development of appropriate MDM cyber security systems are proposed and justified.

2.1 Ultimate Opportunity of National MSSP/MDR Operators

Currently, a number of leading Russian telecom operators and security integrators are actively adopting the best practices of managed security service providers (MSSP), for example, the best MSSP practice of well-known international telecoms AT&T, Deutsche Telekom, France Telecom, etc. Here, the Managed Security Services (MSS) means ready-made technical solutions as the special services for home users and commercial and governmental entities including services for antivirus and spam protection, intrusion prevention, firewalling, secure remote access based on VPN technologies, etc. At the same time, MSSP guarantees the specified Service Level Agreement (SLA), and users manage quality of the provided security services by monitoring these parameters. Let us consider what MSSP models and corresponding technical solutions are in demand in the Russian security sector today [1–5, 7–9, 13, 18–20, 22, 39, 56, 57].

2.1.1 Relevance of MSSP/MDR Cyber Security Services

The relevance of security services is obvious. The number of threats increases every day, and we need to protect our assets – from smartphones and laptops to corporate sites and networks. At the same time, modern cyber-attacks are becoming more complex and sophisticated (Figure 2.1), and insufficient attention to security issues leads to significant financial losses (Figure 2.2).

The annual report data on the protection state of critical information systems (Kaspersky Lab, NIST, SANS, etc.) convincingly indicate the demand for technical information security tool and related security services. At the same time, a minimum basic security package is required for home users. There are:

- Mobile device protection (Mobile Device Management – MDM);
- Antivirus and spam protection;
- Parental control;
- Secure remote access using VPN technologies, etc.

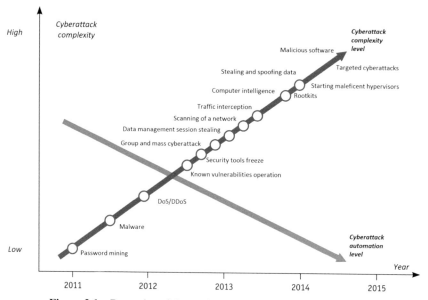

Figure 2.1 Dynamics of the modern cyber-attack complexity growth.

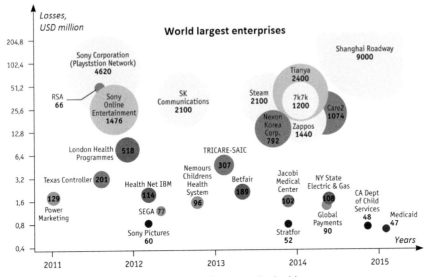

Figure 2.2 Losses from security incidents.

For the enterprises, a much wider range of security services is required (Figures 2.3, 2.4 and 2.5), including:

- Adaptive information security management based on SEM/SIM/SIEM;
- Firewalling and perimeter protection from external intrusions;
- Virtual private networking (VPN);
- Antivirus and spam protection;
- Corporate email filter (antispam, antivirus, etc.);
- Advanced persistent threat (APT) detection and neutralization;
- Detection and prevention of both known and previously unknown mass and group cyber-attacks (IPS/IDS);
- Strong authentication;
- Organization of trusted mobile services, MDM;
- Vulnerability analysis of system and application software;
- Control over the safe software update;
- Private data leak prevention (DLP);
- Security risk management;
- Enterprise policies, regulations, and safety standard efficiency assessing based on SOC;
- Maintaining the business continuity and fail safety, BC and DR, etc.

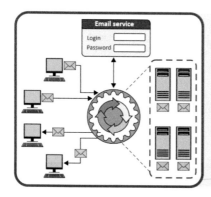

Figure 2.3   Example of a corporate e-mail protection service.

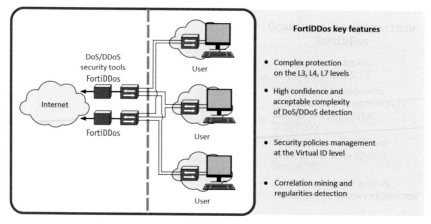

Figure 2.4   Example of a cyber-attack (DoS/DdoS) protection service.

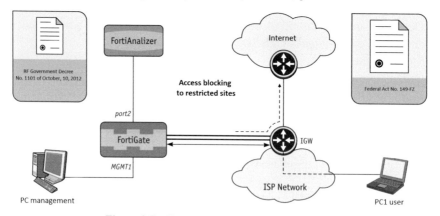

Figure 2.5   Example of a Web-filtering service.

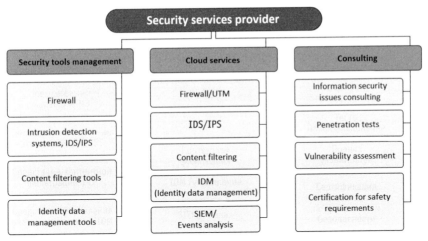

Figure 2.6 Feasible portfolio of security services.

A number of leading Russian security companies, large-scale telecom carriers, and security integrators have begun to form a service portfolio to meet these needs (Figure 2.6). At the same time, there are a number of requirements for such a security services portfolio. First, it is necessary to ensure a high-quality technical security services implementation. To do so, all necessary "cloud" (SaaS, IaaS, and PaaS) and/or classical engineering infrastructure of security services must meet high requirements for functionality, manageability, reliability, and performance. Second, it is important to set the required SLA parameters for each security service and ensure monitoring of compliance these parameters. For this, it is also necessary to line up the qualified afterhours support with the required service quality parameters throughout the service area. Third, we need to provide flexibility and adaptability of the security services portfolio for changing user requirements. Security services should be accessible and convenient for both security services users (fast adding and regulating service parameters, paying for them, viewing reports, etc.) and the MSSP itself (fast user connection, a high degree of operational process automation, as well as service management flexibility and transparency and the feasibility to monitor parameters and control quality).

2.1.2 MSSP/MDR Best Organization Practice

The MSSP market dynamics in North America (Figure 2.7) show the steady growth of the MSS security services. According to IDC, the world MSS

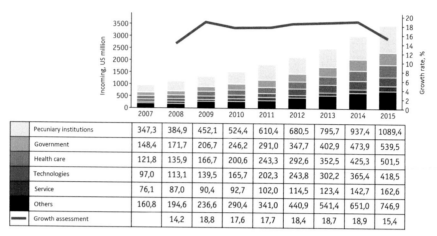

Figure 2.7   MSSP market dynamics in North America (2007–2015).

	2007	2008	2009	2010	2011	2012	2013	2014	2015
Pecuniary institutions	347,3	384,9	452,1	524,4	610,4	680,5	795,7	937,4	1089,4
Government	148,4	171,7	206,7	246,2	291,0	347,7	402,9	473,9	539,5
Health care	121,8	135,9	166,7	200,6	243,3	292,6	352,5	425,3	501,5
Technologies	97,0	113,1	139,5	165,7	202,3	243,8	302,2	365,4	418,5
Service	76,1	87,0	90,4	92,7	102,0	114,5	123,4	142,7	162,6
Others	160,8	194,6	236,6	290,4	341,0	440,9	541,4	651,0	746,9
Growth assessment		14,2	18,8	17,6	17,7	18,4	18,7	18,9	15,4

market amounted to 9 billion dollars in 2013 and continues to grow (in 2017 is 15 billion). According to analyst forecasts, the SMB sector can take 50% of MSS market.

According to Gartner [18–20], the technological leaders of MSSP are companies like AT&T, Verizon, Symantec, Dell SecureWorks, IBM, Solutionary, etc. At the same time, new players are expected to start competing with each other. The main reasons for this include:

- Ability to attract new customers and increase the loyalty of existing ones;
- Feasibility of increasing profits by implementing new differentiated security services;
- Reduction of OPEX costs by 60–70% and CAPEX by 30–40%;
- High return on ROSI investment, etc.

From the technological point of view, there are three main MSS models or MSSP organizations: the client CPE model, cloud, and combined (Figure 2.8) [18–20, 39].

First is the most common, the so-called client CPE model (Figure 2.9), which assumes that the engineering infrastructure of security services (MSS) is primarily located at the end user. Note that this MSSP model is optimal for distributed branch networks of small companies (SMB).

The second the most promising cloud model (Figure 2.10) is characterized by low operating costs and allows offering a wider range of SaaS, PaaS,

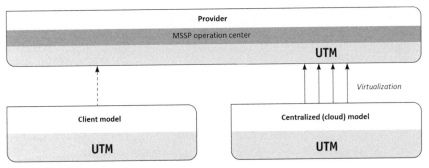

Figure 2.8 Typical MSSP models.

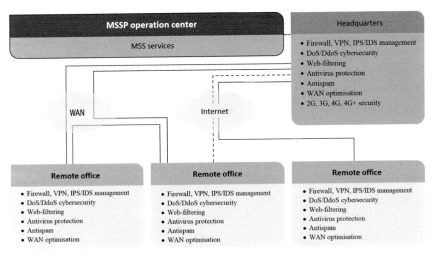

Figure 2.9 Traditional CPE model MSSP.

and IaaS security services for medium-size and large companies. In addition, the cloud model significantly reduces the time to prepare and start providing security services and reduces the cost for deploying the engineering MSSP infrastructure.

It is significant that the cloud model has the properties of adaptability and manageability and allows security service users to timely block malicious traffic, maintaining high centralization of resources (service, management, reporting), Single Pane of Glass, rate services by traffic volume or type of security services, etc.

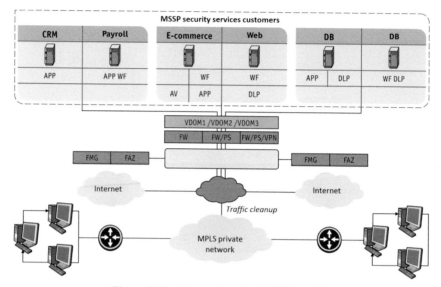

Figure 2.10   Perspective cloud MSSP model.

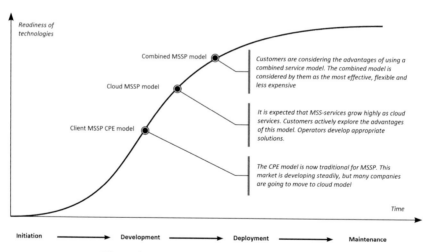

Figure 2.11   Dynamics of MSSP development.

The third model is a combination of the first two [18–20, 39]. The evolution dynamics of the above MSSP models are shown in Figures 2.11 and 2.12.

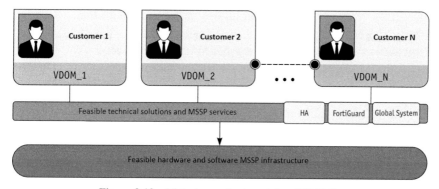

Figure 2.12 Virtual security domaining (VDOM).

2.1.3 Sample of MSSP by AT&T

http://www.att.com/

It represents [39] the largest distributed model of the security service provider (Figure 2.13) in more than 180 countries. It consists of eight global customer support centers and six operating centers, more than 40 data centers, and more than 3,800 local nodes. The model was created to monitor security threats and provide security services for both AT&T and its subsidiaries, as well as for its numerous corporate clients. The AT&T MSSP model

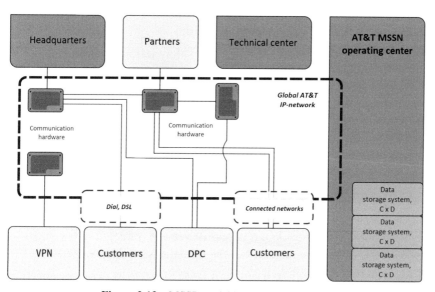

Figure 2.13 MSSP model by AT&T Company.

operates in $24 \times 7 \times 365$ mode and allows monitoring more than 18.7 PB of IP traffic daily.

Typical security services for AT&T MSSP model include:

- Safety management services:
  - Adaptive security management based on SEM/SIM/SIEM;
  - Management of Firewalls AT&T, FW;
  - AT&T systems management for the prevention, detection, and neutralization of cyber-attacks, IDS/IDA/IPS;
  - AT&T segment protection on the Internet;
  - AT&T segment protection from distributed Ddos attacks;
  - AT&T Web-resources security;
  - Traffic analysis and cleanup, DPI;
  - Corporate AT&T email protection;
  - AT&T IP services security management;

- Services for the protected system engineering:
  - Safety requirements development;
  - Sketch, technical and technical-economic design;

- Security consulting services:
  - Certification and accreditation for safety requirements (C&A);
  - Operational security services provision, SOC;
  - Advice on the security issues;

- Services for the security system integration:
  - Implementation of the engineering safety infrastructure;
  - Information security systems operation;
  - Technical and scientific support of security systems;
  - Environmental assessment of security projects, Tco/Tvo/Rosi;

- Security assessment services:
  - Static and dynamic analysis of application code (including NIST, PCI DSS, SANS, OWASP Orizon Code Rewiew, etc.);
  - Internet services vulnerability assessment (in particular, VoIP and Wi-Fi penetration testing);
  - Security threats analysis (susceptibility to social engineering techniques, cyber-attacks on denial of service, targeted cyber-attacks APT, security assessment of the virtual environments, etc.).

For the proper provision of the listed AT&T MSSP security services, special information security facilities and tools, IT, and technical support services are required, including more than 1,500 in cyber security. At the same time, the AT&T MSSP model is constantly improving and developing. In particular, in 2010, the AT&T MSSP security management system was accredited for compliance with the requirements of ISO 27001: 2005.

2.1.4 Sample of MSSP Model by UBIqube

It represents [39] a fairly large MSSP model, developed in 2000. UBIqube is separated from France Telecom and currently holds a leading position in the traditional and special security services providing in Europe. The engineering infrastructure of UBIqube MSSP is a multi-vendor platform (Figure 2.14) with the core is the well-known solutions of Cisco Systems.

The typical UBIqube MSSP security services include the following:

- Information security management services:

    - Intelligent management of infrastructure and security services based on the Cisco IP NGN concept (Cisco NetFlow, Cisco NFP, etc.);
    - Management of firewalls (monitoring the state of packages, proxies, combined);

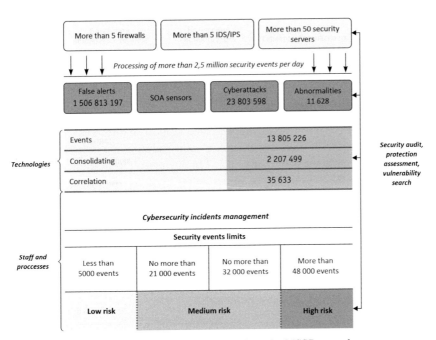

Figure 2.14 System appearance of the UBIqube MSSP control system.

- Virtual private networking, VPN (Ipsec, SSL);
- Management of the cyber-attacks detection and prevention systems, IDS/IDA;
- Management of the antivirus and spam protection systems;
- Management of Web-filtering systems, keeping of reference information and the reporting system support for the current state security services in the $365 \times 24 \times 7$ mode, etc.;

- Services for the mobile corporate services protection:

  - Trusted voice communication, GSM;
  - GPRS 2G, 3G, 4G, and 4G+ data traffic;
  - Trusted LTE data network;
  - In-depth analysis and traffic cleanup, DPI;
  - Assessment of mobile corporate services vulnerabilities, etc.;

- Special security services:

  - Corporate bank of trusted applications (Appstore);
  - Mobile devices and applications protection, Mdm (protection from installing "untrusted" applications, connecting to trusted Wi-Fi network points, blocking lost and stolen mobile devices, etc.);
  - Personal account, personal cloud storage;
  - Trusted payment services (mobile bank, payment for goods and services, NFC services, etc.) with strong authentication;
  - Trusted information services;
  - M2M (location, "alarm button", "mobile doctor", etc.).

2.1.5 Feasible Technical Solutions

To reduce costs for the building, operation, and support of engineering infrastructure, most modern MSSP use complex (multifunctional) security tools, the so-called unified threat management (UTM devices). These solutions have emerged as a result of the modern information security evolution. So, for protection of the digital perimeter of the enterprise, traditionally the firewalls are used. Over time, solutions for VPN were added for ensuring the secure remote access to the enterprise information assets. Then, IDS/IPS systems were added to prevent, detect, and neutralize cyber-attacks. Next, a standard arsenal of information security tools replenished with tools to antispam and antiviruses, as well as Web filters and content-filtering systems, and finally, DLP systems and WAN optimizers.

Further, to solve the problem of compatibility and management of the listed protection measures, the concept of a multifunctional complex UTM device or gateway was proposed and implemented [18–20, 39].

Modern standard security gateways are single multifunctional devices, which include firewall modules, prevention and detection of cyber-attacks (IDS/IPS), VPN, antivirus and spam protection, etc. Currently, the technological leaders in the UTM devices production are Fortinet, Dell (acquired SonicWALL), WatchGuard and Sophos (Astaro), Cisco, Juniper Networks, and Check Point.

Consider the feasible infrastructure MSSP solutions in more detail on solutions by Fortinet company (www.fortinet.com).

Fortinet's basic UTM device is FortiGate (Figure 2.15), which began to ship in the security market in May 2002.

Modern FortiGate is a multifunctional hardware–software complex that combines firewall, IDS, IPS, antivirus gateway, Web-filtering, antispam, and other functions. Currently, more than 1.5 million FortiGate UTM devices are used all over the world. In this case, FortiGate devices of the older models, 5000 series, and/or FortiGate-VM are traditionally used for MSSP.

Here, the firewall monitors and filters all packets, passing through it, in accordance with the established security policies. We can configure policies

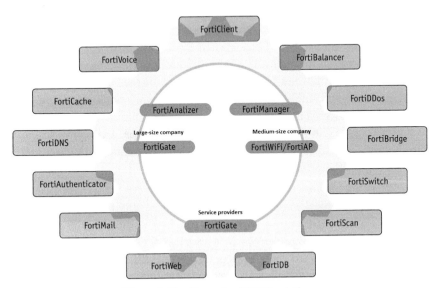

Figure 2.15 Example of UTM solutions.

for the entire local network as a whole and for each of its segments or a specific user. It is possible to build secure VPN connections using IPSec VPN, SSL VPN, and PPTP. Traffic shaping systems are used to control traffic flow and limit downloading information. The intrusion detection and prevention system (IPS) monitors activity on the network to detect unauthorized access to the network and block it.

This component allows conducting the signature traffic analysis, traffic anomaly analysis, implements automatic signature updates, and allows creating own signatures. Systems to protect against malicious program and spyware verify in real time control the incoming traffic on the local network. Web-traffic (HTTP), FTP, e-mail (SMTP, POP3, IMAP), instant messaging protocols (ICQ, AIM, MSN, Yahoo!), and the Network News Transfer Protocol (NNTP) are checked. Antispam checks e-mail (SMTP, POP3, IMAP) for spam, forms "white" and "black" lists of IP and e-mail senders/recipients addresses, signature correspondence analysis. Application control checks the traffic for the specified applications: instant messaging systems (ICQ, AIM, MSN, Yahoo!), P2P networks, voice over IP (H.323, SIP, SCCP), etc., regardless of the ports used. VoIP (IP telephony) support ensures the voice traffic security over IP networks. Web-filtering restricts access to websites based on enterprise security policy.

The filter uses the classification basis of the FortiGuard Web-filtering service. Virtual domains (VDOM) allow creating the multiple virtual devices on the same physical device with security policy management and routing tables.

FortiGate devices run under a FortiOS operating system, data on device operation can be obtained via CLI console, Web-console, or FortiManager device (e.g., 2,000 and 4,000 V series). FortiGate can integrate with other Fortinet security solutions to provide comprehensive security.

At present, FortiGate's UTM device line includes more than 30 different models (Figures 2.16, 2.17 and 2.18), which differ according to two general criteria – throughput in different operation modes and the number of available physical interfaces:

- Firewall bandwidth throughput (from 30 Mbit/s to 480 Gb/s);
- IPSec VPN bandwidth throughput (from 5 Mbps to 204 Gbps);
- IPS bandwidth throughput (from 10 Mbit/s to 36 Gb/s);
- Antivirus bandwidth throughput (from 5 Mb/s to 4.2 Gbit/s);
- Number of concurrent sessions (from 25,000 to 14 million);
- Maximum number of VDOMs in one device (10–250);

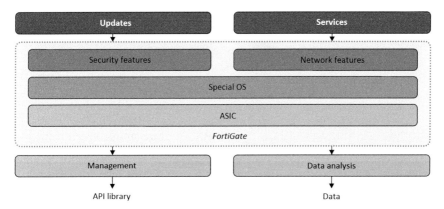

Figure 2.16 Typical components of the Fortinet network security platform.

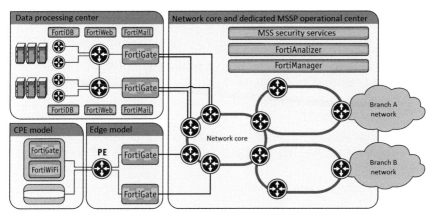

Figure 2.17 Feasible MSSP scheme using Fortinet solutions.

- Different number of Ethernet and SFP ports;
- Different number of Switch/LAN and WAN ports.

All models are divided into four large classes, depending on the scope and number of protected computers in the local network:

- Models for small businesses and "home offices" (FortiGate 20-111);
- Models for medium-size business (FortiGate 200-800);
- Models for large companies (FortiGate 1000-3950);
- Models for telecom operators (Forti Gate 5000).

In addition, Fortinet provides a number of functional security gateways for the MSSP.

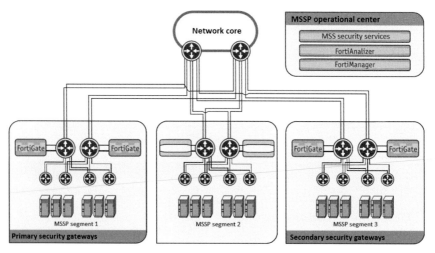

Figure 2.18   Feasible scheme for balancing MSSP performance using Fortinet solutions.

FortiMail is designed to protect e-mail from DoS attacks and Directory harvest attacks, to detect and block malicious programs in e-mail messages, and to block spam. FortiMail allows real-time signature and heuristic analysis of e-mail (SMTP, POP3, IMAP4) for viruses and spam, forming black and white lists of addresses, archiving data and placing suspicious objects in quarantine. The basic MSSP models are FortiMail-5001 A and FortiMail-VM.

FortiWeb performs firewall functions for working with Web applications. It provides load balancing to the server (Figure 2.18) and accelerates the work with Web applications. FortiWeb includes a vulnerability scanner. FortiWeb based on signatures and threat patterns allows blocking attacks against vulnerabilities in Web applications, for example, XSS cross-site scripting, SQL injection, buffer overflow, files inclusion, denial of service attacks, cookie poisoning, schema poisoning attacks, etc. Basic MSSP models are FortiWeb-3000 C and FortiWeb-VM.

FortiDB is designed to protect databases. Its functions are vulnerability management, database activity monitoring (DAM), data loss prevention (DLP), audit and compliance automation, change control, and virtualization. The ability to perform a continuous audit allows identifying at the early stages not only "external" attacks aimed at the database (FortiGate performs this function), but also more "subtle" anomalies in its work, for example, atypical or suspicious transactions performed on behalf of legally registered accounts. The basic MSSP model is FortiDB-2000B.

FortiCache is designed to cache content when working with external information sources (another local network, a LAN segment within an enterprise or the Internet). When caching is used, the data that are being processed are stored in the device memory, and when they are accessed again, the stored data are given. FortiCache recognizes the format of the main content delivery networks and can cache dynamic, distributed, and streamed video content, as well as identify the embedded advertisement. FortiCache can be used in three modes: transparent proxy, partially direct proxy (through policy routing), and a routed direct proxy.

Using FortiCache allows speeding up access to resources and applications, as well as reduces the load on external connections.

FortiBalancer is designed to increase the performance and processing speed of the network by aggregating traffic from various sources. When using FortiBalancer, the data transfer speed is significantly increased due to the fact that the speeds of all data channels are summed up. Devices can work with corporate, mobile, and cloud applications. In FortiBalancer, there are built-in caching and data compression, SSL encryption, and TCP sessions multiplexing. The basic MSSP model is FortiBalancer 2000.

FortiSwitch represents a set of switches for connecting personal computers and other devices to the network. FortiSwitch allows high-speed networking, as well as parallel and cloud computing. The older device models are built according to a modular principle and can be scaled easily if necessary. The basic MSSP models are FortiSwitch 5203B. The number of ports 10/100/1000 Ethernet is from 8 to 48, the number of MAC addresses is from 8 to 128,000, the bandwidth throughput is from 2 to 960 Gbps.

FortiScan is designed to audit vulnerabilities, manage updates about vulnerabilities, and generate reports on the vulnerability audit results on workstations and servers. It allows eliminating mismatches to corporate policies and standards, to perform continuous monitoring and audit, and to assess compliance with internal, industrial, and legislative standards at the operating system level. The scanner works with a lot of constantly updated policies, such as NIST SCAP, FDCC, PCI-DSS, SOX, GLBA, HIPAA, ISO 17799, and FISMA, received through FortiGuard service. FortiScan series currently includes only one device 3000 C, which supports up to 20,000 devices connected simultaneously.

FortiBridge Gateway is designed to redirect traffic of other Fortinet devices. In the event of power failure, a failure in the network equipment, or a critical error, the devices begin to act as a bridge and redirect network traffic around the damaged network segment. As a result, the local network does not stop in the emergency situations. The series includes three types of gateway: 2002, 2002F, and 2002X, which differ in the types of ports available.

Generally speaking, multifunctional Fortinet UTM devices differ from traditional narrow-focused information security tools as follows; first, the improved functionality, including firewalling, intrusion detection and prevention (IDS/IPS), antivirus protection, VPN (Virtual Private Network), spam protection, Web-filtering, private data loss prevention (DLP), application control, traffic optimization and protection, etc; and second, the availability of its own hardware and software solutions optimized for information security tasks. To increase Fortinet UTM-devices performance, Fortinet System-on-a-Chip architecture is used, which allows combining on a single chip a traditional multi-core processor and specialized coprocessors. This allows us to significantly increase the speed of processing network traffic and its verification. The new architecture includes two specialized coprocessors – FortiASIC-CP and FortiASIC-NP. FortiASIC-CP (FortiASIC Content Processor) is designed to accelerate the work of technologies related to content inspection, such as intrusion prevention system (IPS) and antivirus.

Also, the encryption time and authentication procedures are reduced. The FortiASIC network processor (FortiASIC-Net) is designed to accelerate the network services operation. Its use speeds up the firewall operation, encryption procedures, signature searches, and heuristic technologies. As a result of transfer, the processing and analyzing traffic functions to the hardware level, detection, and elimination of threats can be performed in real time and without loss of network bandwidth, and network processors are unloaded for other tasks. A specialized operating system FortiOS is applied to manage all components of the UTM-device. In FortiOS, a management on all the functionality of FortiGate devices is established. We can also manage FortiAP wireless access points with FortiGate or FortiWiFi. Management FortiOS can be carried out through the Web interface, CLI, console, or FortiManager device (e.g., series 3000 C, 5001 A, 4000 B, etc.). The operating system implements the role-based management of several administrators, the access rights differentiation, and the VDOM use to manage virtual devices. Supported protocols are syslog, SNMP, possibly informing about events via e-mail.

Third is the presence of its own expert laboratory FortiGuard Labs, designed to collect and analyze data on security threats, as well as to build and maintain the current status of the cyber-attack signature database. FortiGuard subscription services include updates to subscriptions for antivirus, IPS, antispam, and Web filtering. New updates through the network of Fortinet

Distribution Network servers are downloaded to UTM devices automatically or in the automated mode (push updates).

***

According to Gartner, the world market MSSP volume annually increases by 12–15%. More than 60% companies of Fortune 500 list use MSSP services [18–20, 39]. The MSSP services relevance is confirmed by the savings achieved by the operation and maintenance of expensive network information security tools. The fact is that the end user of security services pays only for providing the appropriate security service in accordance with a predetermined level of service quality (SLA).

In practice, in order to successfully implement the MSSP, it is first necessary to develop a strategy for introducing new security services to the market and then the corresponding technical policy. These documents should define the main MSSP goals and objectives, a plan of priority actions, including a list of the current security services, their SLA quality parameters, and requirements for providing infrastructure solutions. It is also necessary to determine the schemes for the installation and integration of these infrastructure solutions with existing platforms and clouds, the operation scope, technical and scientific support, service, and technical support in $24 \times 7 \times 365$ mode.

2.2 Possible Ways of Providing Cyber Security Services

Prior to the organizational changes, most CISOs in Russian companies certainly asked a number of questions. How to develop an effective IS sourcing strategy that is relevant to the business goals and objectives? Which IS sourcing model is more preferable: internal provider, full outsourcing or selective outsourcing (multisoring)? The importance of the mentioned questions is explained by the fact that the high-quality IS sourcing strategy makes it possible to optimize the company business. In practice, national CISO faced a number of methodological problems and a lack of the necessary practical experience. How to avoid possible mistakes in the selection and implementation of the IS sourcing strategy? Let us try to find an answer to this and other questions [18–20, 39].

As a rule, the IS service activities are strictly regulated by the relevant IS process rules that are performed by the company IS service employees or any third-party organization in accordance with the company previously developed and approved IS policies (Figure 2.19). At the same time, the IS sourcing

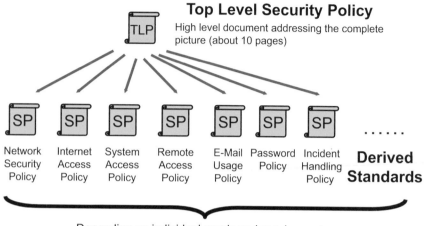

Figure 2.19   Typical set of company IS policies and regulations.

notion plays rather important role, which is usually understood as a way to optimize the company business by combining the internal and external IS services in an optimal way. The IS sourcing involves identification, selection, and use of various IS service providers (internal and external) in the optimal combination. Depending on the different IS service provider selection, there are two opposite IS process variants: insourcing and outsourcing. In the case of insourcing, the IS service provision is performed by internal services (IT service, IS) (internal delivery). In the case of outsourcing, the IS services provision is performed by some external third-party organizations, the so-called IS service providers, security service provider (SSP).

Table 68

Policy in information security (Management statement)		
Regulation on the protection of commercial secrets and personal data		
Information Security. IS standards		
Internet usage policy	Internet use regulation	
Email policy	E-mail use term regulation	
Malware protection policy	Malware protection regulation	

Table 68 (Continued)

Password policy	Password use regulation	
Workplace policy	Workplace organization regulations	
Maintenance policy	Maintenance schedule	
IS audit policy	IS audit regulation	
IS events monitoring policy	IS events monitoring regulation	
Document circulation policy	Document circulation regulation	
IS incident response policy	IS incident response regulation	
Physical security policy of corporate information system	Physical protection regulations of corporate information system	
Policy on work with staff	Regulation on work with staff	
IS roles management policy	Regulation on managing the IS roles	
Information classification policy	Information classification regulation	
Information resources inventory policy	Information resources inventory regulation	
Risk management policy	Risk management regulation	
Software use policies	Software usage regulation	
Policy on work with third parties	Regulations on work with third parties	
Information resources access control policy	Information resources access control regulation	
Automation system development and maintenance policy	Automation system development and maintenance regulation	
Negotiation policy (including the communication means use)	Negotiation regulation (including the communication means use)	Plans to ensure durability and recovery of corporate information system (DRP)
Mobile device policy	Mobile usage regulation	List of protected information resources
Policy on corporate information system structuring	Regulation on corporate information system structuring	Instructions for adapting the regulation templates
Cryptographic information security system policy	Cryptographic information security system regulation	IS checklist

2.2.1 Typical MSSR/MDR Services

The IS service is usually understood as a set of IS processes and supporting assets aimed at performing business tasks in the part of information security that arise in the company business course. The IS service provision result has a clear value for the business and is formulated in the business user terms. The IS service place and composition is shown in Figure 2.20.

For example, the following typical IS service groups can be defined for large companies.

Security services against unauthorized access:

- Software update in case of new versions or product updates;
- Ensuring information protection against the unauthorized access of IP network resources;
- Ensuring users identification, authentication, and authorization;
- Ensuring access to separation and management of the information system resources;
- Ensuring the Internet work communications protection, including the Internet connection.

Antivirus protection services:

- Antivirus mail traffic protection on the gateway;
- Antivirus database protection on the mail gateway;
- Antivirus file servers protection;

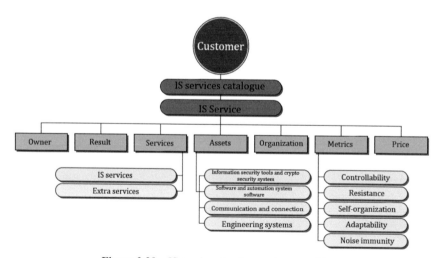

Figure 2.20   IS service structure and composition.

- Antivirus protection provision for user workstations;
- Ensuring timely prevention and response to virus incidents;
- Centralized antivirus protection management;
- Virus infections consequences removal.

Services to control information flows:

- Ensuring the unauthorized transfer detection and prevention of confidential information;
- Ensuring control over the Internet resources and e-mail misuse;
- Ensuring control over the document circulation use;
- Ensuring the blocking and prevention of mass unauthorized mailings (spam).

Security analysis services:

- Detection and analysis of the information system vulnerabilities that allow the use of various attack types on information resources;
- Errors detection in the configuration, settings, and implementation of a network equipment;
- Recommendations development to remove detected threats.

Cryptographic information security services:

- Ensuring the mail system database confidentiality;
- Ensuring information confidentiality when using e-mail;
- Digital certificates management (generation, suspension/termination of the validity period) of cryptographic keys and key information carriers.

Services to provide centralized information security monitoring and management:

- Detection and prevention of various attack types on a network equipment and IP end devices;
- Detection and timely response to incidents.

Application system and database security services:

- Database and application system security analysis;
- Security monitoring and attack detection at the database level in real time.

Security monitoring services:

- Organizational and regulatory documentation development on IS;
- Information risk assessment and analysis, taking into account the tools used and the measures taken to ensure information security;

- Requirement formation for information security at the technical and organizational levels necessary to ensure the required security level;
- Support for a given information risk level;
- Notification and consultation of the company employees on information security issues.

Services to ensure the technical information leakage channel security:

- Measuring the enclosing structure sound insulation of the protected premises;
- Measuring side electromagnetic radiation of the protected premises;
- Carrying out special checks, research, surveys, etc.

In practice, the above-mentioned service set can be expanded depending on the company business specifics.

Inset 2.1 Basic terms and abbreviations

Main terms and abbreviations	
IS	Information security
ISMS	IS management system
SLA	Service level agreement is a written agreement on the IS service provision level. The agreement between the IS service consumer and the party providing it, which sets out the metric values for IS service provision level
SPA	Service provision agreement is an agreement on the IS services provision between the company and an external IS service provider
SSP	Security services provider is an IS service provider
IS infrastructure	Hardware–software tools complex of information security (information security tools from unauthorized access, firewall, VPN, antiviruses, IDS and IDA, cryptographic information security systems)
IS process	Recurring-related action sequence aimed at performing a certain task of the company IS security
IS service	Some IS process sets and supporting assets aimed at performing business tasks in the information security part arising in the company business course. The IS service provision result has a clear value for business and is formulated in terms of the business user
IS sourcing	Way to optimize the company business by the best combination of internal and external IS services. It assumes identification, selection, and use of various IS service providers (internal and external) in optimal combination
IS insourcing	Way to provide the IS services by internal divisions (IT service and IS service) (internal delivery)

IS outsourcing	Way to optimize and improve the efficiency of the company core business by transferring the functions of providing certain IS services types (often noncore) to external specialized organizations and/or individual entrepreneurs on a contractual basis.
	Continuous work model with contractors based on long-term agreement execution (for a period of 2–5 years) that determine how the company will receive services to support current information security processes under a certain quality level (SLA)
IS outsourcer	Organizations that specialize in certain IS service provision and possess all the necessary tools, knowledge, and skills in the information security domain

Criteria for giving IS services to outsourcing

To classify IS services for outsourcing purposes, the following criteria can be used:

- IS service importance for business;
- Internal IS services effectiveness;
- IS service criticality;
- IS services availability on the market;
- IS security service criticality by reliability;
- IS service novelty for the company;
- IS service quality offered on the market;
- IS service cost offered on the market;
- Business requirements stability for the IS service, etc.

Table 2.1 describes the mentioned criteria.

In general, it is possible to separate and rank the key company IS services by such criteria as "importance for business" and "internal IS service effectiveness" (Figure 2.21).

In the future, for more detailed IS services classification and ranking, a company can use expert multicriteria evaluation methods [18–20, 39].

IS sourcing models

At present, the following basic IS sourcing models are known in the IS market (Figure 2.22).

- *Internal provider.* The most common model to provide IS services when the company own department (e.g., IT department or IS department) implement new IS services and/or final architectural solutions for information protection by performing internal projects. This model is the most flexible, since it allows dynamically changing the IS survives requirements, rules, and terms. At the same time, this IS sourcing

Table 2.1   Criteria description for the IS services classification

No.	Criterion	Which question does the criterion answer to?	Possible values	Impact on classification
1	IS service importance for business	To what extent is the provided IS service important for business?	An expert evaluation is given from 1 to 5 points based on a survey and interview	Unimportant for business IS services can potentially be given to outsourcing
2	IS service internal effectiveness	To what extent is the IS service effectively provided by internal company services?	An expert assessment of 1 to 5 points is given based on analysis	If the internal IS service inside the company is provided inefficiently, it should be given to outsourcing
3	Key\nonkey IS service	Is the IS service key important for business and internal efficiency?	I is an insourcing O is an outsourcing IO is an insourcing or outsourcing DO is an internal efficiency development of IS services or outsourcing	The criterion allows potentially dividing IS services into four categories. Nonkey IS services can be outsourced
4	IS services availability on the market	Is the IS service available on the external market? (An IS service is considered to be available if there is a sufficient number of service providers and an acceptable quality of the IS services market)	Y – IS service is available N – IS service is unavailable (IS service availability is determined based on the IS service market analysis)	Available on the external market IS services potentially should be given to outsourcing
5	IS service criticality	Is the IS service reliability uncritical for business? (recovery time, availability, support)	Y is noncritical N is critical (determined expertly or based on the business user requirements)	Noncritical IS services can be potentially outsourced
6	IS services novelty for the company	Is the IS service new for the company?	Y is new N is not new	New IS services are potentially to be outsourced

Table 2.1 Continued

7	The IS service quality offered on the market	Does the IS service quality increase when outsourced?	Y – increases N – does not increase (determined expertly based on the IS services market analysis)	With the IS service quality improvement, it is reasonable to give the service for outsourcing
8	IS service cost offered on the market	Does the IS service cost decrease when outsourced?	Y – decreases N – does not decrease (determined expertly based on the IS services market analysis)	With a tendency to reduce the IS services cost, it is rational to give the service to outsourcing
9	Stability business requirements for the IS service	Are the business requirements to the IS service stable?	Y is stable N is unstable (determined expertly based on interviews and business requirements analysis)	With stable business requirements to IS services, the service can be potentially given to outsourcing

model is also the most limited in relation to both the IS services provision scale (operational activities size) and the expertise and knowledge (experience, innovation, and available additional resources).

- *Multisourcing.* The model is also known as a random or selective IS outsourcing. It implies having a number of separate outsourcing contracts to provide IS services and tactical selection of the best outsourcer for each IS service.
- *Insourcing.* It assumes the IS service allocation to a separate business unit (BU), which provides IS services based on some implemented business rules (with respect to formal contracts, SLA, service cost determination).
- *Joint venture.* It allows removing the limitations inherent in the internal provider models and insourcing. A company creates a separate service company and brings it to the IS service market. As a rule, in this case, the company owns a controlling interest in the joint venture (usually from 50 to 80%) and manages the operating activities of the joint venture.

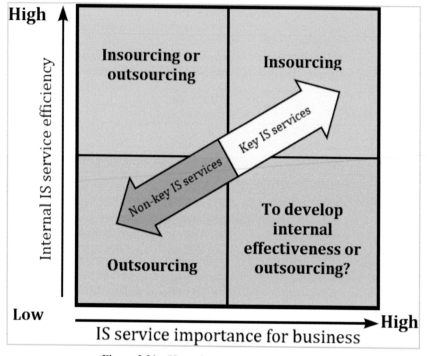

Figure 2.21   IS services classification example.

- *Full outsourcing*. The classical IS outsourcing model, which implies a contract execution with a single provider (ESP) for all IS services. The above IS sourcing model implies:
  - ○ Single provider and single contract approach;
  - ○ Cover most of company needs for IS services;
  - ○ Strategic company top management partnership with the provider;
  - ○ Long-term contract execution (for a period of 5–10 years).
- *Consortium*. The model appeared on the market as a further development of the full IS outsourcing model for sufficiently large state contracts and projects of individual international companies. It is used in case when it is impossible to satisfy the customer requests by involving the single, even very large, IS service provider. The consortium is created based on the customer requirements, and one of the IS service providers takes on the head role.

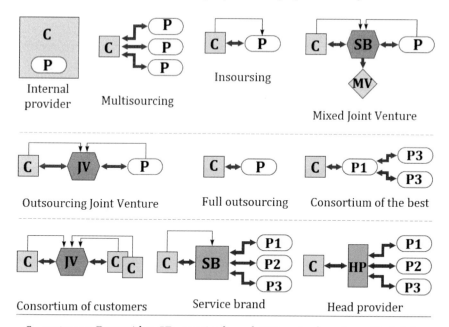

C - customer, **P**-provider, **SB** - service brand, **MV** - mixed venture, **JV** - joint venture, **HP** - head provider.

Figure 2.22 Possible IS sourcing models.

- *Service brand.* It is a model similar to insourcing. The company is created to provide services to a large organization or a company group. The provided IS services are carefully compared with those available on the market; the service company capabilities are taken into account, and the outsourcing IS services part is selectively determined. In this case, the service company, as a rule, aims at providing IS services in the external market in the future.
- *Head provider.* It is a company that provides management and integration of the various IS service provider capabilities to inherit the ability to provide IS services to the customer.

Each model has a right to exist and can reflect a reasonable relationship balance between the consumer and the IS service provider. At the same time, each IS sourcing model has its strengths and weaknesses. Let us just note that if earlier the overwhelming majority of national companies favored the internal sourcing model and extremely rarely the IS outsourcing, then the model is a selective IS outsourcing (multisourcing) [18–20, 39].

2.2.2 IS Sourcing Model Analysis

There are a number of approaches to the development of an IS sourcing strategy. Therefore, the Gartner Group company approach defines five main factors to formulate an IS sourcing strategy.

- *Business goals.* What are the tactical and long-term business goals? How do business goals affect the information security role in a company, what are the IS objectives from the business point of view?
- *Internal effectiveness* of information security. How effective is the IS service provision by internal company divisions?
- *External market opportunities.* What IS services are available on the market with a satisfactory price/quality ratio?
- *IS sourcing models.* What are the best practices to apply IS models? Which of the existing IS sourcing models are applicable?
- *IS services management.* How to effectively organize the management system of internal and external IS services? How to optimally combine IS insourcing and outsourcing?

The IBM approach defines the directions for improving the company information security service performance – ISM-PI (Information Security Management – Performance Improvement) and is based on the following principles:

- Identify the main driving forces of transformations and related factors as the basis for optimization and reorganization in the field of information security services management and building an appropriate model of the IS management system (ISMS);
- Change the principles of the IS service organization from those aiming only at the operational objective fulfillment to the business goal fulfillment;
- More completely cover all IS service activity areas (the IS service interaction with the company management, business units and users, IS management processes, IT security, personnel security, applied systems security);
- Comply the IS services management model with the management principles adopted by the company;
- Use the best experience in the field of information security services management.

The Hewlett-Packard Company approach is based on the ITSM model, which determines the main development directions of the IS management system

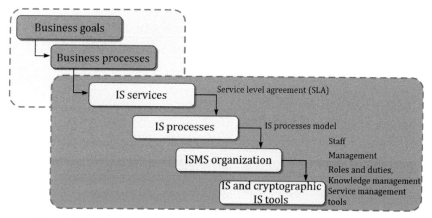

Figure 2.23   The step sequence to design the IS service management model.

based on the IS services provision. This approach defines the IS process as a mechanism to implement the IS services. At the same time, the direct objective of the IS sourcing concept is to study a conceptual model of the IS services management using these approaches.

The step sequence to design the company information security management model is shown in Figure 2.23.

Let us consider that in this case the process of designing and implementing the information security management system is cyclical. The process life cycle of designing and implementing the service model of the IS service management system is shown in Figure 2.24.

In the Gartner Group opinion, the IS sourcing strategy space can look like the following (Figure 2.25).

2.2.3 The IS Sourcing Practice

According to KPMG, the main reasons for the company to address an outsourcer are:

- Service advantages for a fixed fee;
- Difficulties in ensuring the regime $24 \times 7 \times 365$;
- Difficulties in finding/retaining staff;
- Cost reduction;
- Improved service;
- Lack of expertise within the company.

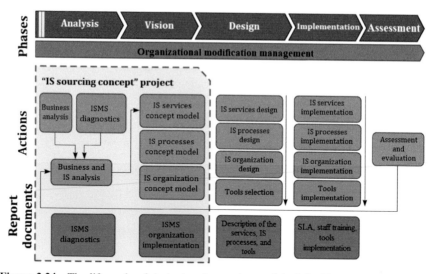

Figure 2.24   The life cycle of designing the service model of the IS service management system.

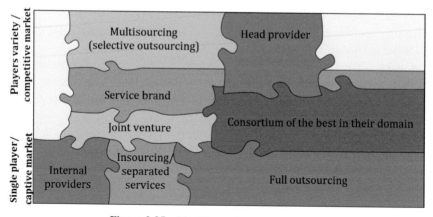

Figure 2.25   The IS sourcing strategy space.

When involving an outsourcer, the following conditions must be met:

- Outsourcer selection must be carried out only on a competitive basis;
- Costs required to be paid for services, provided by an outsourcer should be less than the cost of this service provision by a company staff;
- Services quality performed by an outsourcer must be higher than the same service quality provided by a company staff;

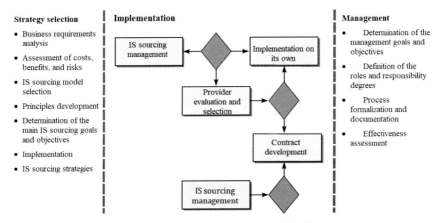

Figure 2.26 The resources usage algorithm.

- Positions reduction of the company staff previously engaged in the service provision transferred to an outsourcer.

The proposal preparation plan for an IS sourcing strategy can include the following steps (Figure 2.26):

- Current state analysis of material, technical, and human resources, as well as the availability and qualifications of the company staff, performing information security (services) that are to be developed independently or outsourced;
- Current level assessment of the financing quality and volume of information security services;
- Analysis of the need in additional funding, required to improve the service provision efficiency and quality on its own;
- Possible risk assessment of service provision on its own (political, financial, technological, technogenic, environmental, level reduction, quality reduction, delayed service provision) and assessment of the associated possible damage for the company;
- Evaluation of service quality performed by outsourcers, and their price dynamics;
- Applicability analysis of the known IS sourcing models;
- Verification of the outsourcers legal status, work experience, commercial reputation, credit history;
- Verification of the necessary licenses and certificates availability from outsourcers.

Proposals on the outsourcing use can be prepared by the company structural units responsible for the IS service provision or the main IS services consumers. Preparing proposals, an official note is to be prepared on the company head name, which gives reason to involve an outsourcer to provide certain IS services, outlines the current problem in the daily company operations, the further resolution of which by its own means is not possible or economically irrational. The official note indicates that the reasons with it is impossible or irrational to provide IS services on its own and considers two options for the company business: with and without an outsourcer. It is important that the official note includes all necessary calculations and analytical materials on involving outsourcers, conclusions on the outsourcing use effectiveness, taking into account possible risks, and, if possible, proposals to reduce the number of company staff performing IS service provision to be outsourced. The official note is approved by the interested structural company departments (security department, legal department, financial department, economic department, etc.) and is signed by CEO.

When the necessary approvals are made, the relevant budgets are adjusted in the established order. After agreeing on the decision draft on the outsourcing use and adjusting budgets, the company head will organize a tender to choose an outsourcer. The tender terms to choose an outsourcer include requirements to have a compliance certificate or a license in cases specified by the Russian Federation legislation. As a competition result, the company head executes a relevant agreement with the tender winner. The following measures are taken to control the outsourced IS service in the company:

- Relevance of prices, volumes, quality, and terms of the information security services performed by the outsourcer is analyzed;
- Annual report is prepared for each work given to the outsourcer;
- Outsourcing contracts accounting is carried out in accordance with the established procedure;
- Acceptance of the IS services performed by the outsourcer is carried out and their payment is ensured.

What determines the success of the selected IS sourcing strategy? Undoubtedly, to a great extent, success will depend on the "maturity" of the management (Figure 2.27) and the actual company business processes. Even with the most favorable external conditions (in relation to IS services), the above-mentioned company business process immaturity can become an insurmountable obstacle for the IS sourcing. Then, the initiatives to use the IS sourcing models will be perceived as a "hindrance to the normal operation"

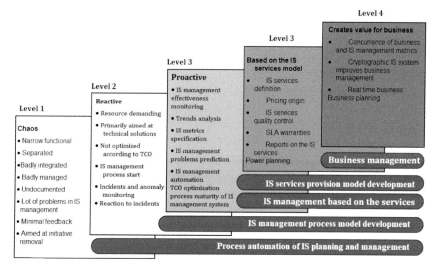

Figure 2.27 The IS management system maturity levels.

of the company. Obviously, with the weak company business processes formalization and the absence of work regulation, the attempts to give the IS functions to outsourcing in the form in which they exist in the organization are doomed to failure. On the other hand, the management based on well-formalized and well-grounded KPIs is an ideal condition to the outsourcing and IS sourcing in general.

Now let us talk about the IS services cost. At present, there are dozens of methods for pricing IS services, reducing the services cost and other tools to survive in a competitive market. Undoubtedly, this is a vital necessity for an independent outsourcer company operating in the free IS services market. However, among the largest Russian companies, there is a special phenomenon, that is, subsidiary outsourcers. In fact, they are completely controlled by the main company firms, providing services only to its "owner". This model provides all the main outsourcing benefits, plus it makes provided services transparent both functionally and financially. As a rule, most of the IS functions are given to such company. At the same time, there are internal settlement payments between their units (the so-called cost allocation) and certain tariffs are established.

Developing an interaction model with IS service providers (Figures 2.28 and 2.29), it is important to take into account the latest changes in the national IS market. Now, not a single, "monolithic" company acts as a customer,

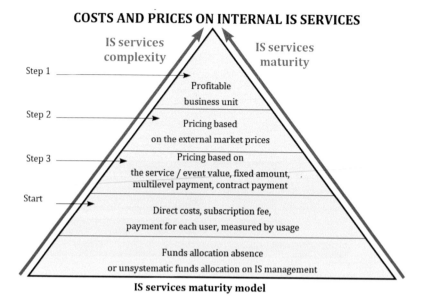

IS services maturity model

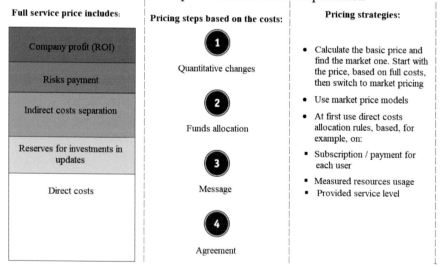

Figure 2.28   Interaction models with IS service providers.

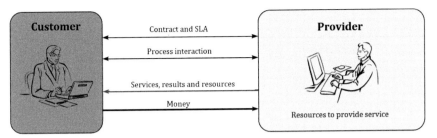

Figure 2.29 Relationships based on the contract.

but a conglomerate of various business units (BUs), divided by domains, geographical location, and sometimes even business culture. In addition, it should be noted that enterprises are becoming increasingly demanding in terms of real monetary profits, cost reductions confirmation, the rationality of innovations, and organizational changes in business. All these requirements are only toughened as the customer increasingly focuses on the core business and switches to the external IS services use. As a consequence, an increasing number of complex enterprises turn to the wide variety use of the "customer – information security services provider" interaction model, previously not used in "monolithic" companies.

From the information security service provider perspective, these relationships can be represented by the following financial factors:

- General income and annual deliveries or current value (NPV);
- Direct costs for the service provision and NPV;
- Gross margin and NPV;
- Initial investments, payments, and costs (for providing resources, for starting a service, and for ensuring the required transformations);
- Risks (penalties, customer business stability, customer obligations, the generic error estimation, uncertainty);
- Possibilities to start additional businesses with this customer (on the vertical market or in the customer location, which should lead to a new product line realization or the business-critical mass reaching).

The relative factor importance changes over time. At the same time, marginality and profits are the most important factors in decision-making, but the investment evaluation and the possible risk calculation should not be forgotten.

Figure 2.30 shows a conceptual scheme to the IS service provision, which can be taken as a basis for the development of procedures for the IS services in national companies.

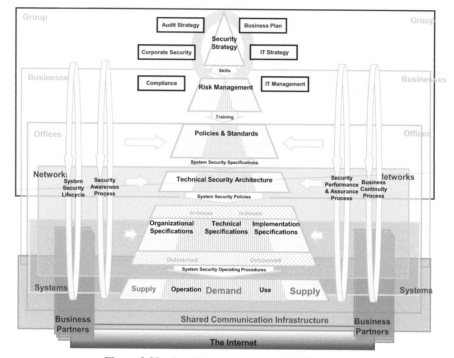

Figure 2.30 Possible scheme to provide IS services.

Let us note that aiming at IS services provision to business users implies a significant company ISMS modification. It is necessary to implement the effective functioning of an internal IS processes number that will be directed to the IS services provision (including the IS services development, the interaction management between the IS department and business users, and the IS services management). To implement these processes, it is required to create special units and staff of relevant analysts and managers who will be engaged in providing new IS service functions.

2.2.4 Sample SLA Content for the Provision of Cyber Security Services

Below there is a structure of the standard agreement on the IS services provision level. The structure contains main agreement sections and a brief sections content description [18–20, 39].

Agreement parties and participants

The definition of the IS service provider and the customer is given, and responsible provider executives and customer are indicated. The procedure for involving subcontractors for the agreement implementation is noted.

Agreement subject

The following definitions are given in the section:

- Agreement subject matter;
- Agreement goals;
- Basic principle formulation of the IS service provision;
- Main advantages (benefits) for the customer.

Agreement terms

This section specifies the agreement validity period, the terms for its validity extension.

Provided IS services

This section describes:

- Service delivery framework;
- Approach to measure the provided service level;
- Parameter description to measure the service level (indicators\metrics);
- Complete provided services list (services groups).

A detailed services specification and a description of the service provision level are given in the relevant annexes.

Reporting on IS services

This section describes the reporting procedure on the provided services and the provided service level, including:

- Reports list and composition (metrics, parameters, management, incidents);
- Reporting interval;
- Reporting periods;
- Reporting terms;
- Reporting analysis interval;
- Description of the order and procedure of holding working group meetings with the customer and provider representatives on the analysis of reporting.

Customer responsibility

This agreement section describes the customer duties and responsibility during providing information security services by the provider under the agreement.

The following subsections are described.

User responsibility describes possible unacceptable user actions that can affect the IS service provision level, under which the customer is responsible for:

- Unauthorized user actions;
- Standards and procedure violation;
- Equipment failure as a result of unauthorized interference;
- Customer power sources failure;
- Refusal or delay of the customer participation in the reporting procedure;
- Other cases of system performance and technologies failure caused by the customer representatives.

Rules of work with third parties discloses the procedure of the implementation delivery and control by the customer representatives of the company IS policies, standards, and procedures.

Parties interaction

This section describes the following procedures.

Satisfying additional customer requests for the IS service expansion or modification (the services provision procedure that is not described in the agreement, or service provision time expansion, or the service provision level improvement).

Interaction procedure. This subsection describes the interaction between the customer representatives and the services provider during IS services provision: the initiation procedure of oral and written instructions and requests, their registration and execution, the problem resolution procedure.

The procedure of the service provision process analysis. The procedure for holding regular meetings on the IS services provision, the meetings format, preliminary agendas, and the meetings entry lists are described.

The downtime removal procedure. The procedures to restore the services in emergency situations are described, including in case of unforeseen circumstances.

IS services cost and payment terms

This section describes the following points.

The *payment procedure* for the provided IS services.

The *cost* of each of the services provided by the provider to the IS service executive for a certain service delivery period. The full cost of all IS services for a specified time period is also given. The IS services cost is detailed in the corresponding annex.

Fines. This subsection describes the penalty system for failure to provide IS services or failure to comply with the services provision level for each of the target parameters. The fine size should depend on IS service provision level deviation from the target level. The penalty system is usually formulated as a credit system. The total fines amount for the reporting period is used to pay for the IS services provision in the future periods. Details on the penalty system for each IS service are given in the corresponding annex.

Agreement terms change

This section describes the procedure of changing the agreement, the procedure of changes and additions approval, and the approved changes application procedure.

Appendix A. Provided IS services and the service provision level.

This appendix contains a detailed description of the IS services and the level of the IS services provided by the provider to the customer and is an integral agreement part.

For each provided service, a description is given according to the following scheme:

- IS service name;
- Service description (result, functionality);
- Responsible service manager (owner);
- Target (planned) service level parameters (availability, response time, delivery time, priority, user number, transaction number, time between failures, etc.);
- Service provision cost;
- Penalties for failure to provide services or failure to meet the parameters of the service provision level;
- Additional or external services;
- Procedure for providing the service;
- Assets used to provide the service;
- Structure and schedule of reporting on the service;

- Security requirements;
- Special requirements to the service provision procedure;
- Existing restrictions on the service provision.

Appendix. Basic definitions

This appendix defines the basic terms used in the agreement. The definitions should be given in the customer business terms and contain the minimum number of the required technical terms, definitions, and descriptions.

2.2.5 Best Practices for Providing Cyber Security Service

Figure 2.31 presents a possible approach to the strategy implementation to source information security in national companies.

To implement the IS sourcing concept, it is necessary to take a number of steps [18–20, 39].

1. *Select a model and develop a plan for the IS sourcing concept implementation.*

   - Choose an information security service provision model.
   - Justify and assess the economic efficiency of the concept implementation.
   - Develop a detailed plan for the concept implementation, to approve the plan, and to initiate the project.

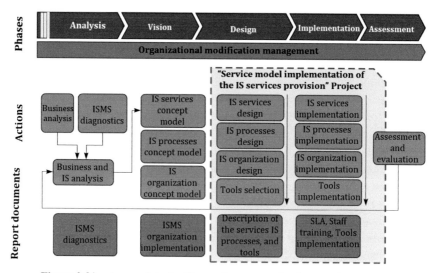

Figure 2.31   Approach to the IS sourcing concept implementation strategy.

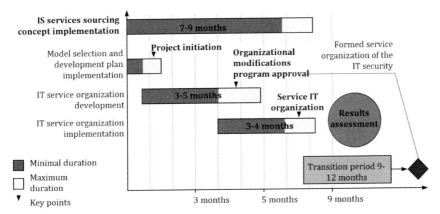

Figure 2.32 The IS sourcing concept implementation. Project time frame.

2. *Design the service organization of the IS service provider.*

- Thoroughly design IS services, included in the service groups, formed earlier. Form the services list.
- Design processes and IS services, as well as procedures for the service organization. Specification is done up to the service element level. Each IS service is completed taking into account the IS services and the required assets necessary to provide the service. The required resources are calculated.
- Design the IS service provider structure (human resources required to provide services, their number).
- Specify the service elements that are supposed to be outsourced (the request for proposal (RFP) to prospective providers for the relevant service elements, the external partner selection).
- Develop and coordinate SLA and SPA for IS services.
- Select specialized software necessary to support the IS services provision.

3. *Implement service organization of the IS service provider.*
The transition to a new model is in progress.

- Implement regulations and procedures for the IS services provision.
- Prepare and approve orders, instructions, and constituent documents to implement a new model.
- Implement new processes in accordance with the regulations developed at the previous stage.

- Execute contracts with external IS services providers based on SPA for the necessary IS services and to implement SLA for each IS service provided to business users.
- Implement relevant software supporting the IS service provision.

Note. After the implementation phase is completed, the transition results of a new model are evaluated by analyzing the satisfaction level of business users and external audit of the information security state, and then the actions improving the organization and next steps are defined and planned. The services cost calculation and the economic organization evaluation of the IS service provider are carried out. These works are recurrent and are carried out regularly, for example, once a year.

During the transition period (Figures 2.33 and 2.34), a periodic assessment of the current IS state is carried out and the necessary steps are taken to improve a model. Budgeting of the information security provision is carried out as before, proceeding from the current costs of technologies and security systems maintenance and development. The service cost calculation in accordance with real indicators, tested in practice, is being clarified. The transition to the calculation by the provided IS services volume is performed only after achieving the stability of the IS service provider.

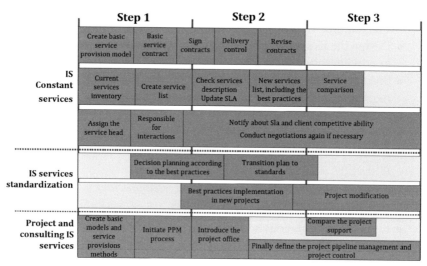

Figure 2.33　The IS services package development.

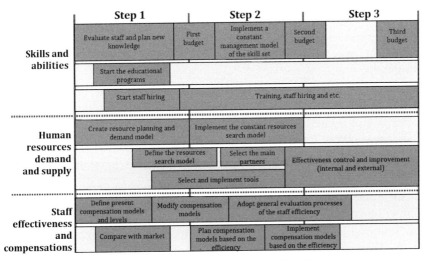

Figure 2.34   Management development of staff and resources.

Critical success factors

Critical project success factors are:

- Management will and increased attention to the information security domain during the transition period, organizational changes support;
- Project approach to the concept implementation;
- Effective problem-solving procedures;
- Key business user involvement in the project implementation process in the center and in the regions;
- Partner involvement in the project implementation who have sufficient experience of such changes;
- Project existence of the target information system architecture (infrastructure, application systems, communications);
- Clearly formulated directions for the IS strategy development;
- Availability of the transition plan to the target architecture to provide information security;
- Projects approved plan/program availability in the information security domain.

***

At present, the majority of the national companies are suspicious about initiatives in the IS sourcing domain. The IS function transition projects to a third party are quite difficult. The main reasons for such state of affairs are:

- Unwillingness of national companies to fully transfer all noncore functions;
- Underdevelopment of the Russian IS outsourcing market;
- Complexity of SLA contract development and KPI calculation;
- Presence of a number of stereotypes, mainly the psychological unwillingness to allow a third-party company to solve security problems, as well as a lack of understanding of the outsourcing project tasks essence and advantages.

At the same time, specialists expect that initiatives in the IS sourcing domain will take their rightful place in the process of improving and increasing the "maturity" of the national company corporate management in the next 3–5 years. National IS services select the IS sourcing initiatives as one of their main developments and improvement trends [18–20, 39].

2.3 Development of National MSSP/MDR Based on Big Data

The known solutions of national and foreign production can be used to modernize national MSSP/MDR Technology Platforms in terms of the effective work organization with Big Data. However, in practice, these solutions often do not have sufficient performance, cyber security, adaptability, and fault tolerance. Moreover, there is a certain compatibility problem of the mentioned solutions between each other. In other words, the best MSSP/MDR practice has not yet developed universal common approaches to determine the optimal structure and functionality of the corresponding Big Data collection and processing subsystems. Therefore, a reasonable selection and implementation of the MSSP/MDR subsystem architecture to work with Big Data is not a trivial task.

2.3.1 Big Data Processing Requirements Analysis

The known solutions for processing ultra-large data flow from IBM, LORD MicroStrain Sensing Systems, Thales, Airbus DS Communications, etc. are expensive and often closed and functionally restricted. Moreover, they are often characterized by insufficient performance, cyber security, adaptability, and fault tolerance. Therefore, the development of some universal monitoring

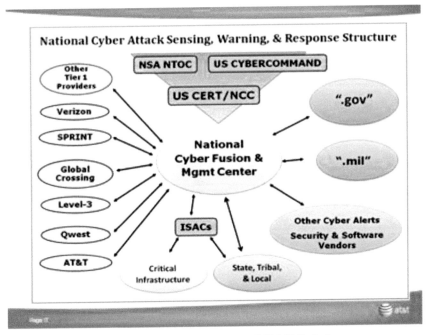

Figure 2.35 National monitoring MSSP/MDR platform.

MSSP/MDR platform to work with Big Data with the required indicator values of the specified system features is an important problem. Such platform could become the MSSP/MDR technology core backbone of a new generation on behalf of the federal and regional executive authorities, as well as big business. There is an urgent need for cloud scalable distributed cyber security monitoring subsystems based on Big Data collection, processing, storage, and analysis technologies capable of operating at high data rates and ensuring super-large data flow processing in real time with minimal delay. In this case, extremely large data means data that occupy an extremely large volume on a physical storage device (the term means the maximum possible volumes determined by the latest advances in the field of physical data storage and software operation by them).

The basic requirements for such scalable MSSP/MDR subsystems to work with Big Data are scalability (the ability to easily add additional servers to increase bandwidth) and high availability (the ability to work steadily without human interference under equipment and network system failures). Such MSSP/MDR subsystems tasks have their own specifics. Data from various MSSP/MDR monitoring objects data flows constantly arrive at changing rate,

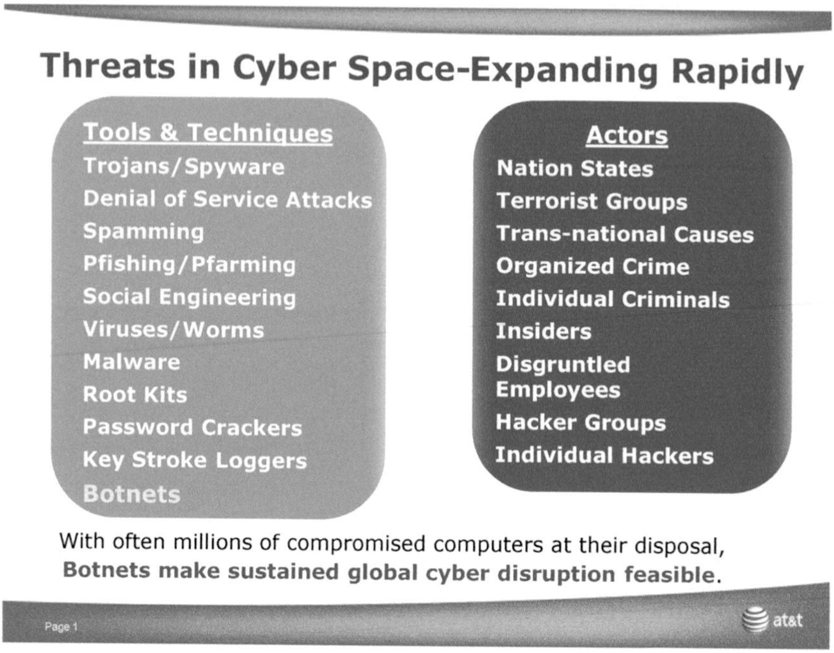

Threats in Cyber Space-Expanding Rapidly

Tools & Techniques
Trojans/Spyware
Denial of Service Attacks
Spamming
Pfishing/Pfarming
Social Engineering
Viruses/Worms
Malware
Root Kits
Password Crackers
Key Stroke Loggers
Botnets

Actors
Nation States
Terrorist Groups
Trans-national Causes
Organized Crime
Individual Criminals
Insiders
Disgruntled Employees
Hacker Groups
Individual Hackers

With often millions of compromised computers at their disposal,
Botnets make sustained global cyber disruption feasible.

Page 1   at&t

Figure 2.36   National MSSP/MDR interaction organization example.

which makes Big Data collection, storage, and analysis in real time a difficult task. The standard technologies used to solve these problems on the behalf of MSSP/MDR, when the number of objects reaches several million, becomes increasingly complex, expensive, and often impossible, since modern systems based on relational databases and application servers cannot cope with the tasks of processing Big Data flows, as well as with the storage of archival data (time series) with petabyte or larger data volumes.

As the MSSP/MDR practice shows, data storage and processing systems based on relational databases cannot be a part of monitoring subsystems because of low performance and scalability. They were developed as a universal solution to process a large number of transactions, maintaining data integrity, and performing simple analytical tasks, such as OLAP. The traditional way to manage petabyte and more data using relational databases stored on servers is not properly scaled, and new solutions are needed to overcome the problem of explosive growth in volume and data flows. With low-cost servers, we can efficiently save Big Data by using horizontally scalable distributed systems. The required monitoring subsystems for future

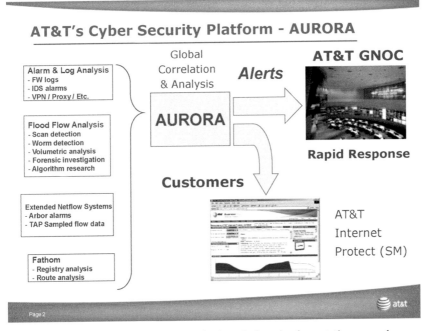

Figure 2.37 The MSSP/MDR monitoring platform implementation example.

MSSP/MDRs must be initially designed, taking into account the requirements of high performance, cyber security, scalability, and fault tolerance. It should be developed primarily on the cloud (foggy) MSPP/MDR infrastructure and meet the specified requirements.

2.3.2 Best Big Data Processing Practice

Nowadays, scalable systems of the NoSQL and NewSQL class, which have successfully implemented the Big Data and Big Data Analytics concept – collection and processing of super-large data flows in real time or close to it – are already known. In particular, they were implemented for a cyber-attack preparation sign identification and for a rapid and relevant response to cyber security incidents and the identified vulnerabilities of the protected information infrastructure. These technologies originated in large electronic commerce companies (Amazon), search engines (Google, Yandex), social networks, and instant messengers (Twitter, Facebook, Skype) and led to a new trend in data management systems and to the so-called NoSQL data storages.

Figure 2.38   Technological reserve for the MSSP/MDR modernization.

The main advantage of NoSQL data management systems is that they are scalable horizontally and can work with thousands of standard servers without a complicated administration process, and the processed data amount can reach tens and hundreds of petabytes or more. A common characteristic of NoSQL systems is the support of a flexible database schema, horizontal scalability, and the denial to support the standard requirements set for relational ACID databases. To achieve scalability and reliability, data are stored and replicated in distributed systems, often in several data centers. The data consistency requirements are reduced to ensure tolerance to network separation and to reduce delays in performing data record operations in these systems. That allows updating data asynchronously and resolving potential conflicts when performing read operations. Therefore, it becomes possible to transfer inconsistent values from distributed data stores, depending on where the read operation is performed. At the same time, data users (administrators, users, and applications) must cope with this potential inconsistency.

Currently, various platforms and frameworks based on the organization of massively parallel computations are known. Most of them use Google MapReduce model, which formed the Google search engine basis. The MapReduce model allows implementing batch processing of static data that are stored in a distributed NoSQL database management system. This model allows organizing a parallel problem solution by decomposition into the same batch tasks that are then executed by accessing various servers and local hard disks, including those without access to the centralized data store. This achieves the necessary performance, flexibility, scalability, and fault tolerance of the Big Data collection and processing system, as well as solves the problem of transferring a simultaneous very large raw data amount from a centralized repository to local computing resources. Today, a large number of open-source projects are implemented based on the MapReduce model. The most famous of them is Hadoop, the Apache Software Foundation project, which is a freely distributed set of libraries and software for developing applications on distributed clusters. Moreover, the MapReduce model constantly improves and becomes more popular. Solutions based on Hadoop allow organizing Big Data processing on clusters from hundreds and thousands of servers with the required performance, scalability, adaptability, and fault tolerance.

At the same time, it should be taken into account that in the MSSP/MDR monitoring subsystems, a big flow of incoming cyber security events (I and II order information features) that come from the cyber security monitoring objects of the protected information infrastructure with uncontrolled data flow intensity is specific. These data must be definitely processed on-the-fly with a minimum delay of several milliseconds. This tasking requires a different data processing architecture from the architecture used for batch processing. There were attempts to implement a stream partitioning strategy of input data into fixed-size segments, which are then processed by MapReduce technology (e.g., the Apache Software Foundation project). The disadvantage of this approach was that the delay in processing data was proportional to the segment length; moreover, there were additional costs necessary for segmentation. Small segments allowed reducing the delay but increased overhead costs and significantly complicated the intersegment dependence management (one segment may require information from others), and large segments led to an increase in delay. Here, the optimal segment size depends on the particular task of Big Data processing and the corresponding application. Therefore, the idea of developing some universal platform for packet and stream processing of cyber security events was not fruitful in the MSSP/MDR construction practice.

Another well-known approach to develop the data processing systems for Big MSSP/MDR Data is based on the actor model that is a mathematical abstraction of the parallel computing organization. This model defines the "actor" concept as a universal parallel numerical calculation primitive: in response to received messages, the actor can make local decisions, create new actors, send messages, and set rules on how to respond to incoming messages. With this approach, the MSSP/MDR monitoring sensors are represented by the finite state machines (FSM), which contain all the current information about the observed object and receive and send messages when the information changes. The processing and analysis of data coming from the MSSP/MDR monitoring sensors is performed by the corresponding FSM, the messages are transmitted in the form of event data. Each object status is not available to other FSM; events transmission and distribution is the only way to communicate between FSM. The main advantage of actors model is the asynchronous interaction of the sender with the recipient: the sender does not wait until his message reaches the recipient and can continue to work. The most popular Big Data processing systems based on the actors model are Twitter company Storm and Yahoo S4. The systems software implementation is performed in the Java programming language by JMS to transfer messages between the FSM. At the same time, the data processing scheme is implemented on the basis of a directed acyclic graph. For example, Storm supports topologies that process continuous data flows entering a system with a changing rate, which, unlike Hadoop, never stop continuing to process the data as they arrive. The system can handle about a million messages per second on a single computing node and has acceptable scalability and fault tolerance.

The next promising way to develop MSSP/MDR monitoring subsystems is based on the Erlang & RIAK Core combination. The functional Erlang programming language was designed by Ericsson to develop software for distributed fault-tolerant computing systems. In this case, Erlang in contrast to the Java language was originally developed for the software actors model implementation, a set of processes interacting with each other by asynchronous messages. The processes here are isolated from each other and do not have a common state. Communication between them is carried out by asynchronous messages about the processes status. Applications execution in the Erlang language is carried out in parallel on several nodes of the distributed computing environment. The macro- and micro-computing modules from the core of one processor, the processor as a whole, the "von Neumann architecture" calculator of 1 to 5 generation, and also cognitive supercomputers on a new cognomorphic and neural-like computing architecture can act as such nodes. Here, the absence of need to block access to the process state to synchronize the interaction nodes greatly simplifies the corresponding applications development for the MSSP/MDR cyber security event monitoring system. In order to organize the processes interaction on different nodes, it is enough to know their names.

Riak Core is an open Basho Technologies library, which was created to develop applications for the Amazon Dynamo distributed computing environment. The difference between this library and the Hadoop project is the ability to create an equal symmetric computing environment (there is no single coordinator like in Hadoop), providing horizontal scalability, high availability, and fault tolerance. In general, the combined Erlang&RIAK Core approach allows creating scalable and fault-tolerant processing subsystems for streaming events and big cyber security data for MSSP/MDR.

There is a conceptual problem when creating distributed processing subsystems for streaming events and big cyber security data for MSSP/MDR. The essence of which is formulated by the CAP theorem. It states that in any distributed system that provides shared data storage, two of the following three properties can be supported simultaneously:

- Consistency means the presence of a single data copy corresponding to the last update operation;
- High availability of these data if there are update operations;
- Tolerance to network partitions.

In practice, this leads to certain trade-offs in justifying and selecting the architecture of the streaming event processing subsystem and big cyber security data for MSSP/MDR, because it is necessary to choose between the cyber security data consistency and availability when partitioning the cluster, and also to implement cluster partition management and restore the cyber security data integrity after the partition. This approach assumes that there are plans to support the operation of the stream event processing subsystems and big cyber security data for MSSP/MDR when sharing and restoring connectivity between clusters.

2.3.3 MSSP/MDR Subsystem Functionality for Big Data Processing

Prospective MSSP/MDR monitoring systems (Figures 2.39 and 2.40) that process large real-time cyber security data flows must meet a wide range of processing requirements for Big Data security. For example, a distributed, fault-tolerant MSSP/MDR application server, developed on the basis of the Elastic Cloud technology of the PaaS class, should provide:

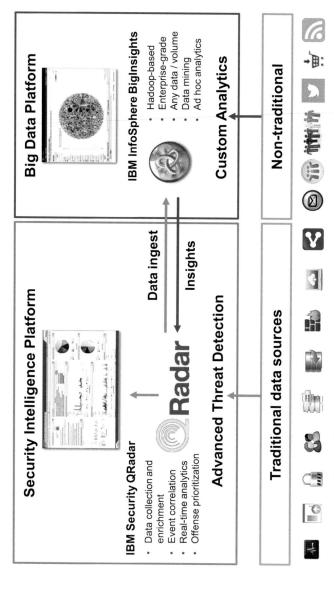

Figure 2.39 Example of Cybersecurity Reporting Using Big Data Technologies.

Security Intelligence From Real-time Processing of Big Data

Behavior monitoring and flow analytics

Activity and data access monitoring

Stealthy malware detection

Network Traffic Doesn't Lie

Attackers can stop logging and erase their tracks, but can't cut off the network (flow data)

Improved Breach Detection

360-degree visibility helps distinguish true breaches from benign activity, in real-time

Irrefutable Botnet Communication

Layer 7 flow data shows botnet command and control instructions

Figure 2.40 MSSP/MDR report examples on the digital enterprises.

- Monitoring of a large number (up to hundreds of millions) of hardware and software detectors, sensors, software agents, robots, video cameras, mobile objects, etc. that collect big cyber security data for MSSP/MDR in "soft" real-time mode;
- Processing of a large number of cyber security events (more than hundreds of thousands per second), taking into account their interrelationships and the MSSP/MDR objects status (complex event processing);
- Small time delay in the cyber security events processing to obtain data on the status of observed MSSP/MDR monitoring objects in real time;
- Intelligent cyber security incidents processing (incident management);
- Collection and storage of various information about a large number (hundreds of thousands and millions) of different MSSP/MDR monitoring objects, collected from different information systems, taking into account their interrelationships;
- Carrying out analytical calculations on large data amounts (petabytes and more), including finding various patterns and hidden relationships on this MSSP/MDR monitoring objects set, as well as the intelligent search possibility, such as taking into account the space-time characteristics of the mentioned objects;
- Closed information security collected by the MSSP/MDR monitoring subsystem to maintain safe and stable operation of monitoring and protection facilities;
- Data visualization on the cyber security state of MSSP/MDR monitoring objects by some Web interfaces, including reports, dashboards, maps, interactive schemes;
- Data display on the cyber security state of MSSP/MDR monitoring objects on mobile devices (smartphones, tablets) running on iOS and Android operating systems, etc.;
- Horizontal scalability (linear performance growth with increasing processing power) so that if the data amount entering the monitoring system significantly increases, the problem of performance shortage could be solved by adding new servers to the system without significantly improving the platform architecture;
- Adding process of the new nodes should not have a significant impact on the operation of the already existing MSSP/MDR cyber security monitoring subsystems; therefore, this process should occur without reloading the corresponding subsystems and should not decrease the overall performance for a long time period, etc.

Let us note that there should be possibilities for carrying out analytical calculations on large data amounts, as well as an appropriate intelligent search for hidden regularities in the cyber security domain, including taking into account the spatial and temporal characteristics of MSSP/MDR monitoring objects. Previously known predictive analytics methods were developed for small data amounts that make them difficult to apply to large volumes of cyber security data generated by large-scale sensor networks and MSSP/MDR subsystems. For these large-scale systems, research on new methods of predictive analytics that are suitable for processing and presenting large cyber security data amounts is required.

2.3.4 Sensor Cloud Architecture Advantages

To develop innovative monitoring systems for super-large MSSP/MDR cyber security data flows, it is proposed to use the Sensor Cloud architecture (Figure 2.41). Here, the Sensor Cloud is a typical cloud platform for collecting, processing, and storing Big Data from sensor networks. Data from sensors enter the cloud through the transport gateway. The information exchange with sensors can take place in both directions, i.e., the platform also provides the ability to control sensory networks. The Sensor Cloud architecture processes the data received from the sensors and ensures that data are stored in the NoSQL storage.

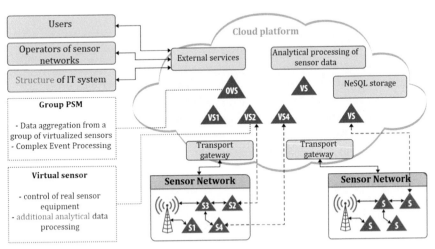

Figure 2.41   Sensor Cloud architecture.

The Sensor Cloud platform requires the client application development through which it becomes possible to manage sensors and access data in real time.

The main advantages of such Sensor Cloud architecture include:

- Possibility of processing large data flows from a wireless sensor network;
- Ability to create scalable MSSP/MDR monitoring solutions, the overall system performance can grow by increasing the server number without requiring large costs;
- High availability of MSSP/MDR monitoring results; access to them through the cloud platform, so responsible persons have the ability to obtain the necessary data regardless of their location;
- API use for the stored data visualization, which makes it possible to represent the parameters values measured by sensors in the form of graphs, tables, and diagrams;
- High automation degree of MSSP/MDR services that significantly reduces the provision time of relevant cyber security services;
- Possibility of sharing resources between different services, etc.

Let us note that the typical Sensor Cloud platform has three architecture levels. On the first level, there are user applications, and on the second level, computing resources are distributed and virtual sensors are created. The third level is the physical sensors level.

Modern cyberspace is filled with extremely large volumes of structured and unstructured data that grow exponentially. The need to obtain qualitatively new and timely information on the quantitative cyber security trends, the state digital economy, and the timely prediction, detection, prevention, and suppression of cyber-attacks of different intruder types becomes increasingly evident. The urgent problem solution is possible based on the considered methods and technologies to process the intensively growing large cyber security data flows in real time.

In particular, the Sensor Cloud architecture allows building distributed fault-tolerant monitoring systems for large cyber security data flows for prospective MSSP/MDR. Despite the heterogeneity of engineering approaches to solve this problem, it can be stated that the general concept has already been formed. On the basis of this concept, the future intelligent monitoring subsystems MSSP/MDR will be built to organize the proper digital enterprises security of the Russian Federation.

2.4 New Methods of Cyber Security Knowledge Management

In early 2016, a number of national companies began working on developing a corporate (institutional) state system segment to detect, prevent, and eliminate the cyber-attack consequences (hereinafter SOPKA). In this regard, the main provisions of the State system concept to detect, prevent, and eliminate the cyber-attack consequences (paragraphs 7, 11, 13 and 14) approved by the President of the Russian Federation on December 12, 2014, No. K 1274, are taken into account, including a new important problem of the semantic cyber security data analysis. The article discusses the possibilities of using Master Data Management (MDM) technology to solve the above problem.

2.4.1 Possible State of the Art

In this section, master data management (MDM) SOPKA is an information system that accumulates input data from various external and internal data sources (Internet/Intranet and IoT/IoT) and ensures centralized data storage and provision in standardized form to make reliable decisions and support the operational SOPKA activities.

The MDM in IT is traditionally used for a data management about products (PIM) and clients (customer data management, CDI). At the same time, MDM is a part of the process-oriented information technology group, and it plays an important role in solving analysis and data processing problems. In practice, the MDM implementation risks are estimated as small, and implementation time as acceptable (Figure 2.42). It is believed that MDM technology is relevant, primarily for large organizations with the number of large-scale distributed applications more than three. According to Gartner (Figure 2.43), a modern MDM solution should:

- Provide opportunities to manage the information quality;
- Provide loading, integration, and synchronization of large arrays and data streams;
- Support workflow management processes and related services;
- Have high performance, availability, and security;
- Support automated modes of data collection, processing, storage, and analysis;
- Meet technological standards and best practices.

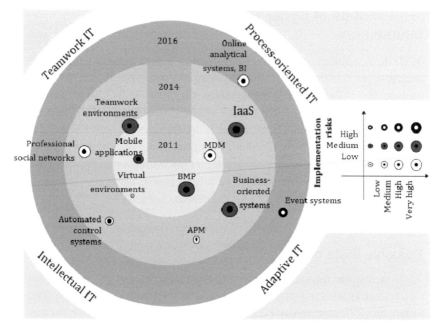

Figure 2.42   The MDM place and role in IT.

The main MDM development trends and prospects are:

- MDM applications shift from the business intelligence and decision support to operational activities that directly affect business results;
- Transition from narrow subject-matter solutions (mainly clients or products) to multidomain solutions (several data types at once: products, customers, finance, security, etc.);
- "Social" MDM uses a modern opportunity of teamwork, social networks, and instant messengers to automate master data processing;

Data governance (unified corporate data management) is an extension of the master data management principles for a wide range of corporate data;

- MDM integration into the corporate business process management system and corporate cyber security systems;
- Including constant information (reference data) along with conditionally constant information (customers, products, etc.) in the MDM sphere;
- MDM adaptation to work with Big Data and streaming data processing;
- MDM implementation in the form of appropriate cloud services: SaaS, PaaS, or IaaS;

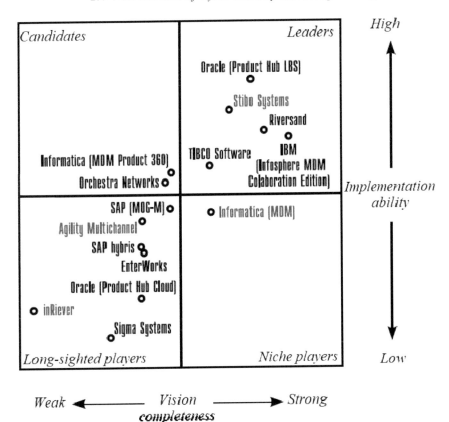

Figure 2.43 The "magic" MDM Gartner quadrant.

• Evolution from syntactic to semantic technology analysis and data processing.

At the same time, the semantic MDM prospects can be estimated from the Gartner "interest curve" (Figure 2.44).

Let us consider that MDM technology shifted from a "lively interest" area to the "mature" technologies field, and the "semantic analysis and data processing" technology is only gaining popularity or is at the meridian of the increased specialists attention in the IT and cyber security domains. The semantic cyber security MDM means a data management system that operates rules of the object behavior and interaction in cyberspace to solve the SOPKA problems in order to prevent the protected critical infrastructure transition to catastrophic states.

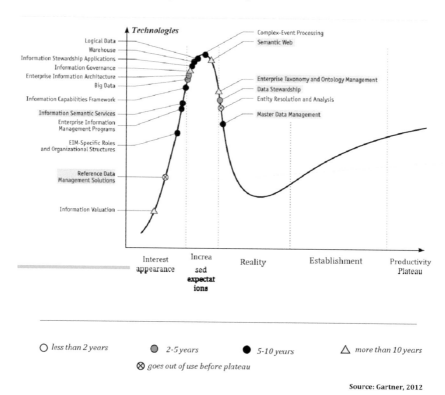

Figure 2.44   The Gartner "Hype Cycle" for information technologies.

Typical objectives of semantic cyber security MDM include:

- Construction and support of the cyber security ontologies that are information confrontation object models, required to solve SOPKA problems;
- Storage model unification of the cyber security data coming from external and internal providing systems, for example, SIEM, IDS/IPS, information security tools from unauthorized access, cryptographic information security systems, etc.;
- Standardization of internal cyber security data exchange protocols;
- Regulating the processes of maintaining cyber security databases and knowledge bases;
- Decision-making support in SOPKA based on the information confrontation semantics, presented in the corresponding cyber security ontology, etc.

Cyber security ontology

As a rule, the knowledge presentation way about qualitative characteristics and quantitative patterns of information confrontation is not clearly indicated in SOPKA. In our opinion, it is reasonable to use cyber security ontology (meta-ontology) for this. According to Thomas Grubber, the cyber ontology is a certain conceptualization specification of the information confrontation domain. This specification can be given analytically, for example, based on the algorithms theory and mathematical logic, as well as graphically, using some schemes that reflect traditional linguistic methods and textual information process methods[58]. The key here is an ontology specification language selection, which allows processing the domain machine-interpreted semantics.

Currently, there are three main ontology description language classes:

- Traditional languages based on the specification (Ontolingua, CycL), descriptive logic (LOOM), frames (OKBC, OCML, Flogic);
- More Web-based languages (XOL, SHOE, UPML);
- Special languages (RDF, RDF SPARQL, SWRL, DAML, OIL, OWL). For example, SPARQL can be used to execute logical queries; Semantics-SDK and Owlim can be used to implement rule-based logic output; and Pellet, FaCT ++, and Hermi to implement semantic tableau based logic output. At the same time, the OWL API framework integration is possible to introduce the ontology in memory, to modify the ontology at the object level, etc.

Two approaches to ontology are also distinguished in accordance with the IDEF5 standard of ontological analysis (IICE, 1994) and the recommendations of the World Wide Web Consortium [2009].

In the literature, the following ontology construction methods are presented (Table 2.2):

- Uscold and King method [Uschold et al., 1998];
- Grninger and Fox approach [Gruninger et al., 1995];
- CycL method [Lenat et al., 1989];
- Kactus method [Schreibe et al., 1995];
- Sensus method [Swartout et al., 1997];
- On-To-Knowledge approach [Staab et al., 2001];
- Methontology method [Ferndndez et al., 2006], etc.

In the works of Russian authors, ontologies are presented on the basis of:

- Finite automata [Kryvyi, 2008], [Beniaminov, 2003];
- Lexical syntactic patterns [Anisimov, 2002], [Rabchevsky, 2009];

Table 2.2 Distinguished ways of ontology construction

Method Name	Completion Degree	Implementation Complexity	Method Flexibility	Software Dependence	Life Cycle Support	Scientific Support	Detailed Elaboration	Method Compatibility
TOVE	Demonstration prototype	Average	Yes	Does not depend	No	Yes	Yes	No
Enterprise model approach		Average	Yes	Depends	No		Yes	No
METHONTOLOGY		Average	Yes	Depends	Yes		No	No
KBSI IDEF5		Average	Yes	Depends	No		Yes	No
Ontolingua		High	Yes	Depends	No		Yes	Yes
Common KADS and KAKTUS		Average	Yes	Depends	No		No	No
PLINIUS		Average	No	Depends	No		Yes	No
ONIONS		Average	No	Depends	No		No	Yes
Mikrokosmos		Average	No	Depends	No		Yes	No
MENELAS		Average	No	Depends	No		No	No
SENSUS		High	Yes	Does not depend	No		Yes	Yes
Cye methodology		Average	Yes	Depends	No		Yes	No
UPON		Average	Yes	Depends	Yes		Yes	No
101 method		Average	Yes	Depends	No		Yes	No
On-To-Knowledge		Average	No	Depends	Yes		Yes	No

- Product systems [Nayhanova, 2008];
- Linguistic methods [Mozzherina, 2011];
- Information granularity [Tarasov, 2012], etc.

For example, in the work of V.B. Tarasov [59], the research results in the field of the cognitive agents and mobile robots theory are presented, which, in contrast to reactive agents functioning according to the "stimulus-reaction" scheme, are given a well-developed, dynamic external environment model. Here, the cognitive function provides a robot with the learning processes of the outside world, other agents, as well as its self-knowledge. Cognitive processes cover an agent environment perception, generalized internal representation formation, interaction and behavior principles understanding, and training. In fact, the "data-information-knowledge" transitions [Gergey, 2004],which are necessary for the efficient agent operation, are implemented in these processes. V.B. Tarasov cognitive robots are able to receive and process heterogeneous information from a human operator in a limited natural language in the form of target designations and instructions, from sensor system actuators and from their own knowledge base.

For the interactive cognitive robot management, it is proposed to develop common ontologies that ensure effective communication between a person and a robot when the latter performs complex tasks in an inaccurate and incompletely defined environment. The main attention is paid to the meta-ontologies formation by granular information representations (in the form of intervals, fuzzy sets, linguistic variables, fuzzy algebraic systems), as well as the space ontology development in the robotics field based on the G. Leibniz and S. Lesnevsky ideas.

Let us note that the granulation meta-ontology and space ontology developed by V.B. Tarasov and his students can be applied to cognitive hardware–software SOPKA agents. The possible architecture of such an artificial cognitive agent is shown in Figures 2.45 and 2.46. In addition to such known artificial agent features as intentionality, activity, reactivity, autonomy, and communication skills, an important ability to granulate incoming information will be added.

The analysis shows that by combining cyber security data from various external and internal information sources and corresponding rules to detect, prevent, and eliminate the cyber-attacks consequences into a single semantic domain model, it is possible to build the required intellectual (and, in the future, cognitive) information space, then to develop the appropriate artificial cognitive agents and the corresponding intelligent "semantic cyber security MDM" software and hardware complex to support SOPKA operations as a whole.

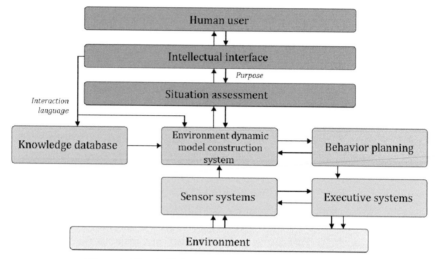

Figure 2.45   Artificial cognitive a, b agents architecture.

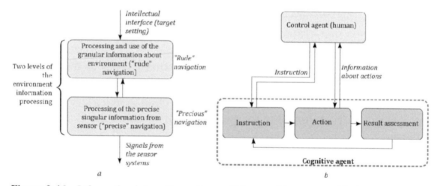

Figure 2.46   Information interaction scheme of the cognitive agent with the environment.

2.4.2 Cyber Security MDM Principles

We formulate the basic development principles of the "semantic cyber security MDM" software and hardware complex [1, 3, 5, 6, 8, 9, 22], designed to manage the main (master) SOPKA data:

1. Cyber security data consolidation from external information sources in the connected Internet and IIoT networks and internal providing information security tools from unauthorized access, cryptographic information security, SIEM class systems, and the SOPKA itself into a single information environment;

2. Presentation of the consolidated cyber security data from the mentioned external and internal systems and means in the form of a single object-oriented cyber security data model;
3. Ontological representation of information confrontation objects, which is a semantic models and submodels of various domains use to store the mentioned information objects;
4. Vision contextuality of the information confrontation objects, which is the objects representation solely in connection with the goals and objectives of the state and corporate SOPKA;
5. Knowledge orientation, which is a knowledge transfer (object behavior and interconnection rules) from the logic of providing application-based cyber-security applications to a single object database.

We will comment on a number of above-mentioned principles to establish the "semantic cybersecurity MDM" software and hardware complex. A consolidated cyber security data repository for each SOPKA segment or center should be the only place where data will be added, modified, or deleted. In other words, the "semantic cyber security MDM" software and hardware complex should be an independent class of technical systems that is not subordinate to any cyber security application system, for example, IDS/IPS or SIEM. The decision rule transfer to the cyber security data models level will make them available for all SOPKA systems and tools. Orientation toward the semantic model construction of information confrontation will allow providing the maximum automation level, as the particular solutions, once included in the semantic cyber security database, will be properly formalized and reused in various SOPKA applications.

Thus, the "semantic cyber security MDM" software and hardware complex allows forming and maintaining a consolidated space of aggregated cyber security data to support the operational activities of each corporate or institutional segment or SOPKA center. Cyber security data for each of the above segments are collected from various external and internal provision systems and accumulated in a single permanent storage location. It is inadvisable to transfer a part of the data beyond the mentioned consolidated space, as this will lead to a connection disruption between the cyberspace objects and automatically to the integrity violation of the information confrontation knowledge system that will limit the developing software and hardware complex capabilities.

The domain model presented in the "semantic cyber security MDM" software and hardware complex should be capable of adaptation and self-organization, timely display the appearance of new objects and relationships, changes in the object behavior rules in cyberspace, and their relationships among themselves. In other words, the semantic MDM should be an intellectual decision-making support environment in SOPKA, regardless of the information confrontation nature and specific content in cyberspace. Contextual representation of the internal the information domain objects structure of the information confrontation must dynamically change depending on the decision to detect, prevent, and eliminate the cyber-attacks consequences.

A key characteristic of the "semantic cyber security MDM" software and hardware complex is the ontological information confrontation objects representation in cyberspace. Without the mentioned ontological model construction, it is impossible to formalize the objects interrelationships with other entities since the compatibility rules for the two objects are determined by the combined compatibility of their constituent parts. At the same time, each object meaning is shown in its semantic links with other information confrontation objects. Obviously, when constructing a semantic information confrontation model within the local MDM system of a single segment or the SOPKA center, it will be necessary to operate with terms and definitions from various knowledge fields and then to combine the mentioned local MDM systems together, for example, using semantic Web technology.

In a long term, the semantic cyber security MDM should be considered as:

- Common language for communication between various applied systems to ensure cyber security;
- Method set to maintain specialized databases and cyber security directories;
- Technological solution set that provides the creation of a single information space to support the SOPKA operations.

2.4.3 MDM Cyber Security System Example

Let us consider the prospective system draft of the "Warning-2016" software and hardware complex that is intended for early cyber-attack warning on corporate and institutional information resources of the Russian Federation.

The possible architecture of the "Warning-2016" software and hardware complex is shown in Figure 2.47.

It is based on SAP HANA, and the required semantic MDM is implemented on the basis of SAP NetWeaver Master Data Management (SAP NW MDM). The typical components of the software and hardware complex system architecture are briefly described.

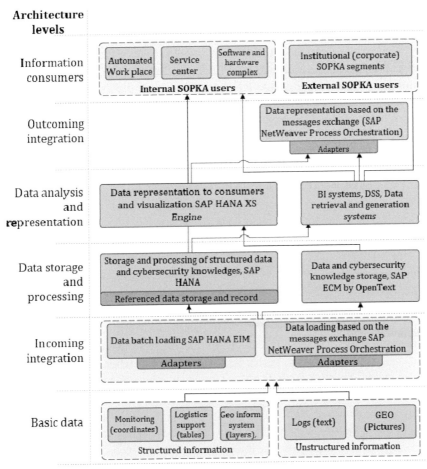

Figure 2.47 The cognitive warning system architecture.

The "Big Data Collection" component is developed based on SAP HANA EIM and NetWeaver Process Orchestration.

Here, SAP HANA EIM "Data Services" (solution of the class Extract/Transform/Leverage, ETL) collects data from various sources. In this case, both standard adapters like Oracle, MS SQL, DB2, Hive, as well as specialized adapters can be used. If necessary, the new adapter development is possible. SAP HANA EIM can perform batch downloading of large data amounts on a schedule up to the online mode (for different data types, different loading time intervals can be configured). When implementing new systems, it will be preferable to organize direct access to databases via the ODBC interface.

SAP NetWeaver Process Orchestration is used to integrate with systems that support queuing and messaging engines. Exchange can occur in either synchronous or asynchronous mode. Process orchestration (PO) provides mechanisms for guaranteed message delivery, including from/to systems in networks with irregular communications. The messages that cannot be immediately processed are saved in the queue for further sending when the connection is resumed. In this case, ordered message delivery is supported (EO – Exactly_Once, EOIO – Exactly_Once_In_Order). The PO ready-made adapters: File/FTP (S), JDBC, JMS, SOAP, WS (WS Reliable Messaging), HTTP (S), Mail (Mail Servers via SMTP, IMAP4, POP3), SFTP, OData, REST, etc. are used for this integration. If necessary, it is possible to modify their functionality or create a new adapter. The structured data are then loaded into the SAP HANA data storage, and the unstructured data are loaded into SAP Extended ECM by OpenText. When connecting new systems to upload data via messaging, the use of Web services technology (e.g. SOAP and WS) and XML format will be a more preferable option.

The data storage component is developed on the SAP HANA basis, which can function in RAM. Due to this, SAP HANA overcomes the main drawback of traditional DBMSs, that is, performance degradation when accessing the disk subsystem. Another SAP HANA feature is the use of the so-called "columned data storage", which allows speeding up the analytical query execution to the database many times.

Within the project framework, a single data model is used in SAP HANA to store consolidated data from various information sources. To do this, we analyze the input data, design the logical repository structure, and create the physical structure of tables and DBMS views. Here, the data are stored in specialized formats (on columns, with compression, etc.), but are represented in the form of a classical relational structure from the user/developer point of view.

Thus, the developing storage based on SAP HANA provides the maximum speed of processing and accessing data arrays, as well as data compression. At the same time, due to the relational presentation form, data access will be carried out in the logical structure optimal for the analytics development.

For this, appropriate analytical tools like the SQL query language, which is de facto the industry standard, can be used. To simplify the modeling process, the so-called data "generalized model" can be used (Figure 2.48), whereas SAP HANA Studio is quite suitable to develop a physical data

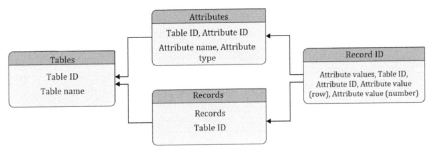

Figure 2.48   The "generalized data model" scheme.

model. Note that with the Graph Engine help, there is a possibility to store data in a graphical form. In this case, data graph access is gained by XS Engine interfaces.

It will be necessary to implement a maintenance system of common directories and classifiers (NSI system) to load consistent data from various sources in SAP HANA. For this purpose, the following data entities are provided in the data model: a directory list, a characteristic list of the objects contained in directories, acceptable characteristic value lists, etc. It is also necessary to have recoding tables between similar directories from different systems. Here, the "semantic cyber security MDM" software and hardware complex is exactly designed to manage the basic (master) cyber security data to support the SOPKA operation as a whole.

Note that SAP HANA supports the developed fault tolerance mechanisms. When one of the working servers failed, the backup server connects to the malfunctioning server data and begins to perform its functions. Thus, the system is fault-tolerant (Figure 2.49).

In addition, it is possible to implement a disaster-tolerant solution by deploying identical SAP HANA installations in two remote data centers with

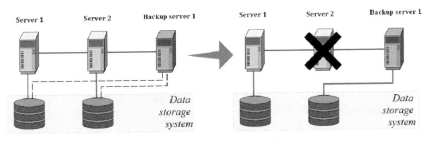

Figure 2.49   Fault tolerance scheme.

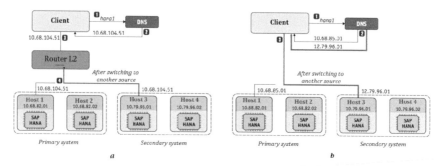

Figure 2.50 Possible switching schemes.

the organization of continuous data replication from the first data center to the second one. When one of the data centers fails, the system continues to function.

Note that, depending on the bandwidth of the existing communication channels, synchronous or asynchronous replication mode can be used.

Switching from one data center to another is also possible by using the so-called virtual IP address or IP spoofing through DNS (Figure 2.50).

To store and manage unstructured information (text documents, audio and video files), it is reasonable to use SAP Extended ECM by Open Text. To scale Open Text, its components (both back-end and front-end) are installed on different servers in such a way that separate servers are responsible for processing user requests and forming interfaces for the latter, while others are in charge of managing the documents storage, building indexes, searching documents, etc. To provide fault tolerance in the 24/7 mode, the Open Text back-end components can be deployed in the HA cluster architecture in Active-Passive mode. For front-end components, constant maintenance is achieved by using backup installations that can be put into balancing at any time when one of the major installation components fails. Switching clients between Open Text installations (if we disconnect one of the installations) are implemented similarly to the case with SAP HANA systems (using a virtual IP address, an IP spoofing).

The "Forecast Analytics" component is developed on the SAP Business Objects BI Platform basis. The named platform supports such functions as generating reports of random complexity, uploading reports in various formats, automatic reports distribution to information consumers, etc. BI Platform complements the SAP HANA XS Engine functionality in terms of data analysis and reporting.

SAP HANA has an accessible library of predictive algorithms (PAL) available for use. If it is necessary to use specific mathematical methods, they can be implemented in SAP HANA in the form of so-called function libraries (AFL).

Users will be able to access the stored information through software or interactive user interfaces implemented using SAP HANA XS Engine, integrated in the SAP HANA application server. To implement interactive user interfaces, it is suggested to use SAP UI 5 interface element libraries. Using UI5, interfaces can be developed with the automatic adaptation function to the size of the device used to output information (wall panel, desktop monitor, tablet screen, etc.). At the same time, information access will be regulated in accordance with user access rights. When accessing data through software interfaces to ensure guaranteed content delivery to consumers, access can be provided using the SAP NetWeaver Process Orchestration component.

Outbound integration is also implemented by SAP NetWeaver Process Orchestration means.

***

The scientific and technical novelty of the "semantic cyber security MDM" software and hardware complex is to apply the process approach in constructing ontological information confrontation models. The proposed solution essence can be briefly expressed as "semantic cyber security management" or, more fully, "the repeated application methodology of knowledge about information confrontation quantitative patterns and qualitative characteristics". The transition to the semantic technologies application in the SOPKA design is a critically important innovation that determines the main development vector in this sphere in the midterm and is a technological advantages source of the systems to detect, prevent, and eliminate the cyber-attack consequences on corporate and institutional information resources of the Russian Federation. At the same time, the appropriate scientific support is, primarily, methods for representing information confrontation knowledge using ontologies, relevant ontological engineering, and semantic search methods.

3

Innovative Methods for Detecting Anomalies

A comparative analysis of the known models and methods of ensuring steadiness (stability, survivability, and fault tolerance) of the Digital Economy technological platforms is given in the current chapter. High vulnerability and insufficient stability of the named platforms are shown in the conditions of rising cyber security threats. The inconsistency of the known methods of information protection and the methods of control and restoration of the computer systems functioning under the destructive effects of intruders have been revealed here. A new approach to ensure the sustainability is substantiated by giving these platforms the ability to develop immunity to cyber-attacks, malware and hardware effects by analogy with the immune system of protecting a living organism, as well as resist the cyberneviation during, not after a hostile impact. A new method for detecting anomalies in the functioning of the critical information infrastructure of the Digital Economy based on the theory of similarity and dimensions has been developed. It is significant that this method allows developing and accumulating the measures to counteract previously unknown cyber-attacks, to detect group and mass impacts that lead to catastrophic states and to partially restore the computing processes. In practice, this new method hinders the degradation of the Digital Economy's technological platforms and allows the reversal of unrestorable or hard-to-recover perturbations against intruders.

3.1 Justification of a New Method for Detecting Anomalies

3.1.1 Analysis of the Existing Approaches to the TCP/IP Network Abnormal Functioning Detection

The problem of the TCP/IP network (hereinafter referred to as the data network) abnormal functioning detection refers to the pattern recognition problems. Therefore, the methodological apparatus of this widely studied science domain can be applied here. Thus, it is reasonable to classify and analyze the approaches used to detect anomalies based on the informative features classification.

All the informative features applied in the pattern recognition are divided into:

- Structural, qualitatively reflecting the nature of the patterns that underlie this or that phenomenon;
- Correlation, revealing quantitative statistical patterns in the features or behavior of the object or phenomenon under study;
- Invariant, defining sets of the attributes or dependencies that are unchanged within each class and are modified when passing to another class.

The methods that apply structural features to detect data network abnormal functioning include the currently widely used signature detection methods (also known as linguistic methods [10, 11]). The general scheme to detect the abnormal data network functioning by signature search is performed by the following algorithm:

- For a set of currently known abnormal functioning types, a model is developed. On its basis, it is possible to obtain the commands or data sequences in the message that clearly identifies the intrusion;
- Software system or hardware–software system consisting of a sensor and a decision circuit is developed. The system analyzes the commands and data flow and then indicates the incident detection fact that matches with the developed model.

This group includes the following methods that are different in the way the intrusion model is specified:

1. *The direct flow scanning method (pattern matching).* A model is formed as a command sequences set or the specific data values. In the modified method version [10], the model elements are specified in the form of regular expressions. The method is characterized by simple model elements development and software implementation.
   The disadvantages of this method are:
   - Impossibility to describe a large intrusion subclass characterized by the parameters variability;
   - Impossibility to set the control dependencies between the model elements;

- Need for the frequent repetition of the model elements corresponding to slightly different abnormal functioning implementations;
- Low information processing rate, caused by a multiple comparison of the target sequence with the model elements.

2. *The anomalies detection method by the expert systems* (e.g., [8, 9, 24, 25]) develops the ideas of the direct scanning method by introducing a control digraph into the model. The digraph vertices are sequences or regular expressions corresponding to the partial intrusion features. The digraph arcs encode the deterministic or probabilistic relationships between the attributes. The decision circuit attainment of the specially labeled notes during the analysis indicates the abnormal network activity detection in the data network.

   Compared to the previous one, the method significantly reduces the model elements number required to describe the same intrusion subclass and reduces the processing time of each data stream implementation simultaneously. However, it practically does not expand the applicability scope and, in the case of the probabilistic processing nature, increases the false response level.

3. *The method of developing the potential active anomalies list* (proposed in [23–25, 40]) is different from the two previous ones in that it does not perform a complete sequence compliance verification for each processed flow element. Instead, the decision circuit maintains the list of actions that possibly begins and is currently unauthorized or unintentional. With each new object arrival from the analyzed traffic:

   - Potential anomaly probabilities in the list are recalculated;
   - Removal from the anomalies list, reliably rejected by the object occurrence fact;
   - List is completed with the records of the anomalies, which are possibly started by this object.

   The method is maximally oriented to work online. The potential anomalies list is maintained in the descending order of their conditional implementation probability based on the already analyzed flow part. This allows maximizing the system computing power resource use under its shortage: in the case of the analyzed objects high arrival rate, the intrusions list is shortened (records with the lowest conditional probability value are removed).

4. *The state transition analysis method* was independently proposed by two research groups [23–26, 40] as a more general variant of the observed command or data flow analysis.

   The method includes a wider range of the intrusion methods into computer systems due to the fact that the model is:

   - completed by the system conditional states set;
   - completed by the relation over the Cartesian product of the system states set and the set of the analyzed command or data flow observed elements;
   - modified by the intrusion signature, controlling the digraph definition not on the set of the observed objects, but on the system state set.

   Obviously, the methods described earlier are special cases of the state transition analysis method in the system. Due to some increase in the model development complexity and its software implementation complication, the method allows describing the unauthorized or unintentional activity methods in the system much more compact and closer to the semantic content. The additional advantage is a possibility to preliminarily classify the anomalies by determining the system state set, mapping to many possible unauthorized action classes.

   The characteristic method properties that use the structural informative features in the decision circuit are:

   - Anomaly model analysis in the data network, however not the model analysis of its normal behavior;
   - Close-to-one probability to detect the anomaly reflected in the model;
   - Close-to-zero false responses level (except for the situations with the boundary parameter adjustment variants in probabilistic models);
   - Impossibility to detect the unauthorized and unintentional activity methods, including new ones that are not described by the model;
   - Qualitative impossibility to describe the certain intrusion classes model, characterized by a wide changes range in their properties.

5. *The correlation features use* for the detection of anomalies is a relatively new approach (the idea was first publicly outlined in [25, 26]), which aims precisely at solving the problem to detect unknown threat implementation ways at the development time. The method class that

uses correlation informative features is characterized by developing in a normal behavior model of data network, which allows separating from the specific threat implementation ways. The most significant disadvantage of the whole class is the probabilistic decision-making nature at the moment. The class is divided into two subclasses: methods of the observed data flow model development and methods of the semantic system model development.

The model development methods of the observed data flow (including statistical) are extensively studied abroad. Within this method subclass, the following models are developed:

- Models of the objects distribution law parameters in the analyzed flow [25, 26];
- Models of the conditional objects occurrence probabilities based on the two [25] and more [26] previous objects analysis.
  The methods modifications differ in:
- Implementation into the model such parameters as "the subject account that generated the objects flow", "time of day", and "day of the week" of the event;
- Resulting space dimension features and, as a consequence, the decision-making based on the threshold values or the cluster analysis [23, 26];
- Model development principles (with the expert help, based on the historical record computation process [23], during the system operational functioning [26]);
- Application of the neural networks [26] and other artificial intelligence theory achievements to develop the model.

The model development methods of the observed data flow are distinguished by the simplicity of the model development and the software implementation. Their disadvantages are:

- High false responses level;
- Impossibility to describe by the model the abnormal data network activity, characterized by the parameters close to the system normal behavior parameters or equal to them.

Table 3.1 shows the comparative characteristics of the described approaches to abnormal functioning detection systems.

The table analysis shows that today there are two actual problems in the issue of detecting the abnormal TCP/IP network functioning. First, it is to effectively detect the new method implementations of unauthorized and

Table 3.1 Informative features selection impact on the abnormal functioning detection system characteristics

Characteristic	Structural Features	Correlation Features
Percentage of the missed known threat method implementations	0%	Close to zero*
Percentage of the missed new threat implementations	100%	Average*
False responses level	Close to zero	High*
Ability to make decisions online	Yes	Yes
Abnormal traffic identification and marking	Yes	Not always

unintentional actions that do not differ from the normal TCP/IP network behavior in the correlation features space. Second, it is a clear identification and marking of the traffic that belongs to such implementations. The methods set used in the modern abnormal functioning detection systems does not allow solving these problems; thus, there is a decrease in the overall security level of the controlled informatization object.

3.1.2 Possible Statement of the Detecting Anomalies' Problem

The current situation analysis in the abnormal functioning detection of the distributed computing system revealed a pragmatic contradiction. It is about the demand for new and promising means to detect the abnormal data network functioning on the one hand and insufficiently developed methods of their application in various directions, on the other. The contradiction raises the scientific problem that is to develop and justify the theoretical provisions of the abnormal data network functioning detection.

The research subject is a process of the abnormal data network functioning detection in the distributed computing system.

The research goal is to improve the process quality of the abnormal data network functioning detection in a distributed computing system based on a method development that takes into account new informative features. Considering the existing method analysis of the abnormal data network functioning detection, the computing system, or the methods, which use the informative anomaly detection features that belong to the invariant class, can become the means to achieve the research goal. In this regard, to achieve the research goal, the following research objectives are formulated:

- Invariant features analysis of the abnormal data network functioning, the selection among them the features, the use of which allows achieving the research goal;

- Model development of the invariant informative features system to design the abnormal data network functioning detection method based on it;
- Model properties analysis;
- Method development of the abnormal data network functioning detection in the distributed computing system, which uses new informative features for decision-making;
- Parameters and algorithm optimization for the method implementation to meet all modern abnormal functioning detection system requirements;
- Abnormal functioning detection system development based on the proposed method and the optimized parameters and algorithms for its implementation.

Results

The modern information threat specificity in the transmission process over the data network of a distributed computer system was analyzed. It was revealed that great part of both unauthorized and unintentional actions implementing information threats generate the abnormal data network functioning incidents, which can be detected by the abnormal functioning detection systems as a part of integrated technical information security system in the computing system.

It is demonstrated that the main quality indicator of the abnormal data network functioning detection process is the incident skipping percentage if the several qualitative requirements are fulfilled. These primarily include the low false response level and the quality indicator stability of the detection process, when the new methods of unauthorized and unintentional actions that were not known at the time of the abnormal functioning detection system, occur, as well as slightly less significant requirements.

The existing approach analysis of the abnormal data network functioning shows that nowadays the methods that implement decision circuit based on the structural and correlation informative features are used. At the same time, the structural informative features separation makes it possible to detect the great majority of the known information threats implementations, but is not able to the detect new methods of unauthorized and unintentional actions. The correlation features separation allows detecting the new information threat methods implementation but it has a high false response level and, in some cases, cannot effectively formulate the rules to determine the message affiliation with the detected incident.

As the analysis result, the research goal was set to improve the abnormal data network functioning detection based on the search for new informative features belonging to the invariant class. The main research objectives were determined, including the method development of the abnormal data network functioning detection, applying the new abnormal functioning informative features in the decision-making process, and the abnormal functioning detection system design based on the method.

3.1.3 Definition of New Informative Features

The structural and correlation informative feature analysis, used in the modern practice of the abnormal data network functioning detection, showed qualitative disadvantages inherent in both the first and second classes as a whole. Therefore, the attempt to use the third informative feature class of the general pattern recognition problem, which is invariant informative features, was made to achieve the stated goal.

The existing invariant informative feature analysis

The invariant characteristics selection of the system classification (in given problem that is classified into two classes: normal and abnormal functioning in the data network) is identical to the two-system isomorphism relating to some mapping. To determine the necessary and sufficient conditions for the systems isomorphism as well as to determine the isomorphism mapping qualitative and quantitative parameters, the similarity theory mathematical apparatus was developed.

L. I. Sedov, V. A. Venikov, A. A. Gukhman, and V. V. Kovalev formulated the basic similarity theory principles. Originally, they were developed to model mechanical [14], electrical, and heat-transfer processes [23–26, 40]. However, in the late 1980s, the results obtained were applied in the modeling with universal digital computers and then transferred to solve a much wider range of problems, including information security, for which S. A. Petrenko formulated the basic concepts [1–3, 5–11, 13–17, 19, 20, 22, 57]

The similarity theory concepts are thoroughly studied in the context of the processes and the systems described by the homogeneous power polynomials systems. Three theorems are fundamental in the similarity theory: direct, inverse, and π-theorems.

Let Φ be two s and S processes, the complete equations of which have the form (3.1) and (3.2), respectively

$$\sum_{s=1}^{q} \varphi_{us} = 0, u = 1, 2, \ldots, r \qquad (3.1)$$

$$\sum_{s=1}^{q} \Phi_{us} = 0, u = 1, 2, \ldots, r; \tag{3.2}$$

where

$$\varphi_u = \prod_{j=1}^{n} x_j^{\alpha_{ul}} \tag{3.3}$$

and

$$\Phi_u = \prod_{j=1}^{n} x_j^{\alpha_{ul}} \tag{3.4}$$

- their parameter homogeneous functions.

A direct similarity theorem asserts that if processes are homogeneously similar, then the relations system takes place:

$$\frac{\varphi_{us}}{\varphi_{uq}} = \frac{\Phi_{us}}{\Phi_{uq}} \quad u = 1, 2, \ldots, r; s = 1, 2, \ldots, (q-1). \tag{3.5}$$

Expressions

$$\pi_{us} = \frac{\varphi_{us}}{\varphi_{uq}} \quad u = 1, 2, \ldots, r; s = 1, 2, \ldots, (q-1). \tag{3.6}$$

are called the similarity criteria or invariants and, as the theorem corollary, are numerically equal for all processes belonging to the same subclass of mutually similar processes.

Thus, the direct theorem formulates the necessary conditions for the studied process correlation with one of the subclasses. The sufficient conditions for the homogeneous similarity of two processes are given in the inverse similarity theorem: if it is possible to reduce the complete processes equations to the isostructural relative form with numerically equal similarity invariants, then such processes are homogeneously similar. A similarity theorem, known as the π-theorem, makes it possible to reveal a functional relationship between the process variables in relative form.

The similarity theory application to control the computing systems stability was first proposed in the V. V. Kovalev scientific school. The direct and π-similarity theorems corollary allowed formulating the invariant informative features for random computational processes in the computing systems in [57] and [5] and developing two methods to verify the necessary criteria for the semantic process correctness. Since the functioning of the data network protocols stack is also a computational process, it is potentially possible to use both methods to achieve the research goal. However, as it will be shown later, the research object specificity leads to the fact that the method proposed in [19] does not satisfy one of the quality requirements of abnormal data network functioning detection.

The analysis method of the computational process dimensions [26, 40] in the computing systems uses the process variables dimensions as the invariant informative features. At the first stage, the computational process algorithm is reduced to the canonical form by introducing the new auxiliary variables in order to derive computational operators from control operators (conditional transitions, ramifications, and cycles). As a result, the process control digraph is reduced to the form where all the computational operators are grouped in the linear process sections' set that is the graph vertex. The digraph arcs are the computation process control operators.

For the linear sections, the error detection problem in the semantic computational process structure is solved as follows. A homogeneous linear equation system with n indeterminate is defined. The value dimension logarithms play the indeterminate functions. For each computational operation, the right-side parsing of the assignment operator to homogeneous summands is performed. As a result, if there are, for example, i summands in the right part, it becomes possible to impose the initial, intermediate, and output variables of the additional i conditions on the dimensional system.

Assume, for example, that the assignment operator has the form (3.7):

$$A = B \cdot C + \frac{D}{E} \tag{3.7}$$

On the right side there are two homogeneous summands that impose the following two conditions (3.8 and 3.9) on the value dimension entering the operator:

$$[A]^1 \cdot [B]^{-1} \cdot [C]^{-1} = 1, \tag{3.8}$$
$$[A]^1 \cdot [D]^{-1} \cdot [E]^1 = 1. \tag{3.9}$$

After taking the logarithms of both (3.8) and (3.9) equations sides, we obtain two homogeneous linear Equations (3.10) and (3.11) for the computational process dimensional system:

$$(1) \cdot \ln[A] + (-1) \cdot \ln[B] + (-1) \cdot \ln[C] = 0, \tag{3.10}$$
$$(1) \cdot \ln[A] + (-1) \cdot \ln[D] + (1) \cdot \ln[E] = 0. \tag{3.11}$$

By combining all the conditions obtained for the given linear program region, a local dimensional system S_i of this region is formed. For example, assuming that six variables are involved in the process, the system containing the conditions (3.10 and 3.11) will have the form:

$$
\begin{Vmatrix}
\cdot & \cdot & \cdot & \cdot & \cdot & \cdot \\
1 & -1 & -1 & 0 & 0 & 0 \\
1 & 0 & 0 & -1 & 1 & 0 \\
\cdot & \cdot & \cdot & \cdot & \cdot & \cdot
\end{Vmatrix} \cdot \bar{L} = 0,
\tag{3.12}
$$

where L is a column variable vector corresponding to the dimensions logarithms of the main process variables.

A criterion for the process semantic structure correctness in the method is the constructed condition system compatibility. In the analytical form, this requirement is transformed to a nontrivial system compatibility S_i.

To construct the S system that is common to the entire computing process, the method proposes to select a random (from the rate optimization point of view, preferably the smallest) implementation combination R_1, R_2, ..., R_w covering the whole process control digraph. In this case, the condition systems combination S_1, S_2, ..., S_w obtained according to the above rule for linear implementation regions gives a general limitation system of the process dimension:

$$
S = \begin{Vmatrix}
S_1 \\
S_2 \\
\cdots \\
S_w
\end{Vmatrix}
\tag{3.13}
$$

The nontrivial S system compatibility, according to the method, is a control criterion of the semantic process correctness.

When developing a program in a procedural programming language for the procedure call points and subprograms, the question of coupling their formal parameters arises. In this case, square permutation matrices are formed T_e, reflecting the correspondence between the formal procedure parameters and the main process variables. In this case, the system (3.13) can have the following form:

$$
S = \begin{Vmatrix}
S_1 \\
S_2.T_1 \\
S_3 \\
\cdots \\
S_{w-2}.T_{e-1} \\
S_{w-1}.T_e \\
S_w
\end{Vmatrix}
\tag{3.14}
$$

The method is characterized by:

- The absence of false positives;
- Sufficiently high work rate for the computing processes fragments of small and medium volume;
- High anomaly detection percentage in the process with the sufficient amount of the computing operators.

The method is not able to detect:

- Semantic errors related to the arithmetic operations ("addition"/"subtraction" operators substitution, etc.);
- Errors in the process control structures (the insufficient or excessive cycle body repetition, incorrect conditional transition direction choice, etc.).

Petrenko [8] proposed the verification method of the semantic process execution structure by controlling the numerical similarity invariant values, the so-called π-complexes. According to him, when forming a complete process equations system at the random execution stage, it is possible to control the numerical similarity invariant values. If the invariant values calculated at the stage are consistent with the reference set, a decision is made as to whether this implementation belongs to the subclass of the acceptable values. If one of the values does not match, a signal is generated about the abnormal computational process flow.

The method has a great potential in terms of abnormal functioning detection in comparison with the dimensional invariant control method. This is due to the processing semantic level information by an order of magnitude. However, this feature significantly limits the economic efficiency of its application to detect abnormal functioning in the data networks. The amount of information processed by the method decision circuit and therefore transmitted over the network is comparable to the amount of protected data flow. That is, the increase in the network traffic generated by the abnormal functioning detection system service data will be 100% in the first approximation. This value, as was shown earlier, is unacceptable for the vast majority of the data networks. Therefore, the use of the π-complexes control method is justified only for local computing processes.

3.1.4 Detection of Anomalies Based on Dimensions

The software processing sequence of the data transmitted via the data networks, being essentially a computational process, can be described in canonical form by the control digraph $G\ (B,\ D)$, where B is the elementary operation set with the selected b' and b'' vertices, and D is the control connection set between them, defined as

$$D \subset B \times B \qquad\qquad (3.15)$$

Any route in the D digraph beginning at the b' vertex and ending at the b'' vertex will be called the p computational process implementation. Obviously, the ordered vertex sequence forming the implementation uniquely depends on the input data set values at the moment when the process passes through the b' vertex.

Let us denote the all subset family of the B set as B and the all subset family of D as D. Let us denote the all possible process implementation set as P (P = $\{p\}$). Two functions are defined on it:

- F_B with the values range equal to B, which matches each implementation with the vertex set from B through which the implementation passes;
- F_D with the values range equal to D, which matches each implementation with control connection set from D involved in the implementation.

Any P_i subset of the P set for which the condition is met

$$F_B(P_i) = F_B(p_0) \cup F_B(p_1) \cup \cdots \cup F_B(p_{NP-1}), \qquad (3.16)$$

where *NP* is an elements number in the P set, let us denote as a representative subset of P. In many cases, it will consist of one element.

Let us match each elementary process operation with the a_i vector of the dimensions matrix row coefficients (3.13) generated by the operation and denote the $\{a\}$ set by A.

Form the algebra $S = <S, \times >$ of similarity invariants (in this case, dimensional invariants) as follows.

The S set is a set of all possible combinations of A:

- Except for the following combinations:
  - Combinations, among which there are at least two mutually proportional ones;
  - Combinations, among which there are elements with nonmutually simple coordinates;
  - Combinations, forming a system for which there are no solutions with all nonzero coordinates.

- With two selected element addition:
  - ○ Zero "0" corresponding to the expression that does not impose any new restrictions on the dimensional equation system;
  - ○ Element of U incompatibility corresponding to the vertex combinations from A forming a dimensional equation system for which there are no solutions with all nonzero coordinates.

The binary operation is defined as the assembly of vectors entering into the operands with several subsequent transformations:

- From each pair of mutually proportional vectors (if such appeared as an assembly result), one is randomly removed;
- If, as a result of the vectors assembly, the dimensional equations system corresponding to the joint set is not solvable with all nonzero coordinates, then the element U is declared to be the operation result.

The algebra introduced above is a groupoid according to the property. The element "0" is a two-sided groupoid unit, since according to the "×" operation instruction:

$$\forall_{s\epsilon S}(s \times 0 = s) \tag{3.17}$$

and

$$\forall_{s\epsilon S}(0 \times s = s), \tag{3.18}$$

including

$$U \times 0 = U \tag{3.19}$$

and

$$0 \times U = U. \tag{3.20}$$

Since

$$\forall_{x,y,z\epsilon S}(x \times (y \times z) = (x \times y) \times z) \tag{3.21}$$

the S algebra is a semigroup, and the "× " operation, having the property

$$\forall_{x,y\epsilon S}(x \times y) = (y \times x), \tag{3.22}$$

allows to define S as an Abelian semigroup in turn.

The S semigroup is idempotent, since

$$\forall_{x\epsilon S}(x \times x = x) \tag{3.23}$$

according to the "×"operation formulation. Including

$$0 \times 0 = 0 \tag{3.24}$$

and

$$U \times U = U \qquad (3.25)$$

Let us denote by F the function defined on the A set with the value range in S corresponding to the initial elements formulation rule in S as described above:

- F(a) = U if the constraint a vector defines solutions only containing zeros;
- F(a) = {a'}, where a' is formed by the reduction of all a vector coordinates by their greatest common divisor;
- F(a) = {a} in all other cases.

Then, the identically defined element equal to F(a_i) matches with each b_i vertex of the control program graph in the S set, and the identically defined element s_b matches each route b = (b_0, b_1, ..., b_{NB}):

$$s_b = F(a_0) \times F(a_1) \times \cdots \times F(a_{NB}), \qquad (3.26)$$

where a_j is the dimensional constraint vector, imposed by the b_j operation.

Within the introduced notation scope, it becomes possible to define a function F with the value set in S on the P set of all program implementations as

$$F(p) = F(a_0) \times F(a_1) \times \cdots \times F(a_{NB}) \qquad (3.27)$$

where a_j is a dimensional constraint vector on the b_j vertex for which the condition is fulfilled

$$b_j \in F_B(p) \qquad (3.28)$$

The F function defines a homomorphic mapping of the computational process execution algebra into similarity invariants algebra S, and the F value equality for the actual p_i implementation to the U value is a sufficient criterion for the abnormal computational process functioning. Accordingly, by the antimplication rule virtue, the condition

$$F(p) \neq U \qquad (3.29)$$

is a necessary criterion for the semantic computational process flow correctness (e.g., the data processing in the data networks).

In the computational process functioning within the same station, the F(p) calculus for any current input data set and, moreover, the value calculus F(p_0) × F(p_1) × ... × F(p_{NP-1}) for a representative computational process implementation sample are not difficult. However, the computational process separation into several workstations with their own address spaces and variables (as is the case for data processing computation processes in the distributed data center) requires additional tools.

Let the computational process implementation p be divided into two implementations "p_K" and "p_L" executable on different K and L workstations in their own address spaces and named variables. In this case, the criterion (3.29) takes the form

$$F(p_K) \times F(p_L) \neq U \qquad (3.30)$$

Let us denote $s_K = F(p_K)$ and $s_L = F(p_L)$. For this case, the semantic correctness criterion verification of the computing process is divided into four stages:

1. Calculate the s_k element at the station K (sender) parallel to the calculation of the p_k implementation;
2. Check the criterion fulfillment

$$s_K \neq U \qquad (3.31)$$

3. Transfer the s_k element parallel to the transmission of the information message with intermediate implementation results to the L station;
4. Check the criterion fulfillment

$$s_K \times F(p_L) \neq U \qquad (3.32)$$

According to the S algebra commutativity property, the calculus of the criterion

$$s_L \times F(p_K) \neq U \qquad (3.33)$$

will lead to the same result, but such option does not correspond to the logic of the processes and data flows in the real domain.

3.1.5 Investigation of Properties of Invariants of Dimension

The graph theory application to the dimensional invariant system properties analysis

The correlations generated by the dimensional constraint system using the graph theory are analyzed. Let G be the set of all nonoriented hypergraphs without loops and α be the isomorphism ratio on it. Define a homomorphic mapping ω of the S set (coefficients matrices of the dimensional equation system) on the quotient set G/α as follows.

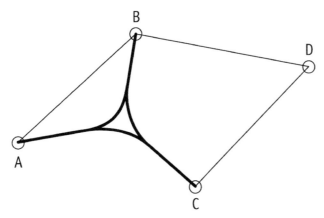

Figure 3.1   The graph family representative is the $\omega(s_X)$ pattern.

Let us compare each computational process variable involved in $s \in S$ unique graph vertex, and match each condition in the system (3.12) with the hypergraph edge that is incident to the vertices associated with those computational process variables that enter this condition with nonzero coefficients. Thus, the ω mapping will transform the s_X system dimensional Equation (3.34) into the graph shown in Figure 3.1.

$$\begin{Vmatrix} 1 & -1 & 2 & 0 \\ 1 & 2 & 0 & 0 \\ 0 & -1 & 0 & -3 \\ 0 & 0 & 1 & -2 \end{Vmatrix} \cdot \bar{L} = 0 \qquad (3.34)$$

The research object specificity, namely the closed variable group's presence of one or related dimensions and a low variable number in one condition, causes certain graph properties corresponding to the computational process dimensional system s in the network protocol stacks:

1. Several connectivity component occurrence (from 2 components for the connection layer protocols to 10–15 for the application level protocols) in the graph;
2. Predominance of the two-and three-degree edges in the hypergraph.

The criterion calculation problem for the semantic network protocol stack correctness according to the proposed method reduces either to verifying the condition of nontrivial system compatibility or to the system solution search with all nonzero components. For both cases, the criterion calculation subproblem is the partition of the general equation system of stage 2 (for F (p_K))

and stage 4 ($F(p_K) \times F(p_L)$) to possibly smaller subsystems, without losing the criterion properties.

Let us call a potentially incompatible matrix formed from the coefficient matrix of the basic homogeneous dimensional constraint system by discarding certain rows and columns and having the following properties:

1. Matrix does not contain zero columns;
2. Matrix rank is equal to the column number.

The $\omega(s)$ mapping for a potentially incompatible matrix s will be a connected hypergraph with at least one simple cycle. Moreover, if the matrix s is square, then the simple cycle will be exactly one.

Not all dimensional invariant system equations are significant to defining the criterion fulfillment. Let us analyze some equation property in the G space. Thus, in the example shown in Figure 3.2, the simple *CDE* and *GIJH* cycles correspond to the potentially incompatible dimensional equation systems. The edges *AB*, *BC*, *EF*, *FG*, and *JK* correspond to insignificant elements of the initial dimensional equation system.

The graph edges, corresponding to the insignificant to the compatibility verification dimensional equation system, are the bridges, since only the

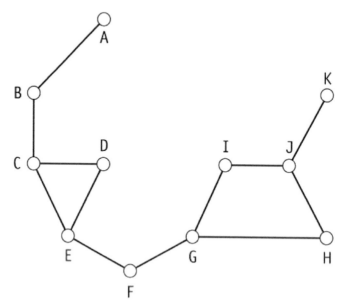

Figure 3.2   The g_{AK} graph is an example of the dimensional system pattern with insignificant elements.

bridge edges in the connected graph are not a part of any simple chain. Let us divide all the bridge edges into two classes:

1. Dividing the graph into the connected components, at least one of which does not contain the simple cycles;
2. Dividing the graph into the connected components, where there is at least one simple cycle in each of them.

In Figure 3.1, the edges *AB*, *BC*, and *JK* belong to the first class, the edges *EF* and *FG* are among the second one.

To detect the insignificant first-class equations is a much simpler computational task. This is due to the fact that if there is no cycle, at least with one of the potentially insignificant edge sides, it is no longer required to verify if another chain connects two cycles, that is, if there is a potentially insignificant edge in any simple cycle. Thus, the insignificant edges analysis and removal of the first class can be performed without connectivity and cycle analysis in the graph.

Define on the G set a unary operation R_1 (R_1:G → G), consisting of the removal from the hypergraph any edge incident to at least one "0" or "1" degree vertex. The edge incident to a "1" degree vertex corresponds to the equation in the dimensional system having one of the variables that does not enter other system equations. This means that when trying to find a solution to a dimensional system, this variable can take any value; therefore, the condition given by this constraint is always possible. The "0" degree vertex in the $w(s)$ graph corresponds to the variable that is not involved in any equation and therefore also does not affect the system compatibility.

Define a unary operation R_1 (R_1:G → G) as the sequential R_1 operation application limit over the given graph:

$$R(g) = R_1^k(g) \qquad (3.35)$$

under the condition

$$R_1^{k+1}(g) = R_1^k(g) \qquad (3.36)$$

and

$$R_1^k(g) \neq R_1^{k-1}(g). \qquad (3.37)$$

Because the R_1 operation reduces either the edge number or the vertex number in the graph, the maximum degree of its possible repetitions (k) is finite and the R operation is defined everywhere. The result of the R operation application over the g_{AK} graph is shown in Figure 3.3.

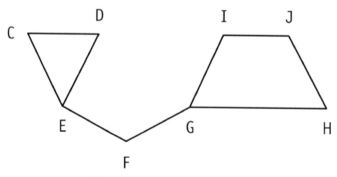

Figure 3.3   The $R(g_{AK})$ graph.

Removal of the insignificant second-class elements from the dimensional system is associated with the need to completely analyze the potentially connected given edge components. In the worst case, the given algorithm has a computational complexity proportional to the product of the graph vertex number by the squared edge number of [56]. Therefore, it seems unjustified to exclude insignificant second-class elements at the first analysis stage of the connectivity invariant graph morphology.

For a more detailed morphology research of the graphs corresponding to the data network dimensional systems in computing systems, the methodology that allows transforming the dimensional system hypergraph to a multigraph without loss of morphological properties is presented.

The potentially incompatible s matrix mapping will be a graph with at least one simple cycle. In relation to the hypergraph edges of the degree higher than two, the term "simple cycle" is not entirely justified. Therefore, define the σ equivalence in terms of the contour number on the G/α set, which allows either to reduce a random hypergraph g to simple graph or multigraph equal to it or to indicate the absence of the potentially incompatible matrices within the $\omega^{-1}(g)$ matrix that generates a hypergraph.

For this purpose, define the unary operation $(T{:}G{\rightarrow}G)$ on the G graph set as follows. Let $g(V,E)$ be the prototype hypergraph with the edges of the random degrees. Choose two (e_i and e_j) edges among its edge E set, each of which is incident to some vertex v_k. The V' vertex set of the g' equivalent hypergraph is equal to the V set, and in this case, the E' edge set is drawn in the following way:

$$E' = (E \backslash E_J) \cup E_{J'} \tag{3.38}$$

and

$$V(e'_j) = (V(e_i) \cup V(e_j)) \backslash v_k. \qquad (3.39)$$

Thus, when fixing the e_i and e_j edges and v_k vertex, with respect to which the equivalent transformation is performed, one of the edges changes the list of vertices incident to it. In addition to conditions (3.38) and (3.39), impose the restriction on the T operation that prohibits this transformation application over the pair of "2" degree edges incident to a common vertices pair.

Consider this transformation from the positions of the g graph vertex degrees. Since the v_k vertex is removed from the list incident to e_j edge, its degree decreases by one. For all the elements of the possibly nonempty vertices subset V

$$V = V(e_i) \backslash V(e_j) \qquad (3.40)$$

the degree is increased by one.

It will be shown that it is always possible to choose e_i, e_j, and v_k parameters of the T transformation at each step in such a way that a finite number of sequential T and R operations applications transforms a random graph to the graph whose connected components consist of two vertices and a certain number of "2" degree edges joining these vertices. Since neither the T operation nor the R operation change the number of the connected graph components without generality loss, the connected graph will be considered from the beginning. If necessary, the result can be extended further to each connected component separately.

Algorithm 3.1

1. Perform the initial transformation

$$g_1 = R(g_0) \qquad (3.41)$$

2. As long as the vertices number in the graph is greater than two, we perform step 3.
3. At each i-th step, select a random graph vertex (as the v_k parameter) and perform the T operation over the graph for two random edges incident to it. This will reduce the selected vertex degree by one.

Repeat inside the step two incident vertices selection from the remaining ones and the T operation application over the graph as parameters. Thus, for a finite number of sequential T operations, it is always possible to reduce the randomly selected vertex degree to "1". After that, at the end of the step, apply the R operation over the graph, which by its definition will remove the edge remaining the last incident to the selected vertex and the selected vertex itself:

$$g_i = R(T(T(\cdots T(g_{i-1})\cdots))) \qquad (3.42)$$

The end of 3.1 algorithm

Since at each step the hypergraph edges set is reduced by one, and the vertices set is reduced at least by one, and this process is finite. The stop occurs when the vertices number is reduced to two or zero. In this case, all the edges remaining in the graph have "2" dimensions and connect the two vertices: a two-vertex multigraph is formed, or the algorithm result is an empty graph $\varnothing(\varnothing,\varnothing)$.

It is assumed that the equivalence relation σ is satisfied between two g_1 and g_2 hypergraphs if and only if the algorithm application over them leads to equal graph accurate to isomorphism. Let us denote the edge number of the *j-th* connected graph component as d(j); in this case, the independent contours number N (corresponding to the potentially inconsistent dimensional system matrices) is calculated by the formula:

$$N = \Sigma_j(d(j)-1). \qquad (3.43)$$

The modified equivalence relation σ', which is different because of the additional restriction introduction on the T operation, is of practical value for optimizing the criterion verification process. According to the restriction, the "2" degree edge cannot be selected as the e_j edge. This leads to the algorithm stopping, not on the two-vertex multigraph state, but when the first graph is equivalent to the original one, the all edge degrees are equal to 2. This kind of equivalence preserves the initial hypergraph contours morphology with respect to the problem of finding potentially incompatible matrices, which allows defining the optimal strategy for verifying their compatibility.

The examples of minimal equivalent graphs for several initial hypergraphs g_0 with three vertices are given (Table 3.2).

It is possible to construct a homomorphism between the algebras $S = \langle S,\times\rangle$ and $G = \langle G,\cup\rangle$. It implements the mapping $\omega:S \rightarrow G$. This fact is due to the fact that the "$\cup$" combining operation result over the graphs set is defined as the graph whose vertices and edges sets are the combination of the operand graph vertices and edges sets [10]. This, in turn, corresponds to the operation of combining two-dimensional constraint system rows: the variables (vertices) and conditions (edges) sets are formed by combining the operand system data sets.

Table 3.2 σ/σ' equivalent graph examples

g_0	$g_c(g_c\sigma'g_0)$	$g_M(g_M\sigma'g_0)$
		$\emptyset$
		$\emptyset$

The graph morphology analysis obtained after combining the operand graphs is essential on the verification step of the semantic correctness criterion on the receiving station L. Here, the F(p_K) and F(p_L) elements are combined by "×" operation, which corresponds to their graph patterns addition:

$$g_{KL} = \omega(s_K) \cup \omega(s_L) \tag{3.44}$$

Dimensional graph of the source station

Dimensional graph of the destination station

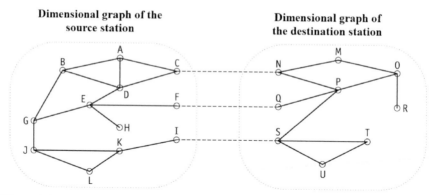

Figure 3.4   An example graph that is a result of adding the dimensional graphs of the source station and the destination station.

The researched object is characterized by the presence of the several closely interrelated variables groups on the source station and on the destination station. Communication between each station groups is realized through a small (usually from 2 to 5) variables set. The variables transmitted over the network usually do not participate in calculations at the stations, but they are only filled in at the transmission time and are processed at the time of the message reception.

- For the $\omega(s_K)$ and $\omega(s_L)$ graphs, the listed domain features are presented by the following properties:
- Graphs contain several high-connectivity subgraphs;
- Graphs contain several low-degree vertices corresponding to the variables transmitted over the network;
- There is at least one edge that goes from each high connectivity subgraph to the corresponding vertices.

The $\omega(s_K) \cup \omega(s_L)$ graph example is shown in Figure 3.4 (the edges corresponding to the variable transmission through the data network are marked with the dashed lines).

Intermediate results

The dimensional invariant applicability as the informative features in the abnormal data network functioning detection systems was analyzed. Two main computational process invariant classes proposed by the similarity theory were considered. It was shown that one of them, that are π-complexes, is not applicable for achieving the stated research goal, since the abnormal functioning detection systems with its use does not meet one of the requirements to the modern abnormal data network functioning detection systems.

The second-class invariants applicability, dimensional invariants, for achieving the stated goal was confirmed by the sufficient criterion development for the abnormal data network functioning of the distributed computing systems. For this purpose, the algebraic dimensional invariants system was constructed, the homomorphic mapping of the data processing computation algebra in the data network protocols stack to the constructed dimensional invariants algebra was defined, and the sufficient condition for the abnormal data network functioning was formulated.

To ensure the ability to construct the algorithms that meet the requirements to the modern abnormal functioning detection systems, the dimensional invariant system properties were analyzed using the graph theory. For the system elements (variables and constraints), significance criteria were determined with respect to their influence on the criterion of the semantic data network operation correctness. The dimensional invariants connectivity property was researched for the research object. The method to reduce the hypergraph corresponding to a dimensional system to the ordinary multigraph was determined, which allowed comparing the multigraph simple cycles to optimize the algorithms for the computation rate by potentially incompatible subsystems.

The developed algebra of the dimensional data processing invariants in the distributed computing system data network and its properties research allow developing a practical way to detect the abnormal data network functioning by the use of the new informative features that satisfy all previously formulated requirements.

3.2 The Main Provisions of the New Method for Detecting Anomalies

The proposed dimensional data network invariant algebra of a distributed computer system is the basis for the development of a new promising methodology to detect the abnormal data network functioning.

3.2.1 The Main Hypotheses for Detecting Anomalies

- Implementation of the software module that builds the dimensional invariants system s_K to the transmitting K station protocol stack of the data network;
- Encoding and transmission of the s_K element parallel to the protected information message to the destination station;

- Software module implementation in the protocol stack of the destination station L, which builds a dimensional invariant system S_L;
- Decision circuit implementation that performs the necessary criterion verification of the semantic data processing correctness in the transmission through the data network

$$s_K \times s_L \neq U \qquad (3.45)$$

To ensure the ability to build the abnormal functioning detection system that meets the equirements to a modern system for detecting abnormal data network functioning on this basis, the main method properties are considered.

The method has a zero false responses level since it checks the qualitative conditions performance and uses the deterministic algorithms. When using this method, all the data network functioning anomalies of the distributed computing systems connected to the dimensional constraint system violation are detected. The criterion fulfillment control is able to detect anomalies that occurred both during the data processing in network protocol stacks of the user stations and during the message transmission at the data network.

The quality characteristics of the abnormal activity detection are unchanged for both already known and new ways of unauthorized or unintentional actions implementations. This property is due to the normal (in terms of dimensional limitations) process modeling of the data network operation and not due to the various anomalous situations modeling.

The method is characterized by the ability to specify the packet, the datagram, and the message where the invariant violation is detected. In the overwhelming case majority, it is possible to determine the station that generated abnormal traffic and the stack network protocol level that performed data conversion with the dimensional invariant violation.

The method development to verify the criterion fulfillment of the necessary semantic correctness

The basic criterion verification procedure is applicable to the whole dimensional constraint systems set *S* and is deterministic.

Basic methodology

The basic criterion verification methodology for the semantic data network functioning correctness is based on the algorithm. The difference is that the computations specificity when processing information in the network protocols stack allows proposing a dimensional control methodology that eliminates one of the critical automatic verification stages that is independent variable selection. The computations that perform information processing (in this case, the network messages) differ from the calculations of theoretical mathematics and physics by the absence of the complex higher-order dependencies involving several (more than four) variables. This fact makes it possible to equate the numerical constants to full-sized dimensional elements in the given domain and to construct the following automatic process verification method in the data network by the dimensional invariant method, on this basis.

The difference from the algorithm considered that all numerical constants are introduced into it as equal variables at the stage of constructing the system. Let us consider an operator different by introducing a numerical constant "1", as an example:

$$A = B \cdot C + \frac{D}{E} + 1. \tag{3.46}$$

According to the methodology described in [6], the operator adds five variables ($A, B, C, D,$ and E) and three-dimensional equations (3.3–3.5) to the dimensional constraint matrix:

$$(1) \cdot \ln[A] + (-1) \cdot \ln[B] + (-1) \cdot \ln[C] = 0, \tag{3.47}$$
$$(1) \cdot \ln[A] + (-1) \cdot \ln[D] + (1) \cdot \ln[E] = 0, \tag{3.48}$$
$$(1) \cdot \ln[A]^1 = 0. \tag{3.49}$$

After substitution of (3.49) into (3.47) and (3.48), the variable that corresponds to the A value dimension disappears from the further analysis, which may entail the skipping of the incorrect code section with A variable. In the methodology [6], it is proposed to control such situations with human participation (most likely, these functions will be performed by the verified program developer).

If we consider a numerical constant as a certain dimensional variable (e.g., with the name $CONST\_1$) in the given example, then the operator (3.46) will already include six variables ($A, B, C, D, E,$ and $CONST\_1$). Equation (3.47) will also change: it will take the following form:

$$(1) \cdot \ln[A]^1 + (-1) \cdot \ln[CONST\_1]^1 = 0 \tag{3.50}$$

As a result, the A variable is stored in a system with a nontrivial value until the calculations are complete.

Such change in the numerical constant status allows determining the semantic computational process correctness criterion in the following form:

Theorem 3.1

"For the semantic computational process correctness, it is required that the dimensional equation system (2.12), built for IT considering the numerical constants, has at least one of the nonzero components among the solution vectors set."

The proof by contradiction will be carried out. The appearance among the variables corresponding to the dimensions, the one that is identically equal to zero at any value of the other variables, means its zero dimension. However, this contradicts with the constraint system development condition, namely the dimension introduction to all process variables and constants.

The theorem is proved

This methodology principal point is the variable creation for each numerical constant in the constraint system. This is a necessary requirement for the correct methodology functioning. Therefore, in the presence of three constants, "1" in the various computational process operators, all three can have different dimensions, despite their identical numerical value.

Thus, the dimensional constraint system and the s_K and s_L elements, accordingly, have a larger column number for the proposed methodology. However, as practice showed, the elements corresponding to the numerical constants increase the requirements of the system computing resources to a negligible level for the research object. This is due to the fact that, in the overwhelming case majority, the variables that have equality in dimensions with the numerical constants in a given domain play the role of either counters or indexes and coefficients when accessing indexed data.

If there is a shortage associated with the increased requirements to the computational resources, then the proposed methodology has a tangible advantage, allowing removal from the verification process the independent variable selection stage that was preformed previously or in advance by the operator (the algorithm developer) or was resolved using any heuristic algorithm. The algorithm that implements the proposed methodology is automatic and deterministic and thus completely solves the variables classification problem into basic and dependent ones.

The determination problem whether a homogeneous linear equation system possesses a solution vector with all nonzero components differs from the traditional problem of the nontrivial system compatibility verification. First of all, this is due to the fact that the requirement that all the solution vector components have to be nonzero is stricter than the solution nontriviality.

For numerical criterion verification on the S coefficients matrix basis, the dimensional equation system of the matrix R was formed. It has a special form:

$$R = \begin{Vmatrix} 1 & 0 & \cdots & 0 & c_{1,1} & \cdots & c_{1,n-k} \\ 0 & 1 & \cdots & 0 & c_{2,1} & \cdots & c_{2,n-k} \\ \cdots & \cdots & \cdots & \cdots & \cdots & \cdots & \cdots \\ 0 & 0 & \cdots & 1 & c_{k,1} & \cdots & c_{k,n-k} \end{Vmatrix}. \tag{3.51}$$

The R matrix in this form can be represented by the following formula:

$$R_{k \times n} = E_{k \times k} | C_{k \times (n-k)}, \tag{3.52}$$

where E is the identity matrix and k and n are the number of the original S matrix rows and columns, respectively.

To form the R matrix, it is sufficient to use three types of operation:

1) Addition of a random matrix row with the other row linear combination;
2) Row permutation;
3) Column permutation.

The main forming process progress (3.51) is identical to the Jordan–Gauss method (see, for example, [30]).

The difference is in the following:

- Double algorithm pass: first one is in the forward direction (from top to bottom) and then in reverse (from bottom to top);
- Column permutation in cases when a nonzero value in any cell within the first k columns, which is not the first nonzero value in succession in a row, cannot be turned into zero due to the absence of other nonzero elements in the column.

With respect to the dimensional constraint system solution, the R matrix is identical to the S matrix except for the possible column permutations. That is, there is the equivalent

$$(S \cdot X = 0) \Leftrightarrow (R \cdot T \cdot X = 0) \tag{3.53}$$

where T is a square permutation matrix of the dimension $n \times n$, corresponding to the performed S column permutations on the R construction step. This result is due to the nature of the transformations performed on the S matrix the R matrix construction.

Formula (3.53) allows using the R matrix instead of the S matrix when checking the semantic correctness. Let us formulate the following theorem for this.

Theorem 3.2

"To have the i-th component identically equal to zero, among the first k values of a dimensional constraint system solution vector (3.12), it is necessary and sufficient that all elements are equal to zero in the C matrix i-th row in formula (3.52)."

Let us prove the condition necessity. The proof by contradiction will be carried out. Let there be at least one nonzero element in the C matrix i-th row (e.g., in position j). Then, setting all $(n-k)$ of the last variables equal to zero with the exception of $(k+j)$th, the following equation will be obtained:

$$\sum_{p=1,p\neq i}^{k} 0 \cdot x_p + x_i + \sum_{q=1,q\neq j}^{n-k} c_{i,q} \cdot 0 + c_{i,j} \cdot x_{k+j} = 0 \qquad (3.54)$$

from which it follows that the variable x_i is not equal to zero in this case. The contradiction was obtained. The condition necessity is proved.

Let us prove the condition sufficiency. If all elements of the C matrix i-th row are equal to zero, the equation is obtained:

$$\sum_{p=1,p\neq i}^{k} 0 \cdot x_p + x_i + \sum_{q=1}^{n-k} 0 \cdot x_{k+q} = 0, \qquad (3.55)$$

from which the desired identity is directly derived.

$$x_i \equiv 0 \qquad (3.56)$$

The theorem is proved

The variables corresponding to the first k columns of the R matrix are the basic (independent) dimensional invariants in the given system. The variables corresponding to the remaining R matrix columns are dependent. Thus, the above theorem determines the relationship between the incident of the abnormal researched computation functioning and the situation when one of the basic variables has the "0" dimension. The reason to this interrelation is that the "0" dimension situation is impossible according to the construction method of a dimensional invariant system.

This methodology (with a complete R matrix construction) is the basis to construct optimized algorithms of the criterion verification [26, 31, 32,

40, 55]. As an input, the algorithm uses a dimension matrix $k \times n$ with Z elements, and the work result is a Boolean variable that has the value "True" if the semantic correctness criterion is satisfied, and "False" otherwise. The intermediate algorithm results are:

- C dimension matrix $k \times (n - k)$ with elements of the rational numbers set Q corresponding to the S matrix notation in the form (3.52);
- The value k_{ERR} is equal to zero if the semantic correctness criterion is satisfied, or equal to the first C matrix row number consisting only zero elements if the criterion violation is detected.

In practice, using the basic methodology, it is possible to vary the R matrix construction algorithm, which consists of its development directly during each operator analysis of the researched computational process. The basic variable selection and the necessary computational transformations over R are done when each new line is added to it. The transformation purpose is to have a dimensional constraint matrix already reduced to the form (3.7) at each analysis step.

This algorithm allows:

- Complete elimination of the computational costs connected with the late (within the Jordan–Gauss algorithm pass) matrix column permutation;
- Reducing the computational operation number while separating the identity matrix in the left part of the R matrix.

The algorithm requires additional permutation T matrix storage during the whole computational process analysis step and slightly slows down access to the matrix elements. However, the effective data structures application makes it possible to reduce the additional costs to a negligible amount.

The direct construction of R matrix while analyzing the computational process makes it possible to detect the incidence of the abnormal functioning before the entire construction end (nevertheless, it is not at all necessary that the semantic correctness criterion will be violated precisely at the moment of adding information about the semantically incorrect computational process operator). This fact is the advantage of the modified methodology in the case if there is a large erroneous packets number in the data network (intentionally or unintentionally generated). In this case, the destination station L, without decoding the message completely, is ready to make a decision to ignore it, thereby freeing up its computing resources. This method ability is very useful when it is included in the echeloned protection system from denial-of-service attacks.

Under normal functioning conditions, the early packet declination possibility does not affect the average statistical workload. This is due to the fact that under such conditions, the proportion of abnormal stack processes implementations of the network protocols should tend to zero.

3.2.2 Control of Semantic Correctness Criteria

The following methodology modification can lead to the criterion verification acceleration for a large dimensional system subclass. The subject domain is characterized by the presence of the several independent variable dimensions groups, such as counters, processed data, network addresses, and protocol parameters. As a result, many variables are split on the subsets with a total number from 2 to 5–10 depending on the protocol specifics for almost all network protocols. These subset analogues in the ω mapping pattern space are the connected graph components. Within each subset, the basic and dependent variables can be distinguished similarly to the basic set.

This property allows reducing the S matrix with minimal computational costs by rows and column permutation to a block-diagonal form:

$$S = \begin{Vmatrix} S_1' & 0 & \dots & 0 \\ 0 & S_2' & \dots & 0 \\ \dots & \dots & \dots & \dots \\ 0 & 0 & \dots & S_g' \end{Vmatrix} \tag{3.57}$$

Further S matrix processing can be performed independently for each of the S_i' matrices. The matrix construction algorithm of the form (3.51) is also modified and applied independently to each of the matrices S_i' in this case. This is possible on the following statements basis:

- All elements outside the R_i' matrix, affected by the operations of rows and column permutations, are equal to zero;
- All elements outside the R_i' matrix, used and assigned during the linear row permutation, are equal to zero.

Therefore, all the operations used in the R_i' matrices construction change the element values only inside themselves.

The general S matrix form after transformation according to the described algorithm has the following form:

$$S = \begin{Vmatrix} E_{n1}|C_1' & 0 & \dots & 0 \\ 0 & E_{n2}|C_2' & \dots & 0 \\ \dots & \dots & \dots & \dots \\ 0 & 0 & \dots & E_{ng}|C_g' \end{Vmatrix}. \tag{3.58}$$

If at least for one of the resulting C'_i matrices the condition similar to the one given above for the general C matrix is not satisfied, then the incidence of the abnormal computational process functioning takes place.

As shown in [10], the total computational complexity of the basic criterion verification under n equation system with n indeterminates is

$$\frac{K_{MUL} \cdot (5n^3 - 8n^2 + 3n) + (3n^3 - 4n^2 + 3n)}{2} \qquad (3.59)$$

processor operations when there are no array samples that store information about rows and column permutations and

$$\frac{K_{MUL} \cdot (5n^3 - 8n^2 + 3n) + (4n^3 - 5n^2 + 3n)}{2} \qquad (3.60)$$

processor operations if there are samples from a similar array. K_{MUL} is a coefficient of computational complexity of the integer multiplication operation in comparison to the integer addition operation for a given hardware platform. For Intel Pentium processors, its value is 6–10 times, for Intel 8086 processors, it can reach 20–40 times [58]. If the caching with a high cache hit rate can be implemented, then K_{MUL} can be reduced two times.

As a consequence, the gain of the computational complexity method basically depends on the S matrix partition proportions into independent components. Thus, if a variable set is bundled into g subsets of the equal cardinality, the K gain in the computational complexity will be calculated in the first approximation by the formula:

$$K = \frac{(5K_{MUL} + 3) \cdot n^3}{(5K_{MUL} + 4) \cdot g \cdot \left(\frac{n}{g}\right)^3} = \frac{5K_{MUL} + 3}{5K_{MUL} + 4} g^2, \qquad (3.61)$$

and will be greater than 1 for any $g \geq 2$ and $K_{MUL} \geq 2$ values. However, for an unequal variable set, the partition formula (3.59) is inapplicable, so the methodology computational gain will be estimated from other considerations.

Let the variable set bundle into the independent variable subsets select the m cardinality subset in it. Then, the K value in the first approximation will be determined as

$$K = \frac{(5K_{MUL} + 3) \cdot n^3}{(5K_{MUL} + 4) \cdot (n - m)^3 + (5K_{MUL} + 4) \cdot m^3}. \qquad (3.62)$$

Let us resolve the inequality

$$K > 1 \qquad (3.63)$$

The equivalent transformations (3.63) considering (3.62) lead to the condition

$$m \cdot (n - m) > \frac{n^2}{3 \cdot (5K_{MUL} + 4)} \qquad (3.64)$$

Bearing in mind that our task is to find the smallest m that translates the inequality (3.64) to true, it is assumed that

$$m \ll n \qquad (3.65)$$

Then, it is possible to take

$$\frac{n}{n - m} \approx 1 \qquad (3.66)$$

and the inequality (3.64) takes the final form

$$m > \frac{n}{3 \cdot (5K_{MUL} + 4)} \qquad (3.67)$$

As already mentioned, the minimum K_{MUL} value with the caching technology use is two times, whereas without caching, it is six times on modern hardware platforms. With this in mind, the coefficient in the inequality denominator (3.67) for cache implementations is 42 or more times, whereas for implementations without caching, it is 100 or more times, which in practice leads to a gain in the methodology computational complexity even at the separation of at least one independent variable.

Thus, this methodology application is justified almost always if the $\omega(S)$ graph has several connected components. The computational costs directly to the search and connected component selection are negligibly low in comparison to the computational complexity of the method. The methodology elements like flowcharts and subroutines algorithms for marking rows and columns and constructing permutation arrays are given in [23–26, 40].

Methodology with the optimization of a controlled cycle set selection

The way to further increase the criterion verification rate lies in the direction of the optimization of the verified subsystem number and their selection law. To solve this problem, a method based on the morphological dimensional system analysis using the graph theory was carried out. The existence verification of a nontrivial or nonzero solution of a linear equation system reduces to checking subsystems having a rank equal to the variable number contained in them. In the $(G/\alpha)/\sigma'$ factor set, the simple cycles correspond to the subsystems with similar properties (hereinafter cycles).

Due to the basic linear equations system properties in order to verify the general system compatibility by the given s matrix, it is sufficient to verify the subsystem subset compatibility satisfying the following criteria:

- Combination of the edges that enter the cycles corresponding to the subset elements must be equal to the $\sigma'(\omega(s))$ graph edges set (the criterion of all system equations covering);
- There must be a route between any two cycles corresponding to the subset elements, that is, a chain of other cycles, where each adjacent cycles pair has at least one common edge.

The possibility to select the subsystems in different ways allows formulating the following optimization problem. It is necessary to develop a methodology to select the subsystem subset that meets the above criteria and requires a minimum operations number to verify a compatibility.

Let us assign the measure proportional to its verification computational complexity to each cycle in the $\sigma'(\omega(s))$ graph. In the first approximation, this parameter can be taken directly proportional (e.g., equal to) to the edges number in the cycle. Define the ψ mapping: $(G/\alpha)/\sigma' \rightarrow H$, where H is the hypergraph set, as follows. Match a vertex in the pattern graph with each edge of the prototype graph, and each cycle with an edge (possibly of the degree higher than 2).

Thus, each constraint system equation matches with a vertex, and each potentially incompatible subsystem matches with the hypergraph edge in the new space. Consider, for example, the g_X graph, shown in Figure 3.5, corresponding to the system with five variables, six connected equations, and two simple cycles. The conditional cycle verification costs A, B, and C are 3, 4, and 5 units, respectively.

The ψ mapping translates this graph into the hypergraph shown in Figure 3.6 (the bipartite hypergraph representation in the form of a vertices set and an edge set is given, according to [33]).

On the H hypergraphs set, the optimization problem formulated above reduces to the problem of a minimal spanning tree construction. It is required to construct a connected graph part containing all its vertices, so that its full measure is minimal. This problem is close to the traveling salesman problem (see, for example, [63]), but in this case, there are no requirements to find the route. The solution can be found in a wider subclass of graphs – trees.

P. S. Prim proposed the problem solution of a minimal spanning tree construction for the ordinary graphs. The Prim algorithm adapted for the hypergraph space is given below.

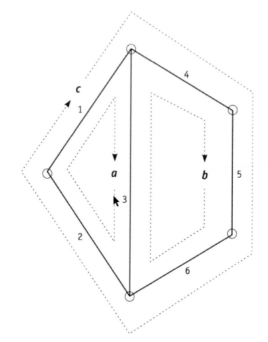

Figure 3.5   The g_X graph.

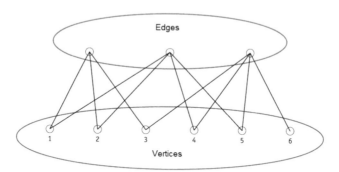

Figure 3.6   The $\psi(g_X)$ hypergraph.

Algorithm 3.1

1. Select an edge with minimal measure among the graph edges set.
2. Repeat the three-step unless the part contains all the vertices.
3. Select an edge and add it to the part. The edge should be with the smallest measure from the edges set, incident to at least one vertex that

enters the already constructed part and at the same time incident to at least one vertex that does not belong to the already constructed part.

3.1 Algorithm end

In the σ' mapping value space, the following selection procedure of the optimal sequence for the interrelated system cycle verification corresponds to this algorithm. At the first step, the smallest length cycle is verified, then the cycles that have at least one common edge with already verified lengths are examined sequentially, unless all the $\sigma'(\omega(s))$ graph edges are covered. This methodology defines a verification plan of potentially incompatible subsystems that is optimal relative to the required computational operations number.

3.2.3 Sufficient Condition for the Criteria Fulfillment

Further acceleration of the criteria verification process of an abnormal operation is possible using probabilistic algorithms. This problem was solved by creating a sufficient condition based on the calculus of the R matrix on a modulo of a prime number.

Lemma 3.1

To have at least one nonzero element in every row of the matrix C, it is sufficient that at least one nonzero element exists in every row of the matrix C_q, obtained from C by calculating the remainder from dividing the corresponding element by the natural number q.

We prove by contradiction. Suppose that in the i-th row of the matrix C all elements are equal to zero, then we have from the matrix C_q the definition:

$$\forall_j(c_{ij} = 0) \Rightarrow_j (c_{ij} \mod q = 0) \Rightarrow \forall_j(cq_{ij} = 0) \qquad (3.68)$$

We have obtained a contradiction.

The lemma is proved.

Theorem 3.3 (sufficient condition for the criterion fulfillment)

"To fulfill the criteria of the semantic correctness of the computational process, it is sufficient that the matrix C in the dimension equation system (3.56) created with account to numerical constants and written as (3.52), calculating modulo an arbitrary natural number q, does not contain zero rows".

Evidence. As a direct consequence of Theorems 3.1 and 3.2, for the correct process operation, it is necessary that the matrix C in the dimension equation system (3.12) written as (3.52) does not have zero rows. In Lemma 3.1, it is shown that to satisfy this condition, it is sufficient that the matrix C_q, obtained from the matrix C by the calculation modulo an arbitrary natural number q has no zero rows. However, because of the distributivity of calculating the remainder from division by an arbitrary number q with respect to operations of addition and multiplication, the application of the calculus modulo q is possible already at the stage of reducing the matrix S to the form (3.52).

The theorem is proved.

This condition is sufficient, but not necessary, because of the sufficiency of the condition formulated in Lemma 3.1. Indeed, if there are zero rows in the matrix C_q, then, some elements in the corresponding rows in the matrix C can be different from zero, namely multiples of q, and therefore the criteria can be true. The above condition is transformed into the necessary one if the search for zero rows in the matrix C is performed for all prime q. Moreover, taking into account the range of initial values in the matrix S and the dimension of the matrix S, we can define the upper bound of the list of primes, the achievement of which ensures the condition necessity. However, the goal set in the section task is to optimize the criteria verification process speed.

A flowchart and an algorithm for verification of the sufficient condition for the fulfillment of the semantic correctness criteria are given in [34]. The set of input and output algorithm values corresponds to the values of the algorithm that implements the basic methodology. The algorithm uses cache tables to store the results of addition, multiplication, and division modulo q, which are filled once in the beginning of the algorithm.

In a general case, the transformation of a system of linear equations modulo of a prime number executes three times faster. This is due to the absence of multiplication operations of rational numbers. Each operation contains three operations for multiplying natural numbers (we assume that the operations of transfer and addition have an order of magnitude less computational complexity compared to the multiplication operation). The only exception is the calculus modulo 2 and modulo 3.

The transformation of a system of linear equations modulo 3 on the set $(-1, 0, +1)$ allows to exclude multiplication of natural numbers from the set of operations performed. All transformations will be done using addition and arithmetic negation. This reduces the computational complexity of checking the condition several times (depending on the architecture parameters of the computing system).

The transformation of a system of linear equations modulo 2 makes it possible to achieve even greater speed. This is achieved due to the ability to process the rows of the R matrix as the bit sequences, thereby consuming the addition and rearrangement of rows by one or two microprocessor instructions. The permutation of the columns is done by cyclic shifts of the binary values.

The disadvantages of using small values q in verification of the sufficient condition are an increase in the probability of failure to fulfill the condition when the value of the criteria is true. Two quantities that critically affect the probability P_{FN} of a similar situation are the value q and the difference between the number of variables and the number of conditions that bind them $(n-k)$. In the first approximation (without taking into account the correlation between the R matrix element values), this dependence is described by the following expression:

$$P_{FN}(q, n - k) = \frac{1}{q^{n-k}}. \tag{3.69}$$

In addition, the specificity of the subject area determines the presence at the beginning of the transformations in the matrix S in the overwhelming number of values $(0, -1, +1)$, since high-degree relations between the variables are quite rare here. This leads to an unacceptably high level of P_{FN} at $(q = 2)$ even in the case of sufficiently large values of the parameter $(n - k)$.

Based on this, it is optimal to verify the criteria:

- Either verify the sufficient condition for $q = 3$, which increases the computational speed in $(3K_{MUL})$ times when the condition is satisfied and insignificantly (in $(1 + (1/3K_{MUL})$ times) reduces the computation speed for its failure K_{MUL} – coefficient of computational complexity of integer multiplication operation in comparison to the operation of integer addition for a given hardware platform;
- Either verify the condition for $q = 3$ and for some simple q greater than 3, which increases the computation speed by about 3 times when the condition is satisfied and reduces the computation speed by 1.33 times when it is not executed.

A greater number of verifications with different values of q does not bring a significant increase in the average test speed of the criteria. The choice of the proposed options is based on practical values of the probability P_{FN} and is most often due to the value of $(n - k)$.

The choice of the q value, greater than 3, for the second variant of the sufficient condition verification is made for reasons of the following considerations. Large q values are more effective in a connection with a decrease in the probability P_{FN} of a sufficient condition false failure. However, the implementation of linear transformations of the R matrix q modulo involves precomputations and storage of multiplication and division tables q modulo, which requires certain resources of the computer system.

The source code for the verification algorithm for a sufficient condition for the case $q = 3$ is given in [24].

Applying the method to the TCP/IP network protocol stack
The TCP/IP stack specificity

De facto since 1980, the TCP/IP stack has become the standard for the interaction of heterogeneous public computer networks. The functions of the network interaction on the stack are made by IP (versions 4 and 6). The transport level protocols are UDP, TCP, and also the more modern SCTP standard [39].

As protocols of the underlying level, IP can use almost all known standards. The most common protocols for link-level protocols are Ethernet, FrameRelay, ATM, PPP, and MPLS [39]. In connection with the special position occupied by the TCP/IP stack in the modern global computer networks, almost all modern application-level protocols are oriented to the use of TCP or UDP as a transport protocol, for example [39]:

- E-mail (SMTP and POP);
- Domain Name Service (DNS);
- WWW – World Wide Web (HTTP);
- File transfer protocol (FTP);
- Electronic news (NNTP);
- Basic network service Microsoft (netbios);
- Interaction of distributed databases (SQL * Net).

To date, the existing system of protocols has the following characteristics:

First, the trend toward the implementation of the functions of the OSI model session level is either a protocol initially designed as a transport (e.g., TCP or SCTP) or an application-level protocol. In some cases, the storage of session information is evenly divided between these protocols. Secondly, in the overwhelming majority of cases, the functions of the 6th level of the OSI model, i.e., data representation, are inextricably integrated with the application level protocol. This trend is due to the fact that at the stage of development of most part of application-level protocols, the format for processing the data presentation level was rigidly fixed. Thirdly, a similar situation is observed with a combination of an approximately equal volume of the functional load physical and link levels.

Thus, most often the stack of network protocols, which the messages pass through, consists of four protocols:

- Application-level protocol;
- Transport-level protocol (tcp or udp);
- IP protocol;
- One or more physical- and link-level protocols.

Accordingly, in the implementation of the method of detecting abnormal operation of the data transmission network (DTN) in distributed CS using the TCP/IP stack, it seems reasonable to independently process invariants at the four specified levels.

Let us consider the approximate operation process of the proposed method in two arbitrary process implementations of TCP and IP.

3.2.4 Implementation of the New Method of TCP Transport Layer Protocol

The two main TCP functions are the confirmation of the message delivery from the sender to the receiver and the control of the data exchange rate. To implement them, the protocol introduced the concept of a window, i.e., the maximum data amount can be transmitted without confirmation from the receiving side. During the data transfer, the window size varies based on the network situation: when the effective network bandwidth is increased, the transmitting side increases the window size, while decreasing it decreases accordingly.

In addition, for guaranteed delivery of messages between the client and the server side, a unique session is established for the message transmission time. During this session, ACK packets are sent. The session start is requested by the client with the service packet SYN with the contents of zero length and is confirmed by the server with the SYN_ACK packet also with contents of zero length. The session end is requested by the service packet FIN and is confirmed by the FIN packet in the reverse direction.

Let us consider a TCP connection model that transmits data in one direction (e.g., from the caller to the called side) based on the TCP symmetry. Describe the TCP client/TCP server system state by the following values:

- Number of SN windows to which the transmitted message will be divided (data type is integer);
- Value of the windows $S_1, S_2, \ldots, S_N$ on the transmitting side (the size of the windows in bits may be not constant, but for this task, it does not matter; data type is the list of string values);
- Buffer of the transmitting SB side in one window size (data type is string);
- Serial number of the current transmitted SI window (data type is integer);
- Value of windows $R_1, R_2, \ldots, R_N$ on the receiving side (data type is the list of string values);
- Buffer of the RB receiving side in the one window size (data type is string);
- Serial number of the RI current received window (data type is integer);
- Number of windows in the message in the opinion of the host RN.

In addition, we introduce the constants responsible for the beginning and end of the session:

- SYN – connection request;
- SYN_ACK – connection establishment confirmation;
- ACK – message window confirmation;
- FIN – message end signal.

The control graph for the selected TCP implementations inside the introduced notations is shown in Figure 3.7. The operations of the assignment SB = RB and RB = SB represent the transmitting of windows over the network using the network-level protocol (in this case, IP). The presence of a cycle and branching in the control graph of the process indicates the possibility of a sufficiently large number of different implementations. Let us choose some of them, reflecting the main entity of the message fragmentation process on Windows.

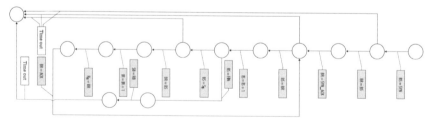

Figure 3.7   Combining the control graphs of some TCP implementations.

In the TCP process model, the following relationships between dimensions exist explicitly:

$$[SB] = [SYN] \qquad [SI] = [1^*] \qquad [R] = [RB]$$
$$[RB] = [SB] \qquad [SB] = [S] \qquad [RB] = [ACK]$$
$$[RB] = [SYN\_ACK] \qquad [RI] = [1^{**}] \qquad [SB] = [FIN]$$
$$[SB] = [RB]$$

In addition, reducing the branching condition of the cycle (SI < N) to the canonical form ($T^* = SI - N; T^* < 0$, where T^* is an auxiliary quantity) adds the dimension equation $[N] = [SI]$.

The S matrix for the implementation sample shown in Figure 3.8 is presented in Table 3.3 (column names contain variable names).

The matrix S has dimension 11×13. The matrix rank is 10, which corresponds to one linearly dependent row (in this case, line 4 is the inverse of line 2). By means of equivalent transformations, we reduce the matrix to the form (3.52) consisting of the basis columns (the matrix C) and the dependent columns (the matrix E) containing one unit element (Table 3.4).

A special matrix form allows us to directly indicate a solution of the model dimension system that does not contain zeros. For this, it is sufficient to choose as a set of values of basis values any nonzero combination that is not orthogonal.

The IP function as a representative of the network level includes the routing of packets in the space of global network names, as well as the fragmentation and assembly of packets (in connection with the variability of the maximum possible packet length in heterogeneous DTSs).

Table 3.3 Matrix for a set of several TCP implementations

SB	SYN	RB	SA	SI	1*	N	S	RI	1**	R	ACK	FIN
1	−1	0	0	0	0	0	0	0	0	0	0	0
−1	0	1	0	0	0	0	0	0	0	0	0	0
0	0	1	−1	0	0	0	0	0	0	0	0	0
1	0	−1	0	0	0	0	0	0	0	0	0	0
0	0	0	0	1	−1	0	0	0	0	0	0	0
0	0	0	0	−1	0	1	0	0	0	0	0	0
1	0	0	0	0	0	0	−1	0	0	0	0	0
0	0	0	0	0	0	0	0	1	−1	0	0	0
0	0	−1	0	0	0	0	0	0	0	1	0	0
0	0	1	0	0	0	0	0	0	0	0	−1	0
1	0	0	0	0	0	0	0	0	0	0	0	−1

Table 3.4   Matrix R for a set of several TCP implementations

SB	SYN	RB	SA	SI	1*	N	S	RI	1**	R	ACK	FIN
1	0	0	0	0	0	0	0	0	0	0	0	−1
0	1	0	0	0	0	0	0	0	0	0	0	−1
0	0	1	0	0	0	0	0	0	0	0	0	−1
0	0	0	1	0	0	0	0	0	0	0	0	−1
0	0	0	0	1	0	−1	0	0	0	0	0	0
0	0	0	0	0	1	−1	0	0	0	0	0	0
0	0	0	0	0	0	0	1	0	0	0	0	−1
0	0	0	0	0	0	0	0	1	−1	0	0	0
0	0	0	0	0	0	0	0	0	0	1	0	−1
0	0	0	0	0	0	0	0	0	0	0	1	−1

In the aspect of detecting anomalies by the method of controlling dimension invariants, the most critical function of the above is precisely the fragmentation and assembly of packets by IP. When building the model, it is necessary to take into account that due to the heterogeneity of network conditions in time, the order of packet fragments arrival to the destination can be arbitrary. This IP is qualitatively different from the TCP with strictly sequential transmitting of windows.

The state of the sender–receiver system when transmitting one IP packet is described by the following values:

- Number of fragments N to which the transmitted message will be divided (data type is integer);
- Value of fragment packets $S_1, S_2, \ldots, S_N$ on the transmitting side (data type is the list of string values);
- Serial number of the current transmitted SI fragment (data type is integer);
- Network buffer B of size equal to the maximum possible length of the IP message (data type is the list of string values);
- Value of $R_1, R_2, \ldots, R_N$ fragment packets on the receiving side (data type is the list of string values);
- Value of buffer RB on the receiving side of the same fragment size (data type is string);
- Value of the received packet number on the PI receiving side (data type is integer).

The control graph of the subsystem for the selected implementations of the network level protocol within the scope of the introduced notations is shown in Figure 3.8. The assignment and reading operations from buffer B are the

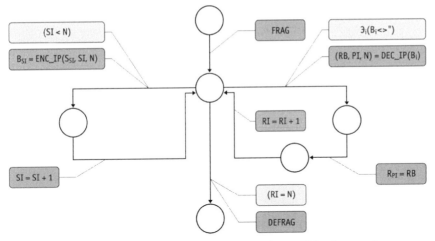

Figure 3.8 Aggregation of the control graphs of some IP implementations.

transfer of windows over the network, using the lower-level protocol. For the reading operation, there are limitations from buffer B: the j-th element of buffer B cannot be considered if the SI counter is less than j.

The procedures listed in the model have the following functions:

- FRAG is the fragmentation of the message into elements S_i;
- DEFRAG is the assembling of the message from the elements R_i and the verifiation of the full reception based on the sequence numbers of the fragments;
- ENC_IP is IP fragment encoding;
- DEC_IP is IP fragment decoding.

This sample of implementations for the IP contains the following associations between the dimensions of the quantities in explicit form:

$$[S] = [RB] \qquad [PI] = [SI] \qquad [RI] = [1^{**}]$$
$$[RB] = [R] \qquad [SI] = [1^*]$$

In addition, the condition operators contain two more restrictions on the dimensions:

$$[SI] = [N] \text{ and } [RI] = [N].$$

The matrix S of the general form for the system of this sample dimension limitations is given in Table 3.5, and the R matrix of a special form is given in Table 3.6. The S matrix has a dimension of 7×9, and the rank of the

Table 3.5 S matrix for a set of several IP implementations

S	RB	R	PI	SI	RI	1*	1**	N
1	−1	0	0	0	0	0	0	0
0	1	−1	0	0	0	0	0	0
0	0	0	1	−1	0	0	0	0
0	0	0	0	1	0	−1	0	0
0	0	0	0	0	1	0	−1	0
0	0	0	0	1	0	0	0	−1
0	0	0	0	0	1	0	0	−1

Table 3.6 R matrix for a set of several IP implementations

S	RB	R	PI	SI	RI	1*	1**	N
1	0	−1	0	0	0	0	0	0
0	1	−1	0	0	0	0	0	0
0	0	0	1	0	0	0	0	−1
0	0	0	0	1	0	0	0	−1
0	0	0	0	0	1	0	0	−1
0	0	0	0	0	0	1	0	−1
0	0	0	0	0	0	0	1	−1

matrix is 7. Similar to the previous section, a special form of the matrix confirms the presence of a zero-free solution and, as a consequence, the fulfillment of the semantic correctness criteria of the model.

Intermediate results

A new promising method and an appropriate technique for detecting the abnormal DTS operation of the distributed CS based on invariant informative features were developed. The method is based on the algebraic system of invariants of the computational process dimension.

It is shown that the development ensures the fulfillment of all requirements to modern methods for detecting abnormal DTS operation; in particular, it has zero false positives, stability of quality indicators applied to new ways of realizing threats to information, and detection of DTS subjects that caused an incident of abnormal DTS operation.

The issue of optimizing the process of verification of the criteria of semantic correctness by execution time is considered separately. For this, methods are developed based on the properties of the system of dimensional invariants obtained in Chapter 2. These methods accelerate the process of verification of the criteria for a large dimension system subclass.

In addition, a sufficient condition for the fulfillment of the criteria for semantic correctness is formulated, which allowed proposing a probabilistic algorithm for verification of the criteria, which increases the average statistical speed of the criteria verification process.

The analysis of the properties of the TCP/IP stack as a de facto standard for modern data transmission systems has been made. The method operation examples for two arbitrary implementations of TCP and IP are given.

3.3 Startup of Anomaly Detection Based on Dimensions

The developed method allows implementing a system for detecting the abnormal operation (SDAO) of the DTS of the distributed CS, which meets the formulated requirements for the modern SDAO. We will make an analysis of practical schemes for implementing the proposed method and perform an assessment of the most important SDAO characteristics based on the test scheme for detecting abnormalities in the DTS enterprise.

3.3.1 Possible Architecture Solutions

The efficiency and performance characteristics of the abnormal network operation detection system, as any intrusion detection system, largely depend on the scheme of its implementation.

Analysis of the possible schemes for the system implementation

The most significant contribution to the change in the SDAO characteristics is made by options for placing the sensors and options for creating the additional (service) information channels.

Currently, there are two most common options for placing the sensors in intrusion detection systems [23–26, 40]:

- Sensor with connection to an environment – medium of storage (network-based IDS);
- Sensor placed on stations – DTS subscribers (host-based IDS).

As applied to the proposed method for the abnormal DTS operation detection, the scheme data are converted to the following system implementation options:

1) SDAO sensor placed on a dedicated computer system:
    - Intercepts network traffic between different subscribers of the SAP, passing through the information-transfer medium and being under control;
    - Receives an element s_k on the associated or dedicated channel;

- Calculates the criteria, performing emulation of the network protocol stack operation of the receiving station L;

2) SDAO sensors placed on each receiving station:

- Receive the element s_k on the associated or dedicated (less often) channel;
- Analyze the execution of the local stack of network protocols;
- Verify that criteria are satisfied.

Flowcharts of the system implementation options are shown in Figures 3.9 and 3.10.

The advantages of the scheme with the placement of a dedicated sensor on DTS include:

- Security of the computational process of monitoring the criteria from the information security violation from within;
- Feasibility of analyzing the abnormal situation developing in the controlled dts segment in general;
- Absence of additional load on computing resources of workstations;
- Significantly lower cost level of the building and the dedicated channel for the control vector transmission.

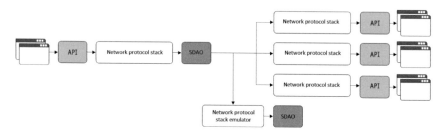

Figure 3.9   Placement of an SDAO sensor on a broadcast network segment.

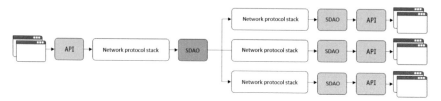

Figure 3.10   Placement of SDAO sensors at receiving stations.

The factors creating the prerequisites for the application of the scheme with the placement of sensors at each workstation of the network are:

- Tendency to reduce the share of broadcasting segments in modern DTS [21];
- Distribution of traffic encryption systems at the network level, including the introduction of the 6th version of the IP network protocol with built-in encryption of the transmitted traffic [61];
- Significant excess of the speed of network traffic growth over the speed of development of the system computing resources (based on [35, 36, 62]);
- Ability to block abnormal network traffic at the stage of transmitting it to the application process;
- Analysis of the real operation of the network protocol stack, and not its emulation. It allows us to analyze all the network protocols involved in this connection instead of only a subclass of protocols known at the system development time.

Issues of portability and scalability [37] may also affect the choice of the SDAO sensor placement. So, in the networks with a permanent set of the used network protocols and a larger range of operating systems installed on workstations, it is more advantageous to use sensors placed on the DTS segment. By contrast, with the same type of workstations and the ever-changing subclass of the used protocols, it is much more efficient (in terms of both the percentage of detected abnormal situations and the costs of material resources) to use sensors placed directly on workstations.

Concerning the additional information channel networking required for SDAO, it is possible to use a dedicated medium for transmitting information (of a similar and other type) and an associated information transfer medium (Figures 3.11 and 3.12). In the second scheme, two modifications are possible, characterized by either the allocation of the element s_K to a single message or its integration with the basic information message.

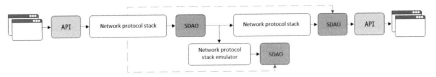

Figure 3.11   Dedicated additional information channel.

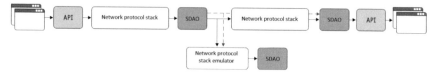

Figure 3.12 Associated additional information channel.

The scheme with the dedicated information channel for vector transmission has the following advantages:

- Ability to build channel, protected from listening and modification, for transmitting information about invariants;
- Ability to adaptively control the bandwidth required to transmit control information (in particular, to use a physical medium with lower bandwidth);
- Minimum delay interval between the receipt by the sensor of the information and control packets.

The scheme with the associated transmission channel of information and control packets is characterized by the following advantages:

- Minimal change in the infrastructure and topology of the network at the stage of system implementation;
- Reliability of the control message delivery (especially for the variant of its integration into the information message), that is, the situation is impossible when, due to the loss of the control message, the recipient station will have to ignore the correct information packet;
- Less physical costs (especially for schemas with sensor placed at a receiving station).

Modification of this scheme by integrating the control information into the main information message has a number of additional advantages.

Among them, it should be specially noted that there is no delay between the arrival of information and messages control, as well as the feasibility of using a unified integrity system for the information and control parts of the message.

However, this modification is not applicable in all cases. This is due to the fact that exceeding the restrictions on the packet length, especially at the channel and physical levels, can be improperly handled by the network equipment. In this case, even if the physical process and the link level are changed (or correspondingly corrected) at the receiving and transmitting workstations, the intermediate network equipment (modems, concentrators, switches) may fail.

For network-level protocols that support additional fragmentation on an arbitrary segment of the packet route (such as IP), this problem can be solved at the sending station K or by setting the largest fragment size, taking into account the maximum possible amount of additional information about invariants, or modification of the network-level computing process to dynamically determine the current fragment length.

For network-level protocols that do not support fragmentation, this method is not applicable. In any case, the use of this modification is possible only in the previously verified data transmission systems and provided the composition of the network equipment used is unchanged.

3.3.2 Features of the Transfer and Control Criteria

In comparison to the subsystems, already developed in CS, in the proposed method, a scheme is newly introduced for transmitting an information invariant element of dimension s_K to the recipient L station. This information must be encoded in some way and transmitted to the receiving side with an acceptable delay with respect to the main packet transmission. Various options for networking the transmission channel of the element s_K are discussed in previous section.

The matrix s_K in practice is rather strongly sparse in the connection with the nature of the interrelations of variables in the investigation object. Thus, the number of nonzero elements in the string does not exceed 4 with a total number of variables from 5 to 30 (the parameters of several protocols are given in [64]). Therefore, the following encoding scheme for the element s_K is proposed in the transmission step.

To the receiving station L, it is necessary to transmit:

- The number of variables involved in the s_k system;
- The positions of variables in network packet headers;
- The number of rows in the s_k matrix;
- The values of elements of the s_k matrix.

The number of variables involved in the system is transmitted in a binary code. Positions of variables are transmitted by two vectors VP and VL. Each element VP_i of the vector VP corresponds to a bit shift from the start of the corresponding level header to the beginning of the field of the corresponding variable and is encoded by a binary code.

Each element VP_i of the vector VL corresponds to the length in bits of the field of the corresponding variable in the header and is encoded by a binary code.

The number of rows in the matrix is encoded in binary code. For effective line coding of the s_K matrix, we propose the procedure, based on the strong matrix sparsity, integer matrix coefficients and grouping of the values of the matrix coefficients near the 0 number.

For each matrix row, vectors VN and VK are written into the packet. Each element of the vector VN corresponds to the nonzero coefficient of the current row of the matrix and stores the ordinal number of the column with such an element in the binary code. Each element of the VK vector stores the value of the nonzero coefficient corresponding to the element of the vector VN with the same index in some effective code (e.g., the Huffman code) [38].

The flowchart and the source code for the encoding algorithm according to this technique, the flowchart and the source code for the complete encoding algorithm of the dimensional invariant system (including zero elements), as well as the flowchart and the source code for the decoding algorithm the system of dimensional invariants are given in [23–26, 40].

The elements of the s_K matrix of the applied and representative OSI model levels can be transmitted in many cases to the receiving station L once at the time of the application-level connection. For protocols, this corresponds to establishing a session-level connection. Matrix elements corresponding to the session and transport levels can in many cases be calculated and transmitted once at the beginning of each message. The network- and link-level elements are transmitted for each message packet.

Ensuring the integrity of the data item transmit scheme

When an s_K element is transmitted over an open-data channel with the modification feasibility, the question of ensuring the message integrity arises. Without additional security tools, the offered method is vulnerable at the stage of transfer of the element s_K for message falsification or an attack by the repetition method (reply attack).

The tasks of ensuring confidentiality and accessibility are not critical for this method.

Message protection from falsification is traditionally performed using either algorithms of electronic signature based on asymmetric cryptography methods or imitation protection algorithms based on methods of symmetric cryptography [14, 23], which are more acceptable for solving this problem, because of the requirement of very-high-speed calculation processes and integrity tags verification.

According to the method of the protection against replay attacks on open messages, symmetric crypto-transformations are performed. In this case, such transformation is the element s_K. In the process, the confidential data block, known only to the sending agent and the SDAO sensor, is involved. This block is a key for the message authentication code generation.

The transformation result is a block of fixed length (message authentication code), added to the open message.

As a practical implementation, an algorithm GOST 28147-89 [25] according to the specification, or any algorithm of a crypto stable hash sum (e.g., MD5 [39] or SHA-1 [40]) can be used for generating the message authentication code. When using hash sums, the message authentication code generation key is added by concatenating to the protected data (the element s_K) before hashing.

The reply attack method consists in changing or replacing an attacker's packet (in this case, the packet, which allows detection the abnormal activity in the network) to a value previously transmitted in the network and leading to the calculation of the positive value of the criteria. When implementing this attack, an attacker is not required to select the message authentication code or a key for its generation, the purpose of which is to hide unauthorized activity from the system. The ability to perform such actions without attacking the crypto algorithm of the message authentication code makes the protection task against this class attacks relevant to the developed method.

In the protection of the reply attack method, either a message serial number unique for a given pair of workstations within a reasonable time period or a high accuracy time stamp (with the need for high-precision synchronization of the machine hours of all station – DTS subscribers) is added to the protected message before the message authentication code is generated. A combination of both methods is possible with a corresponding decrease in the digit number of the serial number and the accuracy of the time synchronization system.

The parameters of the algorithm for ensuring the integrity of the additional information flow are calculated from the following considerations. The key to generation of the message authentication code is long term (it operates unchanged for a sufficiently long period). Because of this, the general requirements for symmetric cryptoalgorithms should be applied, in particular, according to [25] and [41], the key length that is at least 256 bits long is considered to be stable at the present stage of the computer technology development.

The content of the message authentication code is not a subject of any known crypto attacks. However, by generating a chaotic message authentication code for an unauthorized package, an attacker can accidentally create a valid value for it. This attack is decisive for choosing the minimum length of a message authentication code: between two stations for a reasonable time period when using the maximum throughput of the DTS, random generation of a valid message authentication code should be impossible.

The maximum number of packets that can be created on a given physical medium within a certain time interval is determined by the formula

$$N = \frac{BW \cdot 3600 \cdot 24 \cdot 365 \cdot Y}{8L} \qquad (3.70)$$

where BW is a maximum possible network bandwidth between the intruder and the subscriber (in bits per second), Y is time interval length (in years), and L is minimum requirement packet length to attack (in bytes). For determining the upper limit value N_{MAX}, let us take BW = 109, Y = 20, and L = 56. Then, N_{MAX} equals (1.4×10^{15}), which corresponds to the order of 2^{50}, and the length of the message authentication code to achieve the reliability level a = 0.999 must be at least

$$\log_2 \left(2^{50} \cdot \frac{1}{1 - \alpha} \right) = 60 \text{ bit} \qquad (3.71)$$

Taking into account the rounding of this value to the byte boundary required by the hardware platform, we obtain an 8-byte message authentication code length. As the practice of using the SDAO shows, the share of the message authentication code does not exceed 1% of the total traffic volume.

3.3.3 Experiment Results

To verify the research results based on the enterprise data network, a test SDAO implementation was developed. DTS chosen for the test SDAO implementation is a local area network based on Ethernet technology with a capacity of 10 and 100 megabits per second. Network subscribers are 54 stations, including 8 dedicated server stations, 3 active network equipment, and 43 workstations.

As the network-level protocol, IP is used as the transport protocols are TCP and UDP. Among the application-level protocols in the network are NetBIOS, HTTP, FTP, Oracle Net, NNTP, SMTP, POP3, RDP, etc.

The test SDAO was implemented on one server and three workstations with control over the applied and representative levels of traffic exchange via FTP and HTTP. A scheme was used with the SDAO sensor placements at receiving stations. An additional information channel was implemented within the existing Ethernet network, and SDAO information elements were transmitted within the frames of separate packets according to UDP.

The statistics console of the test SDAO is shown in Figure 3.13, and the situation of detecting an incident of abnormal SPD operation is shown in Figure 3.14.

As a result of the SDAO operation in the test mode for 1 month, the following characteristics and quality indicators of the system were obtained.

Statistics of the observed distribution of the message lengths with 7th, 4th, and 3rd levels are given in [23–26, 40].

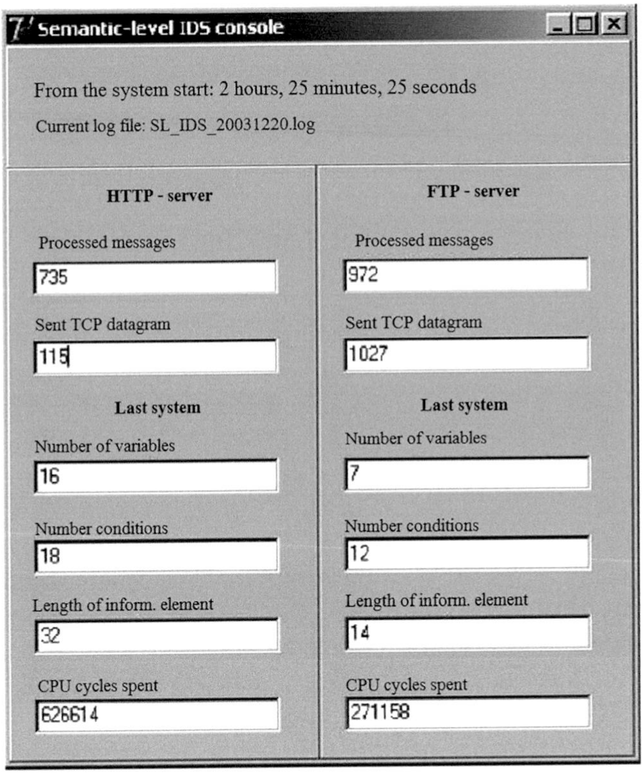

Figure 3.13 Statistics console of the test SDAO.

✓ **Cell of Watchkeepers**
- Service 24/7
- Detection
- Describing events in the tickets
- Basic analyze

✓ **Cell of Analysts / Vulnerability**
- Comprehensive analysis of events
- Technical support for Watchkeepers
- Determination of false positive
- Incident identification
- Recommendation escalate event to cyber security incident

✓ **Cell of Coordination**
- Escalation of events to security incident
- Classification of the incidents
- Cooperation in resolving the incident
- Incident Reporting
- Incident closure

Figure 3.14   Situation of detecting a test SDAO incident.

Specificity of the interrelation between the data of the histograms of the 7th and 4th levels is explained by the fact that the FTP opens a new datagram of the 4th level in addition to the main control session for each application-level message (file, catalog listing, etc.). Thus, the indicated histograms coincide for it.

Unlike FTP, HTTP transfers several objects (usually belonging to the same HTML page) within a single session-level connection. The specificity of packet lengths of the third (network) level is related to the IP partitioning of messages into packets of 1,500 bytes in length (DTS parameter is the maximum transferrable unit), but only for the last packet, there is a length not equal to 1,500 bytes.

The resulting statistics of the HTTP and FTP dimension system parameters and the statistics of the SDAO information elements length distribution for these protocols are given in [23–26, 40]. The observed average value of the information element length was 29.8 bytes; the selective standard deviation is 14.9 bytes. For application-level protocols, the share of additional SDAO traffic in the total information flow was 0.26%. The value application given in [23–26, 40] to the transport and network protocols in the first approximation gives the expected value of the share of additional SDAO traffic at the levels of 0.46% and 2.27%, respectively.

The distribution statistics of the activation of the machine resources of the DTS subscriber stations by SDAO algorithms are also given in [23–26, 40]. The cumulative use of computing resources of the processor for the whole research period did not exceed 1%.

During the SDAO test operation, there were six incidents of operation of the decision scheme, including:

- four incidents corresponding to an error in the network software implementation (the client side of FTP) installed on one of the workstations;
- one incident corresponding to unintentional actions of one of the network operators who mistakenly indicated as the address of the HTTP request a character string from the clipboard containing 420 symbols of a completely different semantic load;
- one incident corresponding to an attempt of unauthorized automatic access to the HTTP server from the side of the auxiliary application, freely distributed on the Internet.

There were no false positives.

3.3.4 Trends and Development Prospects

The proposed method for detecting the abnormal operation of the network protocols stack and the optimization technique for the speed of the algorithms that implement it can be used to solve other problems of ensuring the stability of computing systems.

Distinctive characteristics of the research object in comparison to previous works (e.g., [5, 6]) are:

- Need to transfer, at an intermediate stage, the information criteria to another station of the computer system;
- Set of the domain properties (in this case, the variable stack of network protocols), which determine the coefficient matrix sparsity, the integrality of its elements, and small (in the vast majority of cases, less than 4) absolute values.

The concept of serialization and transfer of numerical elements of the criteria can be applied in CS stability problems, using serialization of the computing context. This method is widely used in object-oriented technologies, for example, in Java and Microsoft.NET interpreter platforms, as well as for storing documents in Microsoft Office programs. If necessary, in the practical implementation of solutions to these problems, the method of efficient coding of the invariant system matrix elements, described, can be also used.

The characteristics of the research object, taken into account when building the criteria of semantic correctness and techniques for optimizing its calculation, allow us to apply the algorithms obtained for a wide range of non-networked CS. In general, the criteria application will be reasonable for data processing that are initially of a digital nature (cycle counters, array indices, line shifts, and large data sets).

Optimization methods for computing speed can be applied as components of the solution to the problem of ensuring the stability of large information structures that process data of various strongly interrelated types. Examples of such CS may be:

- Object-oriented databases;
- Context search systems;
- Rubricators;
- Data visualization subsystems;
- Subsystems for compression, archiving, and data backup.

As the main direction of the further research in this area should be the SDAO integration with systems of control of invariants of a higher level, that is, the same application process. Figure 3.15 shows the most common three-tier CS architecture and feasible control systems schemes for invariants in modern computer technology.

The aggregation of invariant control systems of the entire computer system into a single multilevel echeloned complex has many advantages. First of all, they include high selectivity and, accordingly, the greatest quality of abnormal operation detection and unauthorized actions.

Being built in accordance with the principles of the system approach, such system ensures the greatest convergence of the necessary criteria of semantic correctness with its proper concept.

End-to-end identification of data processed in the computer system at different scheme levels (Figure 3.15) can provide a high-quality and reliable process for backtracking the error or unauthorized exposure. In this case, it becomes possible to uniquely indicate the level of its initial impact and the object that is the impact application point.

A single space of variables or mapping schemes between variables of different levels will help to uniquely determine the minimum set of potentially damaged data as a result of the incident. Moreover, using certain combinations of controlled invariants, it is possible to restore the damaged data in real time with predetermined correcting system ability.

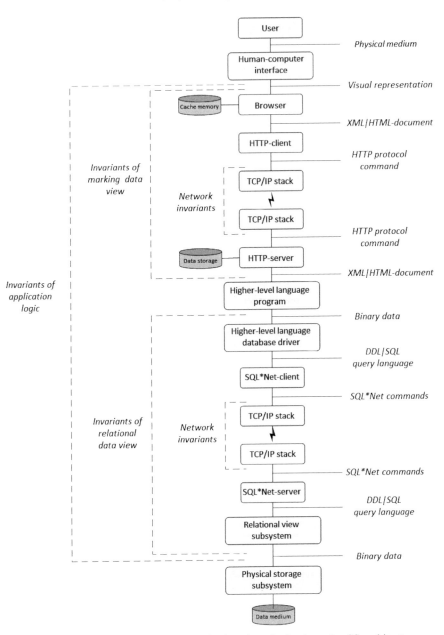

Figure 3.15 Feasible control systems for invariants in the three-tier CS architecture.

Integration of several different levels of anomaly detection systems with the ability to reject or correct data also has additional advantages when processing data using transactions. As early as possible (low-level) detection of data corruption in this case allows us to go to the emergency branch of a transaction of high degree of nesting, in many cases correctly handle the exception and, therefore, complete the main transaction successfully.

Another important direction of searching for abnormalities of network functioning using the method of invariant control is the search for other systems of invariants (except for dimensional equations). Checking several necessary criteria for the semantic correctness of network operation with minimal intersection of truth areas will significantly reduce the level of incident passes. The criteria with scales orthogonal with respect to the set of semantically correct realizations of the process can give particularly good results.

In addition, the goal of searching for other invariants of network operation can be maximizing the corrective capacity of the invariant. The value that is the initial parameter of optimization can be either the amount of additional information transferred or the probability of successful data recovery with the average statistical intentional or unintentional damage.

Results

The main principles of the SDAO construction are outlined based on the proposed new promising method for detecting the abnormal operation of DTS of a distributed CS. The analysis of feasible schemes for the placement of sensors and organization of a channel for the transmission of information elements of the SDAO has been made. It is shown that the main criterion influencing the scheme choice is the protocol heterogeneity degree of the protected SPD and the software installed at subscriber stations.

It was also noted that in cases of high utilization of the main channel of DTS or the presence of active network equipment with specific properties for implementing the information channel of SDAO, additional network capacities are required.

A scheme for the efficient coding of systems of dimensional invariants is presented for their transmission over SPD, taking into account the high sparsity of the system matrices. Based on the specifics of the research object, it is proposed to use for the encoding of the matrix element values of the code variable length dimensions.

The issue of ensuring the integrity of the transmission scheme of information SDAO elements when using packet channels with the modification feasibility of as path sections is considered. The main characteristics of the element ensuring the integrity of this information are calculated.

The test implementation description of SOAP in the enterprise DTS is given. The operation characteristics of SDAO and the quality values for detecting the abnormal operation DTS are given. It is shown that the implemented system based on the proposed method satisfies all the formulated requirements for modern SDAOs and has zero false positives. During the test operation of SDAO, six incidents of abnormal operation of SPD were detected; the causes of their occurrence were discovered and analyzed.

The further perspectives of the way toward integration with multilevel systems of invariant control of computer systems are indicated on the one hand and with echeloned TZI systems on the other. Examples are given of the feasible application of individual research results (techniques for optimizing the verification of the semantic correctness criteria and techniques for efficient coding of strongly sparse matrices with integer coefficients) in related areas of information protection theory.

Result evaluation

Investigation of the current situation in the area of abnormal operation detection in data transmission networks and analysis of the consequences of the similarity theory provided the opportunity to propose the application for the abnormal operation detection of the new informative properties, i.e., dimensional invariants. The proposed method of control of the semantic correctness of the invariant system of the stack size of the network protocols allows improving the quality of abnormality detection in DTS at the zero false positives level significantly.

The main attention was paid to the development of a new promising way of detecting the abnormal operation of DTS of the distributed CS, which makes it possible to build an automatic SDAO that meets all modern requirements for such systems.

The possibility of building based on the new informative signs of a method for detecting abnormal operation that can operate in real time with an acceptable increase in intra-network traffic and a zero level of false positives is proved.

To solve these problems, the following provisions were proposed and substantiated:

- Algebraic system of computational process similarity invariants of the DTS protocol stack is defined and its properties are investigated;

- Method for detecting the abnormal operation of the distributed data based on the analysis of invariant informative features of the dimensional constraint system has been developed;
- Technique for autotesting the semantic correctness criteria of a distributed computational process based on the system control of the dimensional invariants was developed;
- Technique for optimizing the process of criteria verification for the speed of execution by separating the connected components of the dimensional invariant system graph was developed;
- Methodology for optimizing the process of criteria verification for the execution speed with the help of the optimal choice of verifiable subsystems of dimension was developed;
- Sufficient condition for fulfilling the criteria based on the calculation modulo a prime number is formulated, allowing to increase the average statistical test speed of the criteria;
- Test system implementation for detecting the abnormal operation of the data transmission network was carried out, which was carried out in a number of enterprises.

In the course of implementing these provisions, the following scientific and practical results have been achieved:

- Method that allows detecting in automatic mode new ways of implementing information threats with zero false positives was proposed;
- Optimization techniques for the process execution speed of the semantic correctness criteria verification contribute to the implementation based on the proposed SDAO method that meets the requirements for modern SDAO was developed;
- Created test implementation of the SDAO based on the method of dimensional invariant control has successfully passed the tests, and its elements are implemented in the echeloned system of technical information protection of a number of enterprises.

Based on the test implementation of the SDAO DTS, the parameters of the network protocols used in a modern data transmission networks were investigated and the methods and techniques developed for them were used.

The results obtained in practice made it possible to improve the quality of detecting incidents of abnormal DTS operation at zero false positives. This, in turn, led to the economic effect of preventing losses associated with the implementation of unauthorized and unintentional threats to information processed in CS.

The general characteristics of the protected operation were improved in view of the exclusion of service failures associated with the false triggers of the probabilistic solving schemes of SDAO, and the reduction of the total cost of ownership of the computer system by reducing the staff responsible for manual analysis of SDAO triggering.

The application for the abnormalities detection of invariant informative features in general and systems of dimensional invariants, in particular, makes it feasible to approximate the decisive detection scheme as much as possible to the semantics of the controlled process. As a consequence, it is the invariant informative features that can achieve the best correlation between the percentage of incident detection of abnormal operation and the level of false positives.

3.4 New Method of Analytical Verification

The equation analysis and the dimensional analysis are distinguished in the computing process similarity theory [1, 6, 8, 9, 21, 22, 58, 72]. In this regard, there are two possible ways to specify the representation language alphabet of the partially correct computation semantics in the similarity invariants. The first method is based on the equation analysis π-theorem and allows establishing a general scheme (general form) of the semantically correct calculations. Such scheme is represented in the form of dependencies between similarity complexes and simples. Moreover, the dependencies number is equal to the function number to be determined, that is, the number of intermediate and output data. In this case, the final set of these dependencies specifies the representation language alphabet of the partially correct computation semantics in the similarity invariants. The second method is based on the dimensional analysis π-theorem and allows specifying the representation language alphabet of the partially correct calculation semantics in the similarity invariants in the dimensional form. At the same time, the language elements are functional dependences, and usually unique, between the dimensionless data sets obtained from the dimensional formulas. The similarity simples are not included in the indicated dependence. Consider the second method in more detail.

3.4.1 Data Processing Model on the Example of Oracle Solution

Here is a classic ERP Oracle E-Business Suite version according to the scheme "database – application server – thin client". The three main system components are interconnected via a TCP/IP data network (Figure 3.16). In this case, the browser or some other program that acts as a thin client displays data in markup languages HTML or XML for ERP Oracle E-Business Suite users. User-entered queries and commands are sent using the HTTP via the data network to the application server. The ERP Oracle E-Business Suite applications are executed as some computing processes on the application server and, if necessary, interact with the database server (in general case, with the distributed database). In this case, the data exchange process between the applications and the ERP Oracle E-Business Suite database affects the access components to the high-level language relational database, the SQL*Net application protocol level, and the TCP/IP stack. Since the ERP Oracle E-Business Suite is built according to a distributed principle, a cluster of the parallel database servers linked by SQL*Net protocol will be introduced into the scheme (Figure 3.16).

The ERP Oracle E-Business Suite data processing schema analysis presented in Figure 3.16 allows distinguishing the following components responsible for the data transformation, storage, and transfer [26]:

- Browser program that performs visualization of the marked data (conversion "graphical representation – HTML/XML document");
- Marked data transmission and caching subsystem;
- Application program on the application server;
- Data transmission subsystem in relational representation;
- Physical data storage subsystem within the DBMS.

Therefore, in order to control the ERP Oracle E-Business Suite semantic correctness, it is necessary to build some control points or data processing invariants [1–3, 5–11, 13–17, 20–26, 40, 56, 72] and also to develop relevant methods to control these invariants for:

- Marked data and user interface visualization;
- HTTP application layer protocol;
- SQL*Net application layer protocol;
- Stack of the TCP/IP transport, network, and channel levels;
- Application program (application logic);
- Physical representation subsystems of the relational data in DBMS.

Let us thoroughly consider the listed processes of ERP Oracle E-Business Suite data processing.

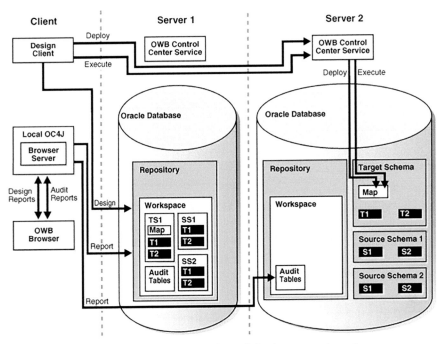

Figure 3.16 ERP Oracle E-Business Suite data processing schema.

3.4.2 Marked Data Visualization

The thin client program control graph is shown in Figure 3.17. Here, the process input parameter is a string value StartURI, which is a unique resource identifier (URI) that initializes a session with the ERP Oracle E-Business Suite user. This parameter can be definitely determined during the ERP system development phase if the user identification and authentication are implemented by the application itself or the StartURI value must be provided by the external identification process in the system.

In general, the branch associated with the ability to display a document copy in the browser cache may depend on the following factors:

- Work logic with the cached documents;
- Cached copy timestamps;
- Presence of the variable component (query) in the resource identifier, etc.

After being downloaded from the server, the document and its parameter copies are cached for later use. In this case, the visualization algorithm

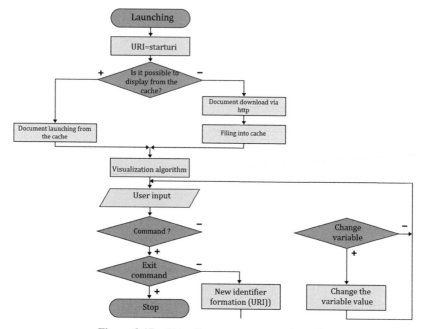

Figure 3.17   Thin client program control graph.

converts the so-called marked data received from the application server into the internal structure of the browser data and then into the graphical representation on the output device. After the document is rendered, the browser switches to the waiting user action loop. There are three variants of the latter:

1. Exit from the program;
2. Variable value editing;
3. Command call to the application server.

When a user attempts to perform an editing operation, the browser should verify the variable availability for modification. When requesting a command for an application server, the browser must:

- Determine the unchangeable part of the unique identifier;
- Define the variable set associated with this query (command);
- Generate a list of pairs "variable name–variable value";
- Generate a full URI resource identifier by concatenation.

Let us describe the internal structure of the data representation in the browser program in the two data set forms:

1. Records about the commands available to the user when working with this document (the "link" and "form" control elements are connected with the commands in HTML), such as:

   - URL_1, URL_2,... URL_{CC} are the permanent components of resource identifiers associated with this command (data type is string);
   - CM1, CM2, ... CMCC is an information about the rules for displaying the control element associated with this command (data type is a structure);

2. Records about the variables (in the HTML, the "text" control element is connected with the constant values, the input control element is connected with the variables), such as:

   - VN_1, VN_2,... VN_{VC} are variable names (data type is string);
   - VV_1, VV_2,... VV_{VC} are variable values (data type is string or, depending on the variable, it is string, integer, real, logical);
   - VW_1, VW_2,... VW_{VC} are the possibility features of the variable value editing in the browser program (data type is logical);
   - VR_1, VR_2, ... VR_{VC} are the order command numbers with which the variable is associated (in HTML and XML marked data representation languages, a strict hierarchical structure of control and visualization elements is observed, therefore each specific variable can be associated only with one command (the data type is integer, the valid values range is [1 ... CC]));
   - VM_1, VM_2,... VM_{VC} are an information about the rules for displaying the control element associated with a given value (data type is a structure).

The two data sets (State) together with the current unique resource identifier (URI) and the marked document in the binary representation (Doc) fully describe the current browser program state. In this case, the information graph of the data processing model is shown in Figure 3.18.

Here, the *Cache* process draws a sample from the document cache by its unique identifier, the *Load* process loads the document using the HTTP, and the *Parse* process parses the marked document and fills it with the *State* data set. The *Request* process assembles a new unique identifier based on the following data:

- URL_j value associated with the selected user (assume that its index is *j*);

- VN_{ki} names and current VV_{ki} values are the variables associated with the user-selected command (i.e., those for which the condition $VR_{ki} = j$ is satisfied).

Note that for the HTML, the rules for building a unique identifier are as follows:

$$URI' = URL_j + \text{``?''} + VN_{k1} + \text{``=''} + VV_{k1} + \text{``\&''} + VN_{k2} + \text{``=''} + VV_{k2} + \text{``\&''} + \ldots + VN_{kn} + \text{``=''} + VV_{kn}.$$

3.4.3 Formalization of HTTP and SQL * Net Protocols

The HTTP and SQL*Net application layer protocols (*Hyper Text Transfer Protocol*) are similar enough (they are on the same classical OSI model level) [26]. The common functions of these protocols are:

- Data transport coding parameters matching (including compression);
- Coding parameters matching of the national language alphabet;
- Application-level name resolution in the resource identifiers.

The differences are obvious in the coding schemes of the information transmitted to the lower layers, as well as in the nature of the meta-information returned from the server.

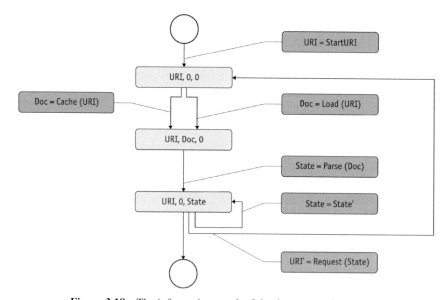

Figure 3.18   The information graph of the data processing model.

The HTTP client process is initialized by specifying the URI parameter, which is the unique identifier of the requested resource. In this case, the process parses the identifier into three parameters:

- Host name;
- TCP port of the HTTP server;
- Relative URI.

If necessary, the host name is represented by the corresponding IP address. In this case, the following are used:

- Domain name resolution service (DNS);
- Windows name resolution service (WINS);
- Local reference book.

Then, the process determines the list of HTTP connection parameters that are acceptable for it, such as transport coding (compression) and the national alphabet coding. In the HTTP request, the entire list is specified, so the HTTP server will be able to select the connection parameters common to the client and server. If there is a document with a similar unique identifier in the local cache, the "Document creation time in the document cache" field is added to the HTTP header. The generated HTTP request text (with the "GET" type indication, "HTTP/1.1" protocol version, relative URI, host name and connection parameters) is transferred to the underlying layer of the network protocols stack.

After the HTTP server receives the request, the process checks the resource availability with the requested identifier on the server. In the requested resource shortage, a message is sent to the client side with error code "404" that is "Document is not found". If there is a resource, three options are possible:

1. If the HTTP request contained the field "Document creation time in the cache" and the time specified therein coincides with the creation time of the resource stored on the server side, a message with the code "304", that is "Document is not modified", is transmitted to the client side. Otherwise, based on its list of possible HTTP connection parameters and the list received in the HTTP request, the server selects the active connection parameters.
2. If the connection parameters cannot be reconciled, an error message with the code "406", that is, "Non-negotiable parameters", is sent to the client side.

3. If the parameters are successfully matched, an HTTP response is generated, including the code "200" ("Resource is transferred"), the content (if necessary, converted according to the active parameters), records of the active connection parameters, as well as information on the date and time of creating the document for caching on the client side.

The full HTTP response text is sent to the lower protocol stack layer. After receiving a response on the client system side, the content length is checked and transport decoding is performed if necessary.

Note that the SQL*Net protocol combines several subprotocols that are responsible for various functions in the organization of client–server and server–server interactions.

The UPIOPI/NPI subprotocol interface provides the ability to execute the following application layer commands in a client–server connection (UPI/OPI) and server–server (NPI/OPI) connection:

- Connection to the server;
- Preliminary SQL query syntax verification by the server;
- SQL query execution;
- Opening the relational data representation cursor;
- Relational data transfer from the server to the client;
- Transmitted data typing based on the records about the data types on the server side;
- Closing the cursor;
- Disconnection from the server.

In this case, the difference between the "client–server" and "server–server" connection schemes appears only at this level, whereas at lower subprotocol levels, there are no differences. The server that responds to the SQL query in both cases uses the OPI interface, which is similar to the HTTP server process. The calling interfaces UPI (client) and NPI (server side when connecting "server–server") differ slightly. From the client side, unified access drivers to relational data (such as, ODBC – Open DataBase Connection) display their interfaces to the UPI interface.

The Two-Task Common (TTC) subprotocol is responsible for selecting a single representation of data types and the national alphabet coding schemes. If necessary, both the client and the server sides of the TTC can perform the transmitted data transformations within the known coding schemes. Unlike the scheme used by the HTTP, Two-Task Common selects the general parameters of the SQL*Net connection once at the opening time of and uses already agreed (actual) parameters when executing all subsequent commands.

The Transparent Network Substrate (TNS) subprotocol implements an interface independent of the underlying network protocol stack with four commands:

- Connect to a remote object;
- Transmit data;
- Receive the data;
- Disconnect from a remote object.

The TNS functions include the object name resolution so that the higher layers can operate with the abstract "alias" notion of the remote process. Names can be resolved using the following protocols:

- Oracle names;
- Local reference book;
- Oracle software adapters for popular name resolution protocols (NetWare Directory Service, StreetTalk, NIS, etc.).

In connection with the uniformity of the functions performed by the protocols, a single process operation model is described for them, extending the ability to cache the results to the entire SQL*Net protocol (only the NPI/OPI interfaces joint has this capability). Next, the term "resource" will be used to the HTML document and data in the relational representation.

Let the state of the client–server system be described by the following variable set:

- RN is a name of the resource requested from the client interface (data type is string);
- RS is a required national alphabet encoding of the requested resource (data type is integer);
- CS is a list of the national alphabet encodings, supported by the client (data type is a list of integer variables);
- CT is a transport encodings list supported by the client (data type is a list of integer variables);
- CC is a resource content on the client side (process output) (data type is string);
- CD is the date and time when the resource was created (data type is a timestamp);
- CACHE_CNT is the resource number allocated in the cache (data type is integer);

- CACHE_NAMES are the resource names allocated in the cache (data type is a list of string values);
- CACHE_DATES are the resource creation date and time placed in the cache (data type is timestamp list);
- CACHE_VALUES are the resources contents allocated in the cache (data type is a list of string values);
- CACHE_CHARSETS is a national language encoding of the resources allocated in the cache (data type is a list of integer values);
- CI is a number of the resource found in the cache (data type is integer);
- CB is a client-side buffer (data type is string);
- SB is a server-side buffer (data type is string);
- TN is a name of the requested resource on the server side (data type is string);
- TS is a list of the national alphabet encodings, supported by the client on the server side (data type is a list of integer variables);
- TT is the transport encodings list supported by the client, on the server side (data type is a list of integer values);
- TDATE is the creation date and time of the resource found in the cache (data type is a timestamp);
- SS is the national alphabet encodings list supported by the server (data type is a list of integer values);
- ST is the transport encodings list supported by the server (data type is a list of integer values);
- VALUE_CNT is the resource number hosted on the server (data type is integer);
- VALUE_NAMES are resources names hosted on the server (data type is a list of string values);
- VALUE_DATES are the resource creation date and time hosted on the server (data type is a timestamp list);
- VALUE_VALUES is the resources content hosted on the server (data type is a list of string variables);
- VALUE_CHARSETS is the national language encoding of the resources hosted on the server (data type is a list of integer variables);
- SI is the resource number found on the server (data type is integer);
- AS is an encoding of the national alphabet chosen by the server as active for this session (data type is integer);
- AT is the transport coding selected by the server as active for this session (data type is integer);

- DS is an encoding of the national alphabet chosen by the server as active for this session on the client side (data type is integer);
- DT is the transport encoding selected by the server as active for this session on the client side (data type is integer);
- CM is a buffer on the client side for the resource content during the conversion (data type is string);
- SM is a buffer on the server side for the resource content during the transcoding (data type is string);
- CE is a server response code on the client side (data type is integer).

In addition, constants are introduced:

- C202 is the response code corresponding to the successful resource sending by the server;
- C304 is the response code corresponding to the unchanged resource since the last download by the client;
- C404 is the response code corresponding to the resource shortage;
- C406 is the response code corresponding to the inconsistency of the client-side and server-side parameters.

The protocol information graph within this model framework is shown in Figure 3.19.

Nominal procedures in the model have the following functions:

- ENC_REQ is a request coding;
- DEC_REQ is a request decoding;
- ENC_RESP is an answer coding;
- DEC_RESP is a reply decoding;
- CHARSET is a coding scheme modification of the national language alphabet;
- ENC_TRANS is a transport coding;
- DEC_TRANS is a transport decoding.

Assignment operations SB = RB and RB = SB represent the message transmission over a TCP/IP network.

3.4.4 Presentation of the Transport Layer Protocol (TCP)

There are two main TCP functions (transmission control protocol) [26]: the messages delivery acknowledgment from the sender to the recipient and the control of the data exchange rate.

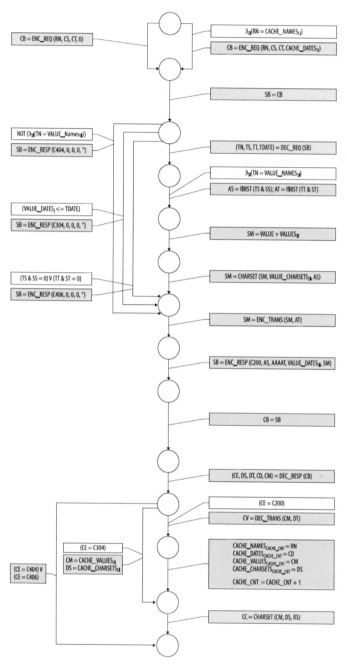

Figure 3.19 Protocols informational graph within the model framework under consideration.

To implement these functions, the protocol introduced the concept of a "window", that is, a maximum data amount that can be transmitted without acknowledgment from the receiver. In the data transfer process, the window size varies based on the network situation: under increase of the effective network bandwidth, the transmitting side increases the window size, when decreasing it decreases.

In addition, to guarantee the message delivery between the client and server sides, a unique session is established for the message transmission time, within the frames of which acknowledgment packets, that is, the delivery confirmation messages, are sent. The session start is requested by the client side by the SYN service package with zero length contents and is confirmed by the server with the SYN_ACK packet also with zero length contents. The session end is requested by the FIN service packet and is confirmed by the FIN packet in the reverse direction.

Based on the TCP symmetry, for the purposes of this paper, it is sufficient to build a TCP connection model that transmits data in one direction (e.g., from the calling party to the called one).

Let us describe the state of the TCP client–TCP server system with the following variables:

- N windows number to which the transmitted message will be divided in (data type is integer);
- Value of windows $S_1, S_2, \ldots S_N$ on the transmitting side (the windows size in bits may not be constant, but for this task it does not matter) (data type is a list of string values);
- SB transmitting side buffer in one window size (data type is string);
- Serial number of the current transmitted window SI (data type is integer);
- Windows value $R_1, R_2, \ldots, R_N$ on the receiving side (data type is a list of string variables);
- RB receiving side buffer in the one window size (data type is string);
- Serial number of the current RI receiving window (data type is integer).

In addition, enter the service values of the receiving and transmitting buffers responsible for the session start and end:

- SYN is a connection request;
- SYN_ACK is a connection establishment acknowledgment;
- ACK is a message window acknowledgment;
- FIN is a message end signal.

The TCP information graph is shown in Figure 3.20. The assignment operations SB = RB and RB = SB represent the window transmission over the network using an IP.

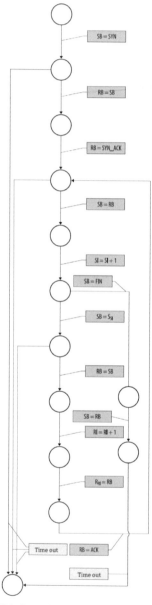

Figure 3.20   The TCP information graph within the entered notation scope.

3.4.5 Presentation of the Networking Layer IP

Network layer protocols are responsible for transmitting messages, relatively independent for the higher layers of the OSI model, over networks with a heterogeneous link layer protocols structure (Ethernet, FrameRealy, ATM, PPP, etc.) [39]. As a consequence, their functions include packet routing in the network layer namespace, as well as fragmentation and assembly of packets (due to the maximum possible packet length variability in various link layer protocols).

In the problem framework, the packet fragmentation and assembly by the IP is in particular the critical function of all above. At the same time, in the process of building a model, it is necessary to take into account that, due to the network conditions heterogeneity in time, the order of the packet fragment arrival to the destination can be random. This IP is qualitatively different from the TCP, which implements a strict transfer windows sequence.

The sender–receiver system state when transmitting one IP packet is described by the following values:

- N fragments number to which the transmitted message will be divided in (data type is integer);
- Fragment packets value $S_1, S_2, \ldots, S_N$ on the transmitting side (data type is a list of string variables);
- Serial number of the current transmitted fragment SI (data type is integer);
- B network buffer size equal to the maximum possible IP message length (data type is a list of string variables);
- Fragment packets value $R_1, R_2, \ldots, R_N$ on the receiving side (data type is a list of string variables);
- RB buffer value on the receiving side of the same fragment size (data type is string);
- Value of the received packet number on the receiving PI side (data type is integer).

In this case, the IP information graph has the following form (Figure 3.21). The assignment and read operations from B buffer are the windows transfer over the network using the lower-layer protocol.

There are restrictions for the read operation from the B buffer: the B buffer j-th element cannot be considered if the SI counter is less than j.

Note that the nominal procedures in the model have the following functions:

- FRAG is a message fragmentation into elements S_i;

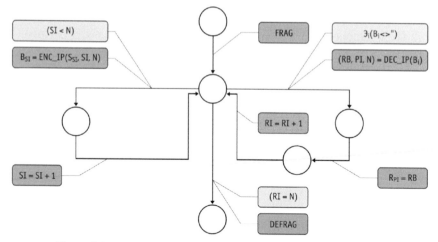

Figure 3.21 The IP information graph in one IP packet transmission.

- DEFRAG is a message assembly from the elements R_i, checking the full reception based on the fragments sequence numbers;
- ENC_IP is the IP fragment coding;
- DEC_IP is the IP fragment decoding.

A characteristic feature of the IP information graph is its behavior nondeterminism under equal initial conditions. A system state graph example for the number of fragments equal to two is shown in Figure 3.22.

3.4.6 Control of the Platform Semantic Correctness

The obtained data processing models are analyzed by the analysis methods of dimensions and similarity. These methods were originally developed by the authors for analytical application verification and are based on verifying the compatibility of the ERP Oracle E-Business Suite data model system [23–26, 40].

Briefly, the proposed method essence is that for the data processing model information graph of the ERP Oracle E-Business Suite, a relationships system between the variables and constants dimensions is constructed. This is implemented as follows. Each model operator containing computational operations and/or an assignment operator and also having a homogeneous form with respect to its arguments is represented as a φ functionals sum of:

$$f_u(x_1, x_2, \ldots, x_n) = 0, \quad u = 1, 2, \ldots, r, \tag{3.72}$$

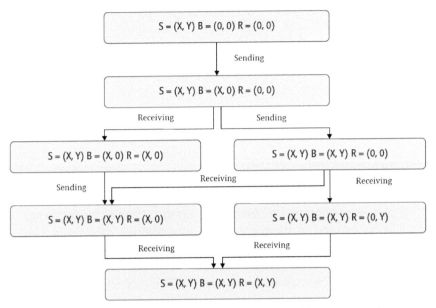

Figure 3.22 A system state graph example for a number of fragments equal to two.

$$f_u(x_1, x_2, \ldots, x_n) = \sum_{s=1}^{q} \varphi_{us}(x_1, x_2, \ldots, x_n) \tag{3.73}$$

where

$$\varphi_{us}(x_1, x_2, \ldots, x_n) = \prod_{j=1}^{n} x_j^{\alpha_{jus}} \tag{3.74}$$

In this case, the provisions of the dimensional and similarity theories make it possible to create a requirements system for the x_j variable dimensions, which implies the following considerations (the notation [X] denotes the "X variable dimensionality"):

$$[\varphi_{us}(x_1, x_2, \ldots, x_n)] = [\varphi_{uq}(x_1, x_2, \ldots, x_n)], \tag{3.75}$$

$$\left[\prod_{j=1}^{n} x_j^{\alpha_{jus}}\right] = \left[\prod_{j=1}^{n} x_j^{\alpha_{juq}}\right], \tag{3.76}$$

$$\prod_{j=1}^{n} [x_j]^{\alpha_{jus}} = \prod_{j=1}^{n} [x_j]^{\alpha_{juq}} \tag{3.77}$$

$$\prod_{j=1}^{n} [x_j]^{\alpha_{jus} - \alpha_{juq}} = 1, \tag{3.78}$$

and after the logarithm:

$$\sum_{j=1}^{n} (\alpha_{jus} - \alpha_{juq}) \cdot \ln[x_j] = 0 \tag{3.79}$$

$u = 1, 2, \ldots, r; s = 1, 2, \ldots, (q-1)$.

A necessary criterion for the semantic correctness of the ERP Oracle E-Business Suite data processing model is the solution existence in the system (3.79), where none of the variables $(\ln[x_j])$ is zero. One of the following sections will be devoted to the problems of this criterion calculation rate increase. In this section, to solve this problem, we confine ourselves to trivial equivalent transformations of the system equations written in matrix form.

Thus, in order to construct the equations (3.73) initial set for each ERP Oracle E-Business Suite data processing model, the following procedure is proposed:

- Randomly take an implementation sample of the process described by the model;
- Verify the sample representativeness by covering all model control graph vertices;
- Perform the model control graph reduction to isolate the computational operators from the operators that verify the conditions and organize the cycles;
- Select from the sample all unique operators that meet the constraints described above for the functional connection type;
- Select the variables and constants involved in the operators, assuming that:

    a) Elements inside the array have equal size,
    b) Numerical constants in pairs have different sizes (their belonging determination to certain classes will occur automatically at the stage of matched dimensional matrix).

The control graph transitions associated with the complex functional dependencies calculation and corresponding to the subprogram call operators supplement the system (2) with formal parameter assignment operations sets. Such additional constraints imply a connecting role between the algorithm main body variables and the subprograms variables. This step must be carried out in the case of building a single computational process model.

3.4.7 Platform Semantic Correctness Control

SQL queries correctness

Typically, an application program is a process that runs on the application server (the "middle link" of the three-tier model) and is responsible for implementing the actual system application logic. The sequence of operations performed by this process on request is shown in Figure 3.23.

Assuming that the calculation algorithm can be reduced to a normalized (divided) form by simple transformations, the absence of the information dimension transformations in the first and last blocks will be taken as their properties. Thus, in the scheme shown in Figure 3.24, the following two system levels are subject to the control [26]:

1) Pre- and postcomputation blocks: they are usually implemented in high-level universal languages such as C++, Pascal, Java, and C# (invariants of arithmetic operations and assignment operators are subject to the control);
2) SQL query: Structured Query Language (SQL) is a high-level application language (based on the dimensional control method, the several group invariants of the typical language operations involving different data types are subject to the control).

Consider the SQL query control abilities.

The SQL clauses, which are extremely widely used today to unify access to relational DBMSs, are commands to the database server to access or edit stored data [26]. Four main clause types are most often used:

- SELECT is a data rows selection;
- INSERT is a new rows addition;
- UPDATE is an editing of the existing rows;
- DELETE is to delete rows.

The context-free grammar elements generating the above clauses in the graphical representation form [23–26, 39, 40] are shown in Figures 3.24, 3.26, and 3.27. The output rules that generate nonterminal symbols that do not affect the dimensional analysis procedure are omitted.

Figure 3.23   Sequence of operations performed by the application program.

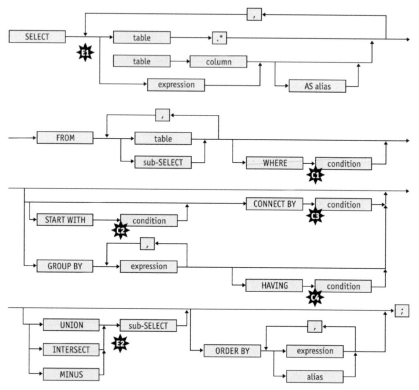

Figure 3.24   SQL clause SELECT.

The SQL clauses consist of two structure types controlled at the dimensional invariants level:

- Condition construction ("condition");
- Implicit dimension matching constructions.

Condition constructions are found in the SQL SELECT, UPDATE, and DELETE clauses (C1, C2, C3, and C4 in Figure 3.24; C1 in Figure 3.25; and position C1 in Figure 3.26). The SQL supports several condition types, but all of them are representable, as in the vast majority of universal programming languages or in a one-place form

$$("condition\ operation"\ "expression - 1") \tag{9}$$

or in a binary one

$$("expression - 1"\ "condition\ operation"\ "expression - 2") \tag{10}$$

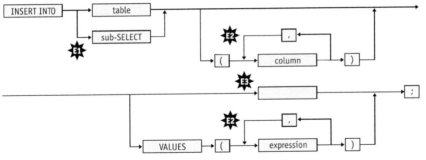

Figure 3.25 SQL clause INSERT.

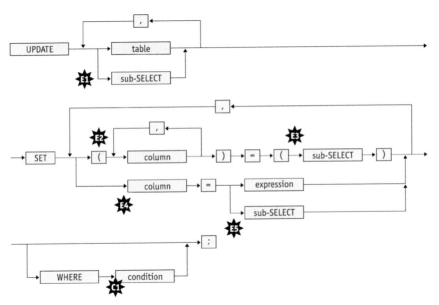

Figure 3.26 SQL clause UPDATE.

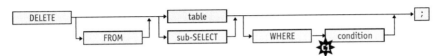

Figure 3.27 SQL clause DELETE.

The exception is the specific ANY, SOME, ALL, IN conditions, involving nested SELECT clauses.

The dimensional analysis of the first two condition classes is analogous to the technique that was considered above:

- Expressions are reduced to homogeneous equations;
- Every equation with q terms is transformed into (q − 1) requirements to homogeneous summand dimensional equality;
- The obtained dimensional equations are logarithmic, resulting in a homogeneous linear equations system (let us call it as A_C) describing the requirements to interrelations between dimensions.

Specific for the SQL is the construction of a dimensional equations system A_E, based on the implicit dimensional compatibility constructions and ANY, SOME, ALL, and IN conditions constructions. The main SQL ideology is working with tuples that is a row processing of the table records and the results of relational relations over them. So, tuples are nonterminal symbols, denoted by E1 and E2 in Figure 3.26 in the SELECT clause, by E1, E2, E3, and E4 in Figure 3.27 in the INSERT, and by E1, E2, E3, E4, and E5 in Figure 3.26 in the UPDATE clause. In any SQL clause, the elements of all the described tuples must have pairwise equal dimensions.

In addition, the similar restrictions are imposed on the elements of special ANY, SOME, ALL, and IN conditions, which in general can be represented in the form:

("tuple − 1" "condition operation" "≪ANY|SOME|ALL|IN≫" "tuple − 2").

Thus, the presence in the SQL statement of the n tuple descriptions from k elements in each

$$\left\|\begin{matrix} < & V_{1,1} & V_{1,2} & \ldots & V_{1,k} & > \\ < & V_{2,1} & V_{2,2} & \ldots & V_{2,k} & > \\ < & \ldots & \ldots & \ldots & \ldots & > \\ < & V_{n,1} & V_{n,2} & \ldots & V_{n,k} & > \end{matrix}\right\| \tag{3.80}$$

introduces a system from $(n − 1) \times k$ additional restrictions on the columns dimensions of tables and expressions containing them. In those cases, when the tuple element is not the actual table column, but the arithmetic expression containing it, an even greater increase of the constraint system is possible (according to the technique for controlling the dimensions of homogeneous equations).

Constructed from the obtained constraints list, the dimensional equations system A_E completes the system A_C obtained earlier. The nontrivial compatibility requirement of the A system, formed by merging the systems A_E and A_C, is a powerful mean to verify the semantic SQL statement correctness before it is executed.

Let us construct an A dimensional equation system and control the correctness of a particular SQL clause in the following example.

Given a DBMS containing the table DETAILS with the fields shown in Table 3.7.

In addition, the constants listed in Table 3.8 are defined in the system.

Suppose an attempt was made to execute the following SQL query:

UPDATE DETAILS
SET (S,VALUE) = (X*Y,
X*Y*PRICE+TRANS)
WHERE ((X,Y) IN (SELECT X+MARGIN,
Y+MARGIN FROM DETAILS)) AND
(PRICE*MAX_SIZE>TRANS).

The main UPDATE clause contains two tuples with two elements each: one nonspecific condition and one specific IN condition. In the IN clause and in the SELECT query associated with them, two more tuples of two elements each are defined.

Thus, verification of the constructions "condition" gives the dimensional equation

$$[PRICE^*MAX\_SIZE] = [TRANS], \qquad (3.81)$$

Table 3.7 Details characteristics table

Field	Description
X	Detail length, m
Y	Detail width, m
S	Detail area, m^2
PRICE	Price per square material meter, rub/m^2
VALUE	Delivery cost, rub

Table 3.8 Table of constants

Constant	Description
TRANS	Shipping cost, rub
MARGIN	Edge width, m
MAX_SIZE	Maximum transportable area, m^2

verification of dimensional matching constructions

$$[S] = [X^*Y] \tag{3.82}$$

and

$$[VALUE] = [X^*Y^*PRICE + TRANS] \tag{3.83}$$

A nested SELECT clause leads to two more dimensional equations:

$$[X] = [X + MARGIN] \tag{3.84}$$

and

$$[Y] = [Y + MARGIN] \tag{3.85}$$

In general, the A_c equations system, generated by Equation (3.81), has the following form (Table 3.9).

The A_E dimensional equations system, generated by (13–16), has the form (Table 3.10).

The matrix corresponding to the A system (formed by A_c and A_E) has a dimension of 6×8 and a rank of 6, which characterizes the presence of two independent dimensions in the system (e.g., X and *VALUE* can be chosen as independent dimensions). This is completely consistent with the subject area, where there are actually two defined independent dimensions: "meters" and "rubles".

Suppose that at the stage of either creating software or parsing a dynamic HTTP request, the error occurred that caused the following SQL clause writing before the execution moment:

Table 3.9   The A_c equations system representation

X	Y	S	PRICE	VALUE	TRANS	MARGIN	MAX_SIZE
0	0	0	1	0	−1	0	1

Table 3.10   The A_E dimensional equations system representation

X	Y	S	PRICE	VALUE	TRANS	MARGIN	MAX_SIZE
−1	−1	1	0	0	0	0	0
−1	−1	0	−1	1	0	0	0
0	0	0	0	1	−1	0	0
1	0	0	0	0	0	−1	0
0	1	0	0	0	0	−1	0

UPDATE DETAILS
SET (S, VALUE) = (X*Y, X*Y*PRICE+TRANS)
WHERE ((X,Y) IN (SELECT X+MARGIN, Y*MARGIN FROM
  DETAILS))
AND
  (PRICE*MAX_SIZE > TRANS)

At the analysis time, Equation (3.85) will be changed. It will take the form

$$[Y]=[Y*MARGIN], \tag{3.86}$$

The A_E matrix, constructed based on equations (3.82, 3.84–86), will take the form of Table 3.11.

The matrix corresponding to the new system has a dimension of 6×8 and a rank of 6, which also characterizes the presence of two independent dimensions in the system.

However, in this case, there are zero values in the decision vector, which correspond to the allegedly existing following relationships:

$$[X] = [MARGIN] = 0 \tag{3.87}$$

This is already fundamentally inconsistent with the subject area and is an unambiguous anomaly indicator in the constructed SQL design.

3.4.8 Verification of Applied Queries

The process implementation sample of the application-level protocol, representative of the control graph vertices covering, explicitly contains the following unique relationships between dimensions (Table 3.12).

In addition, the reduction of condition verification operators allocates one more constraints set on the dimension:

[RN] = [CACHE_NAMES], [TN] = [VALUE_NAMES],
[VALUE_DATES] = [TDATE], [CI] = [CACHE_CNT], [SI] = [VALUE_CNT]

Table 3.11 The A_E matrix representation

X	Y	S	PRICE	VALUE	TRANS	MARGIN	MAX_SIZE
−1	−1	1	0	0	0	0	0
−1	−1	0	−1	1	0	0	0
0	0	0	0	1	−1	0	0
1	0	0	0	0	0	−1	0
0	0	0	0	0	0	1	0

Table 3.12　Relationships between application-layer protocol dimensions

The first group of "client–server" dimensions	The second group of "client–server" dimensions	The third group of "client–server" dimensions
[TN] = [RN]	[AT] = [TT]	[TDATE] = [0*]
[CI] = [1*]	[AT] = [ST]	[TDATE] = [CACHE_DATES]
[TS] = [CS]	[DS] = [AS]	[SM] = [VALUE_VALUES]
[TT] = [CT]	[DT] = [AT]	[CD] = [VALUE_DATES]
[SI] = [1**]	[CE] = [C200]	[CACHE_VALUES] = [CM]
[AS] = [TS]	[CE] = [C404]	[CACHE_CHARSETS] = [DS]
[AS] = [SS]	[CE] = [C304]	[CACHE_DATES] = [CD]
[CM] = [SM]	[CE] = [C406]	[CACHE_CNT] = [1***]

(in all cases, the array elements identifiers are the names of the arrays themselves, starting from the array elements equidimensionality principle, and the signs "*" of the numerical constants mark their potential heterogeneity).

The equation system (3.79) for an application layer protocol, written in a matrix form, is a matrix A_{L7} of 29 × 36 dimension. The matrix rank is 28, which is due to the presence of one linearly dependent equation. The traditional criterion calculation method requires the matrix reduction by equivalent transformations to a special form with the selected base columns. However, for convenience in presenting the results for this model in a book format, we change the initial variable column positions and reduce A_{L7} to the block-diagonal form B'_{L7} (the variables corresponding to x_j in the indeterminate $\ln[x_j]$ of the system (3.79) are written above the columns):

$$B'_{L7} =$$

$$\begin{Vmatrix} C_{L7,1} & 0 & 0 & 0 & 0 & 0 & 0 & 0 \\ 0 & C_{L7,2} & 0 & 0 & 0 & 0 & 0 & 0 \\ 0 & 0 & C_{L7,3} & 0 & 0 & 0 & 0 & 0 \\ 0 & 0 & 0 & C_{L7,4} & 0 & 0 & 0 & 0 \\ 0 & 0 & 0 & 0 & C_{L7,5} & 0 & 0 & 0 \\ 0 & 0 & 0 & 0 & 0 & C_{L7,6} & 0 & 0 \\ 0 & 0 & 0 & 0 & 0 & 0 & C_{L7,7} & 0 \\ 0 & 0 & 0 & 0 & 0 & 0 & 0 & C_{L7,8} \end{Vmatrix} \quad (3.88)$$

where

$$C_{L7,1} = \begin{Vmatrix} CI & 1^* & CACHE\_CNT & 1^{***} \\ 1 & 0 & 0 & -1 \\ 0 & 1 & 0 & -1 \\ 0 & 0 & 1 & -1 \end{Vmatrix} \tag{3.89}$$

$$C_{L7,2} = \begin{Vmatrix} TN & RN & CACHE\_NAMES & VALUE\_NAMES \\ 1 & 0 & 0 & -1 \\ 0 & 1 & 0 & -1 \\ 0 & 0 & 1 & -1 \end{Vmatrix} \tag{3.90}$$

$$C_{L7,3} = \begin{Vmatrix} CM & SM & VALUE\_VALUES & CACHE\_VALUES \\ 1 & 0 & 0 & -1 \\ 0 & 1 & 0 & -1 \\ 0 & 0 & 1 & -1 \end{Vmatrix} \tag{3.91}$$

$$C_{L7,4} = \begin{Vmatrix} SI & 1^{**} & VALUE\_CNT \\ 1 & 0 & -1 \\ 0 & 1 & -1 \end{Vmatrix} \tag{3.92}$$

$$C_{L7,5} = \begin{Vmatrix} TDATE & 0^* & CACHE\_DATES & CD & VALUE\_DATES \\ 1 & 0 & 0 & 0 & -1 \\ 0 & 1 & 0 & 0 & -1 \\ 0 & 0 & 1 & 0 & -1 \\ 0 & 0 & 0 & 1 & -1 \end{Vmatrix} \tag{3.93}$$

$$C_{L7,6} = \begin{Vmatrix} TT & CT & AT & ST & DT \\ 1 & 0 & 0 & 0 & -1 \\ 0 & 1 & 0 & 0 & -1 \\ 0 & 0 & 1 & 0 & -1 \\ 0 & 0 & 0 & 1 & -1 \end{Vmatrix} \tag{3.94}$$

$$C_{L7,7} = \begin{Vmatrix} CE & C200 & C404 & C304 & C406 \\ 1 & 0 & 0 & 0 & -1 \\ 0 & 1 & 0 & 0 & -1 \\ 0 & 0 & 1 & 0 & -1 \\ 0 & 0 & 0 & 1 & -1, \end{Vmatrix} \tag{3.95}$$

$$C_{L7,8} = \begin{Vmatrix} TS & CS & AS & SS & DS & CACHE\_CHARSETS \\ 1 & 0 & 0 & 0 & 0 & -1 \\ 0 & 1 & 0 & 0 & 0 & -1 \\ 0 & 0 & 1 & 0 & 0 & -1 \\ 0 & 0 & 0 & 1 & 0 & -1 \\ 0 & 0 & 0 & 0 & 1 & -1. \end{Vmatrix} \qquad (3.96)$$

The matrix has 8 independent (basic) and 28 dependent columns; the zero line corresponding to the linearly dependent equation is deleted. It is obvious that the homogeneous linear equations system described by the matrix B'_{L7} has a solution that does not contain zeros. As a consequence, the original model dimensional system has a similar solution [23–26, 40].

3.4.9 TCP Verification

The presence of cycle and branching in the process control graph indicates the possibility of a sufficiently large number of different implementations. We will be interested in a selective implementation set covering all control graph vertices set. This condition is satisfied by one implementation set, at least once passing the process cycle, and therefore similar implementations as representative for a given control graph will be considered.

In the explicit form in the TCP process model, the following relationships between dimensions exist:

[SB] = [SYN]	[SI] = [1*]	[R] = [RB]
[RB] = [SB]	[SB] = [S]	[RB] = [ACK]
[RB] = [SYN_ACK]	[RI] = [1**]	[SB] = [FIN]
[SB] = [RB]		

In addition, the reduction of the cycle branching condition (SI < N) to the canonical form (T* = SI – N; T* < 0, where T*is an auxiliary variable) adds the dimensional equation [N] = [SI].

The matrix form A_{L5} of the system record (3.79) for this protocol is shown in Table 3.13. The column headers contain the corresponding variables x_j.

The matrix has the 11×13 dimension. The matrix rank is 10, which corresponds to one linearly dependent row (in this case, row 4 is the inverse of row 2). By means of equivalent transformations, we reduce the matrix to the form B_{L5}, consisting of basic columns and dependent columns containing one unit element (Table 3.14).

Table 3.13 Matrix A_{L5}

SB	SYN	RB	SYN ACK	SI	1*	N	S	RI	1**	R	ACK	FIN
1	-1	0	0	0	0	0	0	0	0	0	0	0
-1	0	1	0	0	0	0	0	0	0	0	0	0
0	0	1	-1	0	0	0	0	0	0	0	0	0
1	0	-1	0	0	0	0	0	0	0	0	0	0
0	0	0	0	1	-1	0	0	0	0	0	0	0
0	0	0	0	-1	0	1	0	0	0	0	0	0
1	0	0	0	0	0	0	-1	0	0	0	0	0
0	0	0	0	0	0	0	0	1	-1	0	0	0
0	0	-1	0	0	0	0	0	0	0	1	0	0
0	0	1	0	0	0	0	0	0	0	0	-1	0
1	0	0	0	0	0	0	0	0	0	0	0	-1

Table 3.14 Matrix B_{L5}

SB	•SYN	RB	•SYN ACK	SI	1*	N	S	RI	1**	R	•ACK	•FIN
1	0	0	0	0	0	0	0	0	0	0	0	-1
0	1	0	0	0	0	0	0	0	0	0	0	-1
0	0	1	0	0	0	0	0	0	0	0	0	-1
0	0	0	1	0	0	0	0	0	0	0	0	-1
0	0	0	0	1	0	-1	0	0	0	0	0	0
0	0	0	0	0	1	-1	0	0	0	0	0	0
0	0	0	0	0	0	0	1	0	0	0	0	-1
0	0	0	0	0	0	0	0	1	-1	0	0	0
0	0	0	0	0	0	0	0	0	0	1	0	-1
0	0	0	0	0	0	0	0	0	0	0	1	-1

A special matrix form allows directly indicating a model dimensional system solution that does not contain zeros. To do this, it is sufficient to choose as a value set of the basis variables any vector that does not contain zero coordinates and is not orthogonal to the vectors compiled row-by-row from the basis column coefficients.

3.4.10 IP Verification

A representative implementation sample for the IP contains the following relationships between the variables dimensions in the explicit form:

$$[S] = [RB] \qquad [PI] = [SI] \qquad [RI] = [1**]$$
$$[RB] = [R] \qquad [SI] = [1*]$$

In addition, the condition operators contain two more restrictions on the dimension:

[SI] = [N] and [RI] = [N].

The matrix record form A_{L3} of the equation system for this model is given in Table 3.15, and the recording format corresponds to the previous paragraph format.

The matrix has a 7×9 dimension, the matrix rank is 7. By means of equivalent transformations, reduce the matrix to the special form B_{L3} (Table 3.16).

Similar to the previous section, a special matrix form confirms the presence of a nonzero solution and, as a consequence, the studied criterion fulfillment for the model semantic correctness.

***

This approach allowed implementing the anomalies detection of the network interaction models in ERP Oracle E-Business Suite based on TCP/IP. Its essential advantage is the necessary criteria development for the semantic correctness of the technological data processing model in ERP Oracle E-Business Suite.

Note that the authors formulated and proved a direct similarity theorem for computational processes, which made it possible to establish the

Table 3.15   Matrix A_{L3}

S	RB	R	PI	SI	RI	1*	1**	N
1	−1	0	0	0	0	0	0	0
0	1	−1	0	0	0	0	0	0
0	0	0	1	−1	0	0	0	0
0	0	0	0	1	0	−1	0	0
0	0	0	0	0	1	0	−1	0
0	0	0	0	1	0	0	0	−1
0	0	0	0	0	1	0	0	−1

Table 3.16   Matrix B_{L3}

S	RB	R	PI	SI	RI	1*	1**	N
1	0	−1	0	0	0	0	0	0
0	1	−1	0	0	0	0	0	0
0	0	0	1	0	0	0	0	−1
0	0	0	0	1	0	0	0	−1
0	0	0	0	0	1	0	0	−1
0	0	0	0	0	0	1	0	−1
0	0	0	0	0	0	0	1	−1

necessary conditions for the semantically correct computations similarity [23–26, 41].

Theorem 1. A direct similarity theorem for computational processes. If the observed computing processes are homogeneously similar, then the required generalized variables satisfy the identical closed equation systems reduced to the relative form.

Proving. Suppose that N is homogeneously similar, semantically correct computations are observed, each of which consists in a modification of the input and output data $x_{1j}, x_{2j}, \ldots, x_{nj}, j = 1, 2, \ldots N$. Let k first data $x_{1j}, x_{2j}, \ldots, x_{kj}$ be independent, and the remaining $m = n - k$ variables $x_{(k+1)j}, x_{(k+2)j} \ldots, x_{nj}$ be dependent. To determine m dependent data, we define the following closed equation system:

$$\mathrm{D}i(x_{1j}, x_{2j}, \ldots x_{nj}) = 0 \qquad (3.97)$$

where the i-th system equation consists of the z_i element sum, and each equation element is the power product of some or all marked data. We call the computation corresponding to the index $j = 1$ as the initial one. Suppose that the variables $X_{1j}, X_{2j}, \ldots, X_{nj}$ are related to each other by the relations $x_{1j} = c_{ij}x_{ij}$

$$x_{ij} = c_{ij}x_{ij} (i = 0, 1 \ldots, n; j = 0, 1, \ldots, m) \qquad (3.98)$$

Let us select N such computations, which are described by the equations system

$$D_i(x_{1j}, x_{2j}, \ldots, x_{nj}) = 0 (i = 0, 1 \ldots, m). \qquad (3.99)$$

For j computation, this system can be represented as:

$$D_i(x_{1j}/c_{1j}, x_{2j}/c_{2j}, \ldots, x_{nj}/c_{nj}; C_{1j}, C_{2j}, \ldots, C_{(zi-1)j}) = 0, \qquad (3.100)$$

For the initial calculation ($j = 1$), this system looks like:

$$D_i(x_{1j}, x_{2j}, \ldots, x_{n1}; i = 1, 1 \ldots, 1) = 0 \qquad (3.101)$$

The relative variables $x_{1j}/c_{1j}, x_{2j}/c_{2j}, \ldots x_{nj}/c_{nj}$ are the part of the equations system $D_i(x_{1j}/c_{1j}, x_{2j}/c_{2j}, \ldots x_{nj}/c_{nj}; C_{1j}, C_{2j}, \ldots C_{(zi-1)j}) = 0$. The transformations indicate that these variables are, respectively, $x_{11}, x_{21}, \ldots, x_{n1}$, and therefore, the given variables will be identical. But the identical variables $x_{1j}/c_{1j}, x_{2j}/c_{2j}, \ldots x_{nj}/c_{nj}$ of a similar computation class must satisfy only the identical equations systems.

Let the class N of similar computations, which are described by the equations system, be

$$D_i(x_{1j}/x_{1j0}, x_{2j}/x_{2j0}, \ldots, x_{nj}/x_{nj0}; \Pi_{1i}, \Pi_{2i}, \ldots, \Pi_{(zi-1)}) = 0,$$

where $i = 1, 2, \ldots, m$; $z_i = z_1, z_2, \ldots, z_m$.

For the initial calculation ($j = 1$), the system takes the form:

$D_i(x_{11}/x_{110}, x_{21}/x_{210}, \ldots, x_{n1}/x_{n10}; \Pi_{11}, \Pi_{21}, \ldots, \Pi_{(zi-1)}) = 0$. The equations system $D_i(x_{1j}/x_{1j0}, x_{2j}/x_{2j0}, \ldots, x_{nj}/x_{nj0}; \Pi_{1i}, \Pi_{2i}, \ldots, \Pi_{(zi-1)}) = 0$ contains the relative variables $x_{1j}/x_{1j0}, x_{2j}/x_{2j0}, \ldots, x_{nj}/x_{nj0}$.

The transformations $x_{ij} = c_{ij} x_{ij} (i = 0, 1 \ldots, n; j = 0, 1, \ldots, m)$ show that these variables are, respectively, $x_{11}/x_{110}, x_{21}/x_{210}, \ldots, x_{n1}/x_{n10}$; therefore, the given variables will be identical. Since the identical variables $x_{1j}/x_{1j0}, x_{2j}/x_{2j0}, \ldots, x_{nj}/x_{nj0}$ of the similar computations class can satisfy only the identical equation system, the equations $D_i(x_{11}/x_{110}, x_{21}/x_{210}, \ldots, x_{n1}/x_{n10}; \Pi_{11}, \Pi_{21}, \ldots, \Pi_{(zi-1)}) = 0$ and $D_i(x_{1j}/x_{1j0}, x_{2j}/x_{2j0}, \ldots, x_{nj}/x_{nj0}; \Pi_{1i}, \Pi_{2i}, \ldots, \Pi_{(zi-1)}) = 0$ are the same.

The theorem is proved.

It follows from the theorem that if the observed computation processes described by the equations system $D_i(x_{1j}, x_{2j}, \ldots, x_{nj}) = 0$ are homogeneously similar, then the required generalized variables satisfy identically closed equation systems reduced to the relative form $D_i(x_{1j}/C, x_{2j}/C_{2j}, \ldots, x_{nj}/C_{nj}; C_{ij}, C_{2j}, \ldots, C_{1j}, C_{2j}, \ldots, C_{(zi-1)j}) = 0$ or $D_i(x_{1j}/x_{1j0}, x_{2j}/x_{2j0}, \ldots, x_{nj}/x_{nj0}; \Pi_{1i}, \Pi_{2i}, \ldots, \Pi_{(zi-1)}) = 0$, where $i = 1, 2, \ldots, m$; $z_i = z_1, z_2, \ldots, z_m$.

The direct similarity theorem corollaries: Let us deduce two main corollaries from the direct similarity theorem of the computational processes.

Corollary 1. Similarity indicators in homogeneously similar correct calculation processes at similar points, as well as in homochronous (similar) time instants are equal to one:

$$C_{1j} = C_{2j} = \cdots = C_{(zi-1)j} = 1. \tag{3.102}$$

In the similarity theory, such relations are called conditioning equations.

Corollary 2. Similarity invariants in homogeneously similar correct computational processes at similar points, as well as in homochronous time instants, are separately the same:

$$\Pi_{1i} = idem; \Pi_{2i} = idem, \ldots; \Pi_{(zi-1)} = idem. \tag{3.103}$$

The following assertion is formulated based on this.

Assertion 1. The similarity computations necessary conditions. For the similarity within a subclass of homogeneous semantically correct computation processes represented by a finite Abelian group G, it is necessary that the elements characteristics of the abstract subgroup G_v nested in G, which specifies the permissible changes in the conditions for program functioning, coincide separately.

An inverse similarity theorem. The authors formulated and proved the theorem, which made it possible to establish sufficient conditions for the correct computational processes similarity [23–26, 41].

Theorem 2. The inverse similarity computational processes. If the generalized variables satisfy the identical closed equations systems represented in relative form, then the observed computational processes are homogeneously similar.

Proving. We take N homogeneously similar semantically correct computations, each of which is a modification of the input and output data $x_{1j}, x_{2j}, \ldots, x_{nj}, j = 1, 2, \ldots N$. Let k first data $x_{1j}, x_{2j}, \ldots, x_{kj}$ be independent, and the remaining $m = n - k$ variables $x_{(k+1)j}, x_{(k+2)j}, \ldots, x_{nj}$ be dependent. To determine m dependent data, the following closed equations system $D_i(x_{1j}, x_{2j}, \ldots, x_{nj}) = 0$ is defined, where this system i-th equation consists of the z_i element sum, and each equation element is the power product of some or all marked data. We call the computation corresponding to the index $j = 1$ as the initial one. Suppose that the quantities $x_{1j}, x_{2j}, \ldots, x_{nj}$ are related by the relations $x_{ij} = c_{ij} x_{ij} (i = 0, 1 \ldots, n; j = 0, 1, \ldots, m)$.

The equations system $D_i(x_{1j}, x_{2j}, \ldots, x_{nj}) = 0 (i = 0, 1 \ldots, m)$ can be represented in the form of an equations system: $D_i(x_{1j}/c_{1j}, x_{2j}/c_{2j}, \ldots, x_{nj}/c_{nj}; C_{1j}, C_{2j}, \ldots, C_{(zi-1)j}) = 0$. Let the equations systems $D_i(x_{1j}/c_{1j}, x_{2j}/c_{2j}, \ldots, x_{nj}/c_{nj}; C_{1j}, C_{2j}, \ldots, C_{(zi-1)}) = 0$, which describe N homogeneously similar semantically correct computations, be identical. For this, it is sufficient in the aggregate $x_{1j0}/c_{1jo} = x_{110}, x_{2j0}/c_{2j} = x_{210}, \ldots, x_{kj0}/c_{kj} = x_{k10}$ to implement the equalities $x_{(k+1)j0}/c_{1j} = x_{(k+1)10}, x_{(k+2)j0}/c_{2j} = x_{(k+2)10}, \ldots, x_{hj0}/c_{nj} = x_{n10}$. In the realization, we choose $x_{1j}/c_{1j}, x_{2j}/c_{2j}, \ldots, x_{nj}/c_{nj}$ so that the relations $C_{1j} = C_{2j} = \cdots = C_{(zi-1)j} - 1$ are valid.

In this case, the equations system will be identical to $D_i(x_{1j}/c_{1j}, x_{2j}/c_{2j}, \ldots, x_{nj}/c_{nj}; C_{1j}, C_{2j}, \ldots, C_{(zi-1)j}) = 0$. At the same time, the N observed computational processes are homogeneously similar. Similarly, the corresponding equations systems are identical to

$$D_i(x_{1j}/x_{1j0}, x_{2j}/x_{2j0}, \dots, x_{nj}/x_{nj0}; \Pi_{1i}, \Pi_{2i}, \dots, \Pi_{(zi-1)}) = 0.$$

The theorem is proved.

From the inverse theorem, it follows that for similarity within a subclass of homogeneous correct computation processes described by formally identical closed equations systems, it is sufficient to implement in a certain set of these processes parametric points such a data transformation that:

1. Similarity indicators, which enter the first relative form of equations systems describing homogeneously similar semantically correct computation processes, were equal to one;
2. Similarity invariants, which enter the second relative form of these equations systems, were equal to each other.

In addition to this, it is necessary to achieve the conditions identity for the calculation operation given in a relative form.

With this in mind, the following provision is formulated.

Assertion 2. Sufficient conditions for the computations similarity. For similarity within a subclass of homogeneous semantically correct computation processes, represented by a finite G abelian group, it is sufficient for some observable computations states to transform their features sets in such a way that the corresponding elements features of the abstract subgroup G_v nested in G that assigns permissible transformations of calculation functioning conditions would coincide with each other.

The equations analysis π-theorem. Finally, the authors formulated and proved a π-theorem for the computation equations analysis, which allows establishing a general solution form of the equation system that defines homogeneously similar semantically correct computational processes [23–26, 40].

Theorem 3. The computation equations analysis π-theorem. The equations system solution defining a class of homogeneously similar computational processes can be represented in the dependencies form on similarity invariants and simplices.

Proof. We take the similar computational processes class determined by the equations system $D_i(x_{1j}, x_{2j}, \dots, x_{nj}) = 0$ $(i = 0, 1 \dots, m)$, which can be represented as an equations system $D_i(x_{1j}/c_{1j}, x_{2j}/c_{2j}, \dots, x_{nj}/c_{nj};$

$C_{1j}, C_{2j}, \ldots, C_{(zi-1)j}) = 0$. Suppose that the solution of the equations system $D_i(x_{1j}/c_{1j}, x_{2j}/c_{2j}, \ldots, x_{nj}/c_{nj}; C_{1j}, C_{2j}, \ldots, C_{(zi-1)j}) = 0$ exists and can be represented as follows:

$$\begin{cases} X_{(k+1)j}/C_{(k+1)j} = \varphi_1(x_{1j}/C_{1j}, \ldots, x_{kj}/C_{kj}; C_{1j}, \ldots C_{(z1-1)j}) \\ X_{(k+2)j}/C_{(k+2)j} = \varphi_2(x_{2j}/C_{2j}, \ldots, x_{kj}/C_{kj}; C_{1j}, \ldots C_{(z2-1)j}) \\ X_{nj}/C_{nj} = \varphi_m(x_{1j}/C_{1j}, \ldots, x_{kj}/C_{kj}; C_{1j}, \ldots C_{(zm-1)j}) \end{cases}$$
$$(3.104)$$

or $F_1^0(x_{1j}/C_{1j}, \ldots, x_{kj}/C_{kj}, \quad x_{(k+1)}/C_{(k+1)}, \ldots, x_{nj}/C_{nj}; C_{lj}, \ldots, C_{(zl-1)j}) = 0$, where $c_{i,j}$ are the multipliers of the similarity relations transformation and φ_i are the functions of all or some relative data. Based on the direct similarity theorem corollary, we have $F_1^0(x_{1j}/C_{1j}, \ldots, x_{nj}/C_{nj}) = 0, l = 1, 2, \ldots, m, j = 1, 2, \ldots, n$.

Since this equations system describes processes from a class of homogeneously similar processes, the functions F_1^0 should not depend on the transformation factors $D_i(x_{1j}/c_{1j}, x_{2j}/C_{2j}, \ldots, x_{nj}/C_{nj}; C_{1j}, C_{2j}, \ldots, C_{(zi-1)j}) = 0$ and, consequently, should not contain them. This requirement can be satisfied if the data $x_{1j}/C_{1j}, x_{2j}/C_{2j}, \ldots, x_{nj}/C_{nj}$ are combined in a certain way into similarity invariants $\Pi_{1i}, \Pi_{2i}, \ldots, \Pi_{(zi} - 1)$. In this case, we obtain a equations system form $F_1^0 = \Pi_{1i}, \Pi_{2i}, \ldots, \Pi_{(zi} - 1) = 0$.

It is clear that the relation $x_{1j}/x_{1j0}, x_{2j}/x_{2j0}, \ldots, x_{nj}/x_{nj0}$ do not depend on the transformation factors of the $C_{1j}, C_{2j}, \ldots, C_{(zi-1)i}$, and therefore are similarity invariants. It is shown that it is possible to represent the equations system as follows: $F_1^0 = \Pi_{1i}, \Pi_{2i}, \ldots, \Pi_{(zi} - 1), i; x_{1j}/x_{1j0}, \ldots, x_{nj}/x_{nj0} = 0, l = 1, 2, \ldots, m$. And this equations system solution with respect to the desired relations (simplexes) is as follows:

$$\begin{cases} X_{(k+1)j}/C_{(k+1)j} = \varphi_1(\Pi_{11}, \ldots, \Pi_{(z1-1)}; x_1/x_{j0}, \ldots, x_{kj}/x_{kj0}) \\ X_{(k+2)j}/C_{(k+2)j0} = \varphi_2(\Pi_{12}, \ldots, \Pi_{(z2-1)}; x_1/x_{j0}, \ldots, x_{kj}/x_{kj0}) \\ X_{nj}/C_{nj} = \varphi_m(\Pi_{1m}, \ldots, \Pi_{(zm-1)j}; x_1/x_{1j0}, \ldots, x_{kj}/x_{kj0}) \end{cases}$$
$$(3.105)$$

The theorem is proved.

The obtained results showed the following [23–26, 40]:

1. Since the dimensional analysis is essentially an analysis of the equations describing the observed computating process in the most general form, both the equations analysis and the dimensional analysis can be used to describe the equations system solution. It can also be used to describe

network interaction models in ERP Oracle E-Business Suite based on TCP/IP.

2. Dimensional analysis should be used only when a closed equation system describing the observed computating process is missing or difficult to obtain.

To optimally construct the invariants trajectory (route) of the network interaction model similarity in ERP Oracle E-Business Suite based on TCP/IP, it is advisable to use the known structural completeness criteria, in particular, the criterion of all vertices cover. To determine the route weight, it was suggested to modify known Halstead metrics. This allowed taking into account the probability of the computational process passing along a certain control graph route, the superpositions number in symbolic route execution, and also the route representativeness determined by the similarity invariants number.

4

Development of Cyber Security
Technologies

This chapter provides an overview of the best practices for the organization and research conduction, R&D in the field of cyber security. The urgency of continuous improvement of the cyber security requirements is shown in the example of GOST R IEC 61508. The development of new ontologies of cyber security for intelligent networks of the new generation, Smart Grid, is grounded. The peculiarities of the development of promising Digital Economy technological platforms based on software-defined computing systems and/or software-configurable networks, software-defined networking (SDN), is revealed. Possible ways of developing a flexible methodology for the development of Agile software with regard to the requirements of cyber security are considered. The addition to the classical Agile methodology of a number of special methodical techniques from the best practices of SDL Microsoft, Cisco and PCI DSS, ISO/IEC 15408, ISO/IEC 27034-1, 7.3.5 STO BR IBBS-1.4-2018 and GOST R 56939 is substantiated; this allowed to form a basic set of requirements for the development of secure software for Digital Economy applications. The development of a next-generation BI platform for predictive cyber security analytics is proposed.

4.1 Cyber Security R&D Best Practice

Recently, there has been a major international political scandal caused by the exposures of E, during which Snowden and several other journalists revealed that the US National Security Agency (NSA) has direct access to the central servers and data centers of the largest Internet companies like Google, Yahoo!, Microsoft, Facebook, Skype, YouTube, and Apple [62, 75, 76].

Information is intercepted in the mobile and landline communication channels all over the world. More than 85,000 of special hardware and software bookmarks are used. Only in Germany, the NSA monitored more than 500 million electronic and computer connections every month, at the end of 2012, and over 70 million such connections were tracked in France in 1 month, and the number of tracked social network accounts was more than 250 million for a year [62, 77, 78]. According to U. Binni, a former NSA employee, this organization has data on 40–50 trillion telephone conversations and e-mail messages from around the world. It was established that the NSA conducts cyber-espionage activities against more than 35 heads of state and government, including German Chancellor, Brazilian and Mexican presidents, as well as authorized representatives of the European Union Council of Ministers and the European Council, 38 European Union embassies and diplomatic missions, and its member countries in the headquarters buildings of the International Atomic Energy Agency (IAEA) and the United Nations (UN) [62, 77, 79].

The European, Asian, and Latin American countries immediately reacted and announced the promotion of their own and US-independent technical policy in the cyber security field: 12 Latin American states belonging to the Union of South American Nations (UNASUR) announced their intention to create their own network that is an Internet analogue. The Brazilian authorities began developing their "secure" e-mail service. The German government banned the Internet traffic transmission between German users through the network nodes located outside the country to eliminate sniffing by foreign special services. The European largest operator Deutsche Telekom called on German companies to unite under the project to create "national routing", etc.

4.1.1 Cyber Security R&D Importance

In the 20th century, a number of technologically advanced countries (more than 120 countries) declared the "cyberwarfare" development. In the USA in December 2011, Congress received the permission to develop "offensive" cyberwarfare. The US 2011 "International Strategy for Cyberspace" determined cyberspace to be a potential battlefield as land, sea, air, and space [62, 80, 81]. In France, in 2008, the "White Paper on Defense and national security" introduced the "cyberwar" concept and defined its components like cyber defense and offensive capabilities for cyberwar. The "Security Strategy in Cyberspace" was adopted in Germany in February 2011. A similar document was introduced in the UK in November 2011. In 2011, the "Internet troops" creation in the People's Liberation Army of China was officially announced. In India, the cyber security strategy was adopted in May 2013, including the creation of the Indian National Coordination Center for Information Control on the Internet to prevent foreign cyber espionage and hacker attacks. The 2013 Japanese Ministry of Defense White Paper noted the exceptional cyber security importance to ensure the security of the state and its armed forces. In particular, attention was drawn to the fact that cyber-attacks on the information and communication networks of the various countries government and military institutions increased. In 2012, there were more than 1,000 cyber-attacks on Japanese institutions recorded in Japan itself.

In 2011, NATO approved a new version of the "Cyber Defense NATO policy" document and "NATO Action Plan for Cyber Defense", containing practical recommendations on actions in the domain. On June 19, 2013, cyber security strategy entered into force in the European Union. In early 2013, a cyber security strategy was developed in Finland.

Since 2010, technologically developed countries have started to create special organizational and staff structures designed to plan and manage cyber operations both in peacetime and in high alert. For example, the NSA already created a specialized unit from "highly skilled" hackers called TAO (Tailored Access Operations), capable of "reaching the inaccessible" in 1997. Since 2011, there has been a separate cyberspace department in the US State Department, whose leader announced cyber security "the US foreign policy imperative". In June 2009, a special US Armed Forces cyber command was created (*USCYBERCOM*), whose main tasks were "conducting operations aimed at ensuring action freedom for the USA and its allies in cyberspace, as well as limiting this freedom to rival countries".

Since June 2010, the Government Communications Headquarters (GCHQ), one of the secret services, has been operating a cyber security operations center in the UK. National cyber security centers were established: in June 2011 in Germany, in January 2012 in the Netherlands, and in January 2013 in Denmark. In France, such unit is a part of the main external security department; in Germany, the Federal Intelligence Service (Bundesnachricht-endienst, BND) has a special department to counter hacker attacks and cyber espionage. In Latvia, there is Center for Countering Cyber Threats, and in Lithuania, there is National Center for the Incidents Prevention in Information Technology. In Estonia, the advanced NATO center, the Cooperative Cyber Defense Center of Excellence, was created, in which a number of countries (Hungary, Great Britain, Germany, Spain, Italy, Latvia, Lithuania, the Netherlands, Poland, Slovakia, USA, Estonia, etc.) participate. At the same time, the special International Secretariat management department on new challenges and threats of the Brussels unit coordinates NATO activities in cyber security. Within the European Union, the European Network and Information Security Agency operates, and the Cyber Crime Center was launched on January 1, 2013.

A whole range of research and development (R&D) activities in the cyber security domain are carried out to develop special software and hardware complexes:

- PRISM (metadata collection and processing);
- Gourmet through and Jet plow (remote "bookmarks" integration in personal computers);
- Quantum Insert (redirecting traffic to false Internet sites);
- Dropout Jeep (remote information retrieval from Apple phones);
- Monkey calendar (SMSs about the mobile phones location);
- Rage master (information interception from computer screens);
- Genie (monitoring the functioning of 85,000 "spyware" around the world, etc.).

Since 2006, large-scale transnational Cyber Storm attacks have been regularly taking place in the USA, and Cyber Europe has been running in the European Union since 2010, during which, based on the "alleged enemy actions", various cyber security aspects are tried out and tested, by modeling of the hacker attacks and "cyber-attacks" on critical objects information systems of state and military control in real time. According to the Washington Post, as E. Snowden said, the US intelligence service conducted 231 cyber-attacks against other countries in 2011, for which more than US $652 million was spent. Three quarters of cyber-attacks, as the publication specified, were directed against Russia, Iran, China, and North Korea, including those against the nuclear programs of these countries. According to Snowden, the USA conducted more than 61,000 hacker cyber-attacks around the world. Thus, it seems that cyber security issues will be among the priority areas to ensure global security and global stability at least until 2020.

R&D in the cyber security domain is also relevant for Russia. In particular, in 2012, the National Advanced Research Foundation (ARF) was established based on the Federal Law of October 16, 2012 No. 174-FZ and the Presidential Decree of the Russian Federation of May 7, 2012 No. 603. The foundation goal is to promote the scientific research and development implementation in the interests of national defense and state security, including for the Russian Digital Economy, associated with a high-risk degree of achieving qualitatively new results in the defense, technological, and socioeconomic spheres.

The main foundation objectives include the development of:

- Most effective methods to determine the most important trends in the scientific and technological sphere and the innovative solution needs;
- System evolutionary models of technological space and corresponding knowledge bases;
- Most effective interaction forms and ways with scientific and expert communities;
- Automated means of scientific and technological prediction and decision support using qualitative and quantitative approaches and operating with large volumes of unstructured and semi-structured data;
- Control technologies in communication systems with random routing based on a dynamic network of heterogeneous repeaters;
- High-performance computing technologies in distributed heterogeneous networks;
- Creation technology of an elemental quantum computers base;
- Teaching and imitation technologies of the human thought process;
- Technologies of meaning and multivalued information context understanding by technical systems;
- Technologies that ensure the technical system ability to generalize and efficiently work with incomplete, inaccurate, or distorted data;
- Technologies to determine the psychoemotional state and to predict people's behavior;
- Demonstration samples of neural interfaces with ability of feedback and personalized cognitive abilities enhancement to control robotic combat, reconnaissance, transport, and other modules, improving the quality of personnel training;
- Digital models of the brain and its work simulation in artificial and hybrid intelligence systems by using in cognitive technical systems (cognitive human–machine interfaces, brain–computer interfaces, and eye–brain–computer, anthropomorphic, and neurromorphic robots), etc.

The most important ARF objective is the creation of a completely new system and model to manage the development of advanced innovative solutions in disturbed production chains. The allocated status allows the ARF to transform strategic challenges and cyber risks into specific scientific and technical projects, implement goal-setting, and conduct the development of priority interspecific, interdisciplinary, and intersectoral scientific and technical research projects. Continuous communication with the scientific community, industry, authorities, and development institutions opens new opportunities in technological breakthroughs in the next decade to properly ensure the cyber security of the Russian Digital Economy at the new technological 4.0. Industry (http://fpi.gov.ru/activities/areas/information).

4.1.2 Cyber Security Project Management

Nowadays, a number of developed countries (DARPA), Russia (Advanced Research Foundation), France (GDA), Israel (MAFAT), China (SASTIND), India (DRDO), and others have been conducting a wide range of research and development in cyber security to gain advantages in international relations and the world economy. At the same time, all the mentioned R&D can be relatively divided into three large classes:

1. Fundamental research in the cyber security domain, conducted by large state universities and academic research institutes;
2. Applied research, confirming a fundamental research in practice; as a rule, this class of research is carried out by enterprises of the defense industry complex, possibly under the scientific supervision of some academic institute or university;
3. Applied research at the stage of weapons and military equipment operation and maintenance, usually carried out by industrial enterprises, sometimes in partnership with the customer operations department.

Let us consider the possible project management organization in the cyber security domain on the agency DARPA (USA) for example, which successfully performs its research activities (Figures 4.1 and 4.2).

The DARPA management believes that in order to achieve the project goals, original scientific ideas and an appropriate team with all the necessary knowledge and skills are required. A possible management scheme of DARPA search programs is shown in Figure 4.3.

According to DARPA executives, all technological projects should be evaluated in the coordinates "technical risk – potential military utility". In this case, the advantage should be given to those that have both high risks and high pay-offs and thus provide breakthrough achievements. As a rule, these are large long-term conceptual projects in which various DARPA research

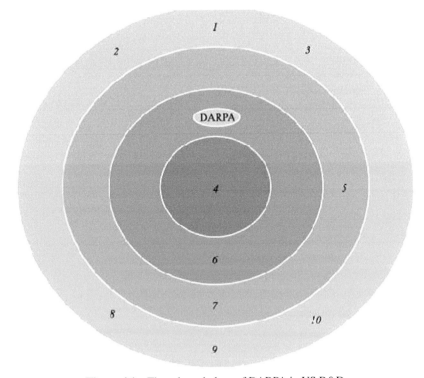

Figure 4.1 The role and place of DARPA in US R&D.

1 – communication forms with the military–industrial complex and business in the aerospace and defense industries; 2 – cooperation programs and projects with universities; 3 – consulting companies (aerospace and defense practice); 4 – national military branch laboratories and scientific centers; 5 – intelligence community contractors of the private military companies; 6 – corporations and universities; 7 – small technological business; 8 – venture companies of the military–industrial complex; 9 – new business and public relations forms; 10 – public associations of the military–industrial complex.

departments are involved. The main investment part about 60% is allocated to them. For projects with low risk and high pay-off (associated with the adaptation and application of ready-made commercial products to the current defense needs lying at the intersection of military departments responsibilities), 20% or more investments are allocated in general. Projects with high risk and low pay-off also receive about 20% of investments, including the dual-use technologies development, the niche occupation in the commercial security market, and the new military technology creation.

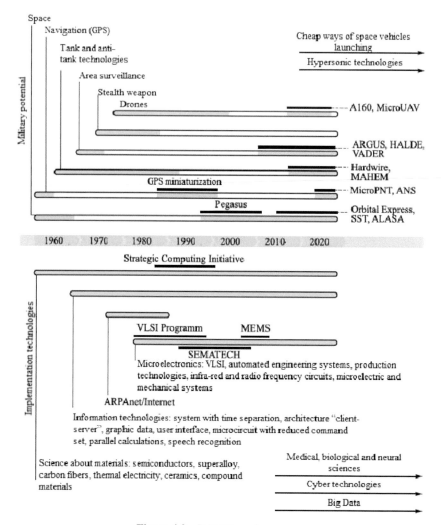

Figure 4.2   DARPA performance.

Let us note that, in 1975, DARPA Director George Heilmeier outlined the following pressing issues for future researchers:

1. What are we trying to do? What is the problem that we are trying to solve?
2. How is it done today? What are the limitations of the existing experience?

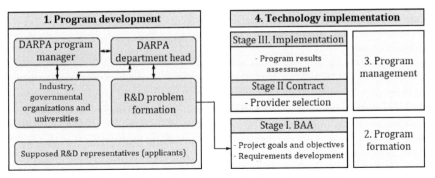

Figure 4.3 Project management scheme in DARPA.

3. What is new in our approach and why do we think that it will be successful? What is the evidence that the new approach will work?
4. If we succeed, what changes will it lead to?
5. How long will it take? How much does it cost? What are the medium-term and target indicators?

The following stages of the project implementation are defined in the DARPA agency:

- From the idea to the concept – the R&D implementation demonstrating the principle possibility of the technology realization;
- From concept to prototype – a development of test or demonstration prototype that might be scaled;
- From the demonstration prototype to the industrial prototype and limited production;
- From limited production to mass production of an industrial prototype for military use.

At each stage, DARPA places solicitations on the implementation of the corresponding development with specified tactical and technical characteristics. Organizations that meet the stated selection criteria of R&D providers receive a certain reward and the right to conclude a further contract. In general, the US government applies 40–50 different types of federal contracts. They can be divided into groups: "fixed price" contracts and "provider cost reimbursement" contracts. At the same time, contracts are divided into "incentive" and "repeatedly incentive" types, which have a complex pricing mechanism and a wide variation in the amount of material incentives for the contractor, depending on the achieved results.

DARPA agency pays great attention to R&D risk management. For each project, an individual approach and risk management methodology is selected. In addition, the regulatory authorities carefully carry out audits. Right before an R&D contract conclusion, research programs are developed. Based on the results of the interaction with all interested parties, as well as studying the existing technologies and achievements in the area, the program manager announces the procedure of requests for competitive proposals (RFP). Then, after completing the proposals collection, he creates an R&D requirements specification (Broad Agency Announcement, BAA). In this case, the BAA terms and requirements are closed until the competition announcement and are not subject to disclosure. The application evaluation procedure can last up to 3–4 months, and the contract determination with the competition winner can last up to 2 months. At the same time, most contracts are concluded for a period of 1–3 years.

4.1.3 New Cyber Security Problems Statement

It should be noted that at least until 2020 the scientific and technical studies in the cyber security domain will be a part of priority research programs, R&D of relevant state and commercial enterprises and organizations, as well as the development programs of the leading universities and academic research institutes. The Russian Advanced Research Foundation program will also include them. It will require a more sophisticated methodology for scientific research, which allows properly protecting the national Digital Economy information infrastructure from catastrophic consequences. Such problem statement requires a substantial revision of the well-known information security concept of the critical Russian Federation objects. The fact is that modern information objects, which are complex distributed heterogeneous computing systems, no longer possess the required stability for targeted operation in the current or supposed conflict in cyberspace. This is due to the high complexity of these construction objects, blurring the concept of "protection perimeter", the potential danger of undeclared equipment operation and system-wide software, including hidden hardware and software intruder hypervisors. The detection and complex neutralization means of information and technical influences connecting the possibilities of joint combined technologies use to obtain unauthorized access, hardware-program bookmarks, and malicious software, in particular, with targeted attacks, are not still sufficiently effective.

Moreover, neither traditional information security means at the levels: Level 4 – ERP; Level 3 – MES; Level 2 – SCADA; Level 1 – PLC/RPA; Level 0 – field devices including traditional means of protection against unauthorized access, firewalling, traffic filtering (Modbus, OPC, IEC 104), cyber-attacks detection and prevention (IDS/IPS), antivirus protection, cryptographic information security, security analysis, integrity control, and cyber security management in general (based on SCIRT/CERT/SOC), nor the known means of ensuring the technical systems resilience, using the capabilities of redundancy, standardization, and reconfiguration no longer ensures the required systems and networks operability of the Internet, Internet of Things (IoT) and Industrial Internet of Things (IIoT) under growing cyber security threats.

The known approaches analysis to ensure the information infrastructure cyber security of the Russian Digital Economy under mass and group cyber-attacks proves the prospects of using theories, models, and methods:

- Multi-level hierarchical systems to design the mentioned intellectual cyber security subsystem;
- Cognitive systems for predictive analytics when working with Big Data (ETL + Big Data + NoSQL) and proactive cyber security systems creating;
- Agent-oriented (and multi-agent) systems to coordinate the control systems using monitoring and self-restoration systems;
- Artificial intelligence, including neural networks to solve identification and management problems; expert systems for learning, training, and early detection and localization of cyber security emergencies and incidents;
- Formal languages and grammars to generate and recognize possible types of mass perturbation structures;
- Catastrophes to analyze the behavior dynamics of the disturbed system functioning processes by analogy with the perturbation simulation in living nature;
- Control, recovery, and self-recovery to form the "immunity" to destructive disturbances, etc.

As a result, there should be an opportunity to create a systemic image of a prospective intellectual subsystem providing the information infrastructure cyber security of the Russian Digital Economy in the theory model types that allow synthesizing the desired self-recovery structure, as well as designing and executing a relevant self-recovery plan under growing cyber security threats.

4.2 Development of the Cyber Security Requirements in Terms of GOST R IEC 61508

At present, the functional safety standard of GOST R IEC 61508-2012 is widely used when building Russian software and hardware systems for cyber security. This standard differs in its requirements from the traditional GOST 34 series standards and is identical to the international standard IEC 61508 "Functional safety of electrical/electronic/programmable electronic safety-related systems". Let us consider the standard application specifics in practice.

4.2.1 Analysis of the Cyber Security Requirements

Nowadays, the development of Russian hardware and software security systems for critical facilities (CF), including the fuel and energy complex (FEC), is based on the following regulatory documents:

1. Federal Act of the Russian Federation of August 21, 2011 No. 256-FZ "On the safety of the fuel and energy complex objects";
2. Order of Russian FSTEC of March 14, 2014 No. 31 "On the approval of the requirements for securing information in the CPCS";
3. Russian FSTEC documents are "Information security measures in state information systems" (2014), "Recommendations for ensuring information security in key information infrastructure systems" (2014), "Basic model of information security threats in key information infrastructure systems" (2007), and "Methodology for determining current information security threats in key information infrastructure systems" (2007);
4. GOST R series "Communication industrial networks. The Network and the System Security (cyber security)" includes GOST R 56205-2014 (IEC/TS 62443-1-1: 2009) "Part 1-1. Terminology, conceptual points and models", GOST R 62443-2-1-2015 (IEC 62443-2-1: 2010) "Part 2-1. Scheduling the security (cyber security) ensuring of the control and industrial automation system", GOST R 56498-2015 (IEC/PAS 62443-3: 2008) "Part 3. Security (cyber security) of the industrial measurement and control process";
5. GOST 34.XXX "Complex of standards for automated systems";
6. GOST R IEC 61508 "Functional safety of electrical/electronic/ programmable electronic safety-related systems", etc.

At the same time, the original IEC 61508 standard is relatively young: its first English version appeared in 1998 (Table 4.1) and has been actively developing (Figures 4.4 and 4.5). Today, more than 200 substantive functional safety standards for 38 directions of ISO (International Organization for Standardization, ISO) and IEC (International Electrotechnical Commission, IEC) standardization are known. In the European Union (EU), CEN and CENELEC have more than 30 standards (the so-called European standards, EH)

Table 4.1   GOST R IEC 61508 development prerequisites

Document No.	Name	IEC 61508:1998	IEC 61508:2012
ISO 12207	Systems and software engineering; software life cycle processes	ν	
ISO 16085	Systems and software engineering; life cycle processes; risk management	ν	ν
ISO 25030	Software engineering; software product quality requirements and evaluation (*SQuaRE*); quality requirements	ν	
ISO 25051	Software engineering; systems and software quality requirements and Evaluation (*SQuaRE*); requirements for quality of ready to use software product (*RUSP*) and instructions for testing		ν
ISO 15504-6	Information technology; process assessment. Part 6: An exemplar system life cycle process assessment model	ν	
ISO 14762	Information technology; functional safety requirements for home and building electronic systems		ν

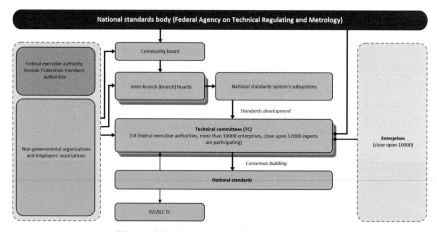

Figure 4.4   Standard development scheme.

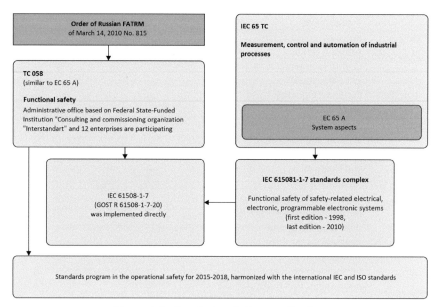

Figure 4.5   Standard evolution coordination.

for functional safety in 11 subject application areas. In 2010, the Russian Federation created a special Technical Committee No. 058 "Functional Safety". It conducted the standard IEC 61 508 translation into Russian and put into operation. In 2012, the second (up-to-date) version of the GOST R standard IEC 61508 was released. GOST R IEC 61508-2012 consists of seven parts:

- Part 1. General requirements;
- Part 2. System requirements;
- Part 3. Software requirements;
- Part 4. Requirements and definitions;
- Part 5. Risk-reduction measures and assessment methods;
- Part 6. GOST R IEC 61508-2 and GOST R IEC 61508-3 Implementation Guidance;
- Part 7. Methods and tools.

GOST R IEC 61508-2012 (Figure 4.6) contains:

- Basic concepts and definitions of functional safety;
- Criteria and indicators of functional safety;
- Requirements for ensuring functional safety;
- Functional safety concept;
- Life cycle stages of the security systems;
- Specification of functional safety requirements;

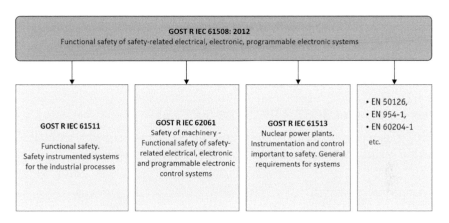

Figure 4.6   GOST R IEC 61508 applications.

- Safety integrity levels, SIL;
- General methodology for risk assessment;
- Methods and tools of ensuring the functional safety;
- Feasible problem descriptions of security systems analysis and synthesis, etc.

According to GOST R IEC 61508-2012 [82], functional safety means the ability of the hardware–software system associated with security, to perform all the security functions provided in the system with the residual risk retention of dangerous events occurrence at an acceptable level. At the same time, functional safety is part of the overall security that relates to managed hardware and control systems and depends on a system or hardware that operates correctly in response to input impacts. Functional safety is ensured at all stages of these systems life cycle.

In the fourth part of GOST R IEC 61508-2012, safety is defined as "freedom from unacceptable risks". At the same time, risk is understood as a combination of the damage occurrence probability and weight. Danger is a potential damage source. A risk is considered acceptable (Figure 4.7), if it is accepted in these circumstances.

GOST R IEC 61508-2012 sets four safety integrity levels from the minimum SIL1 to the maximum SIL4. For each of these security levels, the corresponding PFD (*Probability of failure on demand*) values are defined (Table 4.2).

Other indicators of GOST R IEC 61508-2012 are:

- RRF (risk reduction factor) is the ratio of the incidents frequency without taking security measures and the accepted incidents frequency;

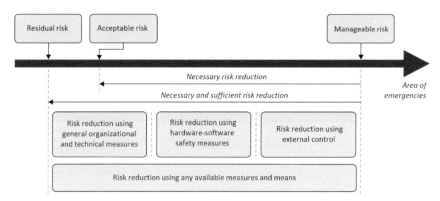

Figure 4.7   The general idea of risk management in IEC 61508.

Table 4.2 Correspondence of the safety standards concepts

ISO EN 13849-1	IEC 62061	IEC 61805 GOST PM∋K 61508	IEC 62061
PL (performance level)	PFHd (probability of failure per hour, dangerous)	SIL (safety integrity level)	SIL CL (safety integrity level, claim level)
a	$10^{-5}-10^{-4}$	–	–
b	$3\times10^{-6}-10^{-5}$	1	1
c	$10^{-6}-3\times10^{-6}$	2	2
d	$10^{-7}-10^{-6}$	3	3
e	$10^{-7}-10^{-8}$ and more	4	4

- MTTF (mean time to failure) – this indicator can be interpreted as the device service life, if it is not subject to recovery or repair);
- MTTR (mean time to repair) – the maintainable devices indicator;
- MTBF (mean time between failures) – usually expressed in years;
- MTBF = MTTF + MTTR,
- MTTF = MTBF – MTTR.

Inverse value MTBF is the component or device failure rate: SFF (safety failure fraction). According to IEC 61508-2012, security systems components are divided into types A and B. Type A components are simple devices whose behavior and failure modes are well known. Type B components are complex components (microprocessors, processors, etc.) with potentially unknown failures types.

In the security systems development, along with GOST R IEC 61508-2012 requirements, ISO, IEC, or EH requirements apply (Figures 4.8 and 4.9), for example, requirements of IEC 61496 (1, 2), IEC 62061, GOST ISO 13849-1, etc. In this case, according to GOST EN 414 (defining five possible types of safety standards: A, B, B1, B2, and C), GOST R IEC 61508-2012 is of type B2 (applies to secured devices).

4.2.2 Need for GOST R IEC 61508 Development

In practice, GOST R MEK 61508-2012 requirements are necessary, but are already insufficient to ensure the stability of CPCS, a more fundamental property that contains the sub-property "functional safety".

Stability can be intuitively defined as some constancy, invariance of a certain structure (static stability), and system behavior (dynamic stability).

ISO 13849-1, EN 954-1, GOST ISO 13849-1 categories   ISO 13849-1 Performance Level

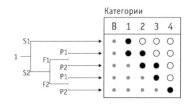
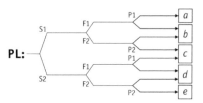

1 - initial point of the risk assessment for the security-related management system element;
S injury severity:
S1 - slight injury (usually reversible);
S2 - serious injury (usually irreversible), including death;
F frequency and/or duration of exposure:
F1 - from rare to very frequent and (or) short duration;
F2 - from frequent to continuous and (or) long duration;
P possibility of danger avoidance:
P1 - possible under certain conditions;
P2 - almost impossible

Figure 4.8   Correlation of basic safety standards concepts.

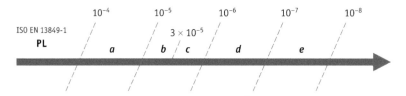

Figure 4.9   Probability of a dangerous malfunction for an hour.

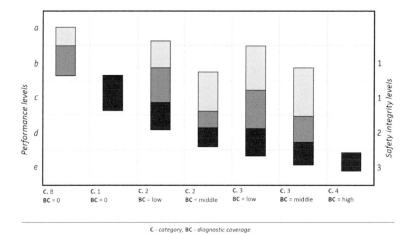

Figure 4.10   Correlation of safety standards concepts.

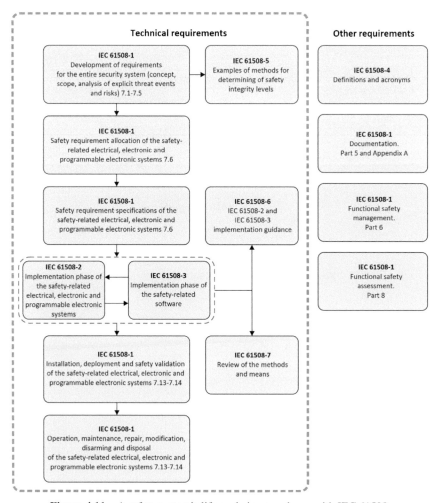

Figure 4.11 A safety system's life cycle in accordance with IEC 61508.

With reference to technical systems, the definition of stability was given by the outstanding Russian mathematician A. M. Lyapunov: "Stability is the system ability to function in states close to equilibrium, under conditions of constant external and internal disturbing influences".

In the context of information confrontation, the concept definition is concretized as follows: "The stability of the CPCS operation of a critical object is the ability of a system operating according to a certain algorithm to achieve the operation goal in conditions of attacker's information

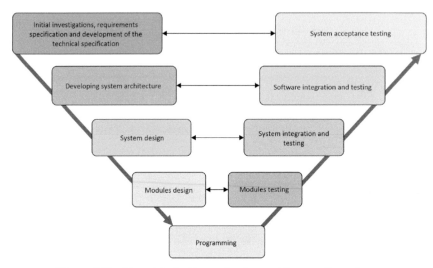

Figure 4.12   V-model of software development for security systems.

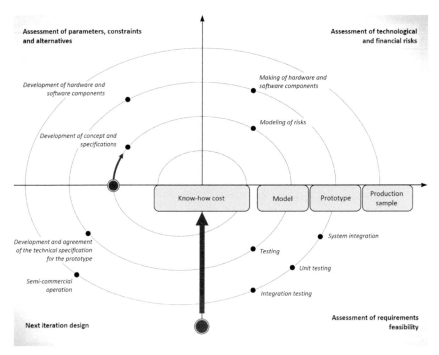

Figure 4.13   Spiral security software development model.

and technical influences" [17, 23, 40]. Indeed, according to Fleischman, B. S., it is necessary to distinguish the active and passive forms of stability. An active form of stability (reliability, fault tolerance, survivability, etc.) is inherent in complex systems whose behavior is based on the decision act. Here, the decision act is defined as the alternative choice, the system trend to achieve a state preferable to it. Passive form (strength, balance, homeostasis) is inherent in simple systems that are not decision-makers. In addition, unlike the classical equilibrium approach, the central element here is the notion of structural-functional stability. The fact is that the regular operating CPCS mode is, as a rule, far from equilibrium. At the same time, external and internal information and technical influences constantly change the equilibrium state itself.

Accordingly, a proximity measure that allows one to decide whether the behavior of a system undergoes a significant change under the influence of a perturbation is here a set of functions performed.

After Glushkov, V. M works, the stability theory development was devoted to Lipaev, V. V, Polovko, A. V, Ryzhikov, Yu. I., Dodonov, A. G, and a number of other Russian scientist-researches. However, the sustainability theory in these works developed only from the point of view of the technical system structure vulnerability without explicit consideration of its behavior vulnerability in conditions of a priori uncertainty of the attacker's information and technical impacts. As a result, the system in most cases is an example of a predetermined changing and the saving of relations and associations. This saving is achieved to save the system integrity for a certain time period in the normal operating conditions. This predetermination is of a dual nature: on the one hand, the system responds better to the negative impact, and on the other hand, the system is incapable of resisting other, a priori, unknown attacker information and technical influences, changing its structure and behavior.

4.2.3 Method for Anomaly Detection in the CF CPCS Behavior

The main co-problems of ensuring the CPCS stability of critical Russian facilities in the information confrontation conditions include:

- Insufficient stability of technical systems operation;
- Increase in complexity of the hardware and software systems component structure and behavior;
- Difficulty in detecting the quantitative regularities that allow one to explore the technical systems stability in the information confrontation conditions.

Let us comment these co-problems.

1. The CF CPCS stability is often lower than required. In many cases, the hardware and software components of these systems are not able to fully perform their functions for a variety of reasons, such as the following:

   - Inconsistency of the real behavior system parameters in software and hardware specifications;
   - Reassessment of the current development level of programming and computer technologies;
   - Destructive information and technical impact of external and internal factors on the systems, especially in the context of the attacker impact;
   - Reassessment of the feasibilities of modern information security methods and tools in CPCS, CS fault tolerance, and software reliability.

   Ignorance or avoidance of the above causes leads to the CS CPCS efficiency decrease. Moreover, this problem is greatly escalated in the context of information confrontation.

2. Modern CPCS, generally, are heterogeneous distributed systems, consisting of many components of different architectures. According to the authors, in the modern CS CPCS structure at Level 4 – ERP, Level 3 – MES, Level 2 – SCADA, Level 1 – PLC/RPA, Level 0 – field devices include more than:

   - 20 types of BI based on Big Data and streaming data processing;
   - 12 types of ERP;
   - 40 types of SCADA systems;
   - 20 operating systems;
   - 80 translators and interpreters;
   - 800 network protocols;
   - 20 types of network hardware;
   - 35 types of controllers;
   - 40 types of field devices;
   - 21 types of information security tools (SOC, CERT/CSIRT/SIEM, IST from unauthorized access, CIPF (VPN, PKI), antivirus software, IDS for SCADA systems based on Modbus networks/TCP, tools for verification and filtering network packets by their contents (Modbus DPI firewall, etc.), security policy monitoring tools, specialized software for penetration testing (for example, VPN Hunter,

Exploit Search, NmapOnline, Metasploit, GLEG SCADA + Pack, OpenVAS), etc.).

The following features are relevant to the CS CPCS operation:

- Slightest system downtime leads to the shutdown of a rigorous technological process and significant costs for disaster recovery;
- Use of proprietary technology protocols of hardware manufacturers, which could have hard-to-detect vulnerabilities;
- Probability of catastrophic consequences as a result of system failure: deaths, environmental disasters, etc.;
- Prohibition of false alarms, resulting in interruptions in ensuring the regular operation of technological processes;
- Use of buffer, demilitarized zones for the organization of MES, ERP, BI, and other systems interaction with the enterprise system;
- Need to provide remote access and management of CPCS by contractors, etc.

The above-mentioned CS CPCS operation features lead to the threat range expansion that can be followed by the attacker as a result of information and technical impact on the named systems and determine their high vulnerability.

3. The third co-problem is the difficulty in identifying quantitative regularities that allow one to research the CF CPCS stability in the context of information confrontation. The fact is that the systems operation processes are significantly influenced by the factors of the external and internal environment. These factors considered within the system are either fundamentally impossible to manage, or management takes place with an unacceptable delay. In addition, the external and internal environments have the property of incomplete certainty of their possible states in future periods. In other words, the factors influencing the system behavior undergo such changes in time that can fundamentally change the operation algorithms or even make the set goals unattainable. The changes that external and internal factors are undergoing occur both naturally and accidentally, so in the general case, they cannot be predicted accurately, as a result of which there is some uncertainty in their values. On the other hand, the CPCS having a specific goal possess a certain "safety margin" – the features that allow achieving the set goals with certain factors deviations affecting the process of the external and internal environment.

Until recently, two main approaches have been used to identify these operation regularities of technical systems: experimental (e.g., mathematical
statistics methods and experimental design methods) and analytical (e.g.,
methods of analytical software algorithms verification). In contrast to the
experimental methods that make it possible to study the single CF CPCS
behavior, the methods of analytical verification allow us to consider the
most general system behavior properties, which are specific for the operation
processes class as a whole. However, these approaches have significant disadvantages. The disadvantage of experimental methods is the inability to extend
the results obtained in this experiment to a different system behavior, which
is different from the one studied. The lack of analytical verification's methods
lies in the difficulty of moving from the class of system operation processes,
characterized by the derivation of universally significant properties, to a single process that is additionally characterized by the corresponding operation
conditions (in particular, specific parameter values of the system behavior in
the information confrontation conditions).

Consequently, each and every approach is not sufficient for an effective
research of the CF CPCS operation stability in the information confrontation
conditions. It seems to us that only by using the strengths of both approaches,
combining them into a whole, one can obtain the mathematical tools that are
necessary to identify the required quantitative regularities.

The practice of creating Russian CF CPCS security systems demonstrates
the following: the information confrontation conditions give technical systems features that exclude the feasibility of modeling the CF CPCS behavior
by traditional methods; the resulting complexity factors and the resulting
difficulties are listed in Table 4.3.

In the above table, factors 1, 4, and 7 are decisive. They exclude the feasibility to take into account only the generally valid CF CPCS properties under
the information confrontation conditions. However, the traditional ways of
ensuring functional safety and stability in general are based on the following
approaches:

- Simplification of the technical system behavior till the resulting of
  significant algorithmic properties;
- Generalization of empirically established particular regularities of
  system behavior.

The use of these approaches leads not only to a significant error in the results,
but also has fundamental disadvantages. The disadvantage of analytical modeling of the system behavior in the information confrontation conditions is

Table 4.3 CF CPCS complexity factors and the resulting difficulties

No.	CF CPCS Complexity Factor	Resulting Difficulties
1.	Complex structure and system behavior	The cumbersomeness and multidimensionality of the solved tasks
2.	Stochastic system behavior	Uncertainty of the system behavior description, complexity in the task formulation
3.	System activity	The determination complexity of the limiting laws of the potential system efficiency
4.	The mutual influence of data structures on each other	Cannot be taken into account by models of the known types
5.	Impact of failures and malfunctions on the system	Uncertainty of the system behavior parameters, complexity in the task formulation
6.	Deviations from the standard operating conditions of the system	Cannot be taken into account by models of the known types
7.	Information and technical impact on the system by attacker	Uncertainty of the system behavior parameters, complexity in the task formulation

the difficulty of transition from the system behavior class, characterized by the conclusion of general algorithmic properties, to single behavior, which is additionally characterized by the operation conditions in information confrontation conditions. The lack of empirical modeling of the behavior system is the inability to extend the results to another system behavior, which differs from the studied one by the operation parameters.

Therefore, in practice, the traditional approaches to ensure the functional safety and stability of the system as a whole can only be used to develop systems for approximate prediction of the CF CPCS operation stability in frames of information confrontation.

We propose an approach based on similarity methods, which is devoid of the noted shortfalls and allows us to realize the so-called CF CPCS decomposition principle in information confrontation conditions, based on the structural and functional characteristics, to resolve these contradictions. In the similarity theory, it is proved that the set of relations between the essential parameters for the considered system behavior is not a natural property of the problems under study. In reality, the influence of certain factors of the technical system's external and internal environment, represented by different quantities, is shown not separately but combined. Therefore, it is proposed to consider not individual quantities, but their aggregates (similarity invariants) that have a definite meaning for the CF CPCS operation.

***

Thus, the application of similarity methods makes it possible to formulate the necessary and sufficient conditions for the isomorphism of two models of the allowed CF CPCS behavior in the information confrontation conditions described by the formal systems of homogeneous power polynomials (posets).

As a consequence, it becomes possible to:

- Perform analytical verification of the system behavior and verify the isomorphism conditions;
- Determine numerically the coefficients of some system behavior model representation in order to achieve the isomorphism conditions.

This, in turn, allows us to:

- Control the semantic correctness of the technical system behavior under the impact conditions by comparing the observed similarity invariants with the invariants of the standard, isomorphic behavior representation;
- Detect, including in real time, the technical system behavior anomalies, which resulted from the information and technical impact of the attacker;
- Restore behavior parameters that significantly affect the stability of the technical system.

A method was developed for detecting the CF CPCS behavior anomalies based on the similarity theory application of π-equation analysis and π-dimension analysis of the feasible stability metrics, and measures and functional security for the mentioned technical systems were substantiated and proposed. This allowed developing engineering methods for modeling, observing, measuring, and comparing the behavior system stability based on similarity invariants, including a new technique for simulating the semantically correct CF CPCS behavior standards, consisting of four stages.

The first stage is the π-analysis of the technical system behavior regularities. The main goal of this stage is to identify the semantically correct system behavior based on similarity invariants. The stage procedure includes the following steps:

1) Identification of structural and functional standards;
2) Detection of temporary standards;

3) Development of control correlations necessary to determine the semantic correctness of the system behavior.

The second stage is the algorithmization of obtaining the standards of semantic correctness system behavior. Its main goal is to obtain, in matrix and graphical form, probabilistic standard algorithms or similarity invariants of the system behavior. The stage procedure consists of the following steps:

1) Standard algorithm development in the tree form;
2) Algorithm implementations enumeration;
3) Weighing the algorithm implementations (probabilistic algorithm development);
4) Rationing the algorithm tree.

The third stage is the synthesis of the standards of the semantic correctness system behavior, adequate to the application's goals and objectives. Its main goal is the algorithmic structures synthesis formed by a set of sequentially executed standard algorithms. This procedure is carried out in the following steps:

1) Synthesis of structural-functional standards;
2) Synthesis of time standards;
3) Symmetrization and ranking of matrices describing standards.

The fourth stage is the simulation of stochastically determined algorithmic structures of the standards of semantic correctness system behavior. The procedure of the stage includes the following steps:

1) Analysis of empirical standards of semantic correctness;
2) Determination of the type of empirical functional dependence;
3) Development of control ratios sufficient to determine the semantic correctness system behavior.

As a result, the applicability of similarity methods for the decomposition of the behavioral CF CPCS algorithms by functional features and the formation's necessary invariants of the semantically correct operation for technical systems was demonstrated. The presence of the self-similarity property of the similarity invariants made possible forming the static and dynamic standards of the semantically correct behavior of these systems and using them for engineering solution of information and technical influences monitoring, detection, and neutralization in the real CF CPCS operating conditions.

4.3 Creation of New Cyber Security Ontologies

At present, Smart Grid technology is widely used in a number of developed countries in the world. It is designed to deliver electricity to the consumer using modern digital technologies, which ensure energy saving, reduce costs, and improve lines reliability and transparency of the management process. To implement programs for electrical lines modification into Smart Grid and to develop relevant standard solutions, the world largest companies have created the Smart Energy Alliance. These standard solutions became as close to known telecommunications solutions by their structure and functionality as possible.

4.3.1 Analysis of New Requirements of Cyber Security

The objective importance is explained by the need to create a smart grid to ensure the stability of "smart" power systems under information confrontation [1, 2, 5–7, 13, 15–26, 39, 57, 61, 64, 67, 73, 74, 82–121].

Today, the most significant projects of power systems development based on the Smart Grid are implemented in the USA and EU countries, in Russia, Canada, Australia, China, and Korea. For example, Miami (USA) implements a large smart energy line project called the Energy Smart Miami, where such well-known companies as General Electric, Cisco Systems, and Silver Spring Networks take part along with the local Florida Power & Light power company. In Denmark, there is a large-scale project EDISON, uniting IBM, Siemens, and DONG Energy companies. For Europe as a whole, the Strategic Energy Technologies Plan (SET_PLAN) that introduces the European electric lines transformation into smart grid within the next 10 years was developed [13, 57, 58].

In 2007, the "Maturity Model" was developed under the IBM leadership to assess the electrical lines readiness to transform to smart grid. It was put into practice by programmers from Carnegie Mellon University, SEI (Software Engineering Institute). In addition, SEI researchers develop a similar model (Capability Maturity Model Integration – CMMI), based on the recommendations and requirements of so-called best practices, including recommendations from the National Institute of Standards USA, NIST-EISA (Title XIII), and relevant international standards. In addition, the US Energy Department together with the SEI develops the global Smart Grid management system model (SGMM).

In Russia, a large-scale project is implemented to develop a Smart Grid system with active-adaptive network (SGS AAN). The expert working groups led by the Architectural Committee of the scientific and technical council of FGC UES and the Russian Academy of Sciences (RAS) developed the main provisions and approaches to вуышпт, a standard architecture of the mentioned smart grid system. As part of the project implementation in the East United Power Systems, the SGS AAN range was created for the period until 2014 over 2020. The range is a complex of software and hardware means that form the environment to support the SGS AAN solutions development. The SGS AAN range is located on the territory of FGC UES research and development center and includes, in particular, the software modeling Power Factory complex and the power grid model of the East united power systems. In the energy "Elga Coal" cluster, a pilot implementation of a multi-agent automated voltage and reactive power management system based on the Power Agents Platform was performed [13, 57, 113].

It is significant that in these projects the key point is to introduce and develop into future Smart Grid systems the following two new capabilities [13, 57, 113]:

- Resistance to negative impacts: the availability of special methods to ensure stability and survivability, to reduce the physical and information vulnerability of all energy system components and contribute to prevention as well as to the rapid recovery from accidents in accordance with energy security requirements;
- Self-recovery in emergency situations: the power system and its elements must be able to maintain their technical condition continuously in the efficient state by identifying, analyzing, and switching from event management after the incident occurs to a preventive (warning) event management.

The self-recovering power system should allow maximum possible minimization of disruptions (disturbances) with the smart control system help, including its most important component that is a cyber security insurance subsystem.

In other words, Smart Grid's intelligent grid system should be proactive in relation to changing functioning conditions and monitoring the pending technical problems before they can affect its safety and stability in general (Figures 4.14 and 4.15). Therefore, the developing smart cyber security subsystems should include the relevant components of containment, prevention, detection, neutralization, and self-recovery.

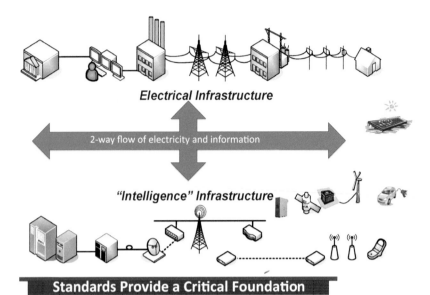

Figure 4.14   Smart Grid cyber security challenges.

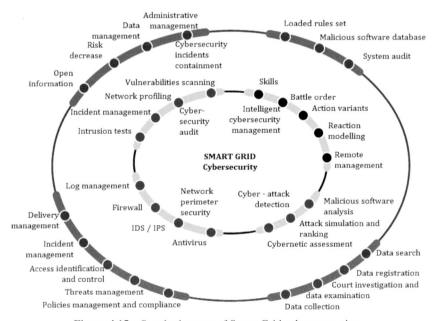

Figure 4.15   Standard means of Smart Grid cyber protection.

The known methods analysis to develop perspective Smart Grid systems demonstrates the relevance of using the following methods and technologies to ensure the required stability:

- Multiagent systems to coordinate the control systems by the transient regimes monitoring system and FACTS devices, self-recovery of district power plants;
- Artificial intelligence, including neural networks to solve identification and management problems, as well as expert systems to learn and conduct training and realize early detection and localization of emergency pre-crash regimes;
- Adaptive vector control of flexible alternative current systems for primary and secondary automatic voltage and reactive power management, power mode optimization;
- Adaptive automatic control for renewable energy sources, including wind, tidal, solar, and, in the future, space solar power plants;
- Intelligent cyber security (Tables 4.4 and 4.5), capable of providing the required stability of the future Smart Grid systems under information confrontation, etc.

4.3.2 Known Cyber Security Ontologies

Previously, the ontological modeling problems were considered by foreign scientists T. Gruber, N. Guarino, and others, and in Russia by G. S. Pospelov, D. A. Pospelov, E. V. Popov, V. F. Khoroshevsky, T. A. Gavrilova, Yu. A. Zagorulko, A. S. Narinyani, A. S. Kleshchev, I. L. Artemyeva, I. V. Kotenko, A. G. Lomako, D. N. Biryukov, L. S. Massel, T. N Vorozhtsova, and many others. Nowadays, knowledge representation models are introduced in the form of frame systems, semantic networks, and production systems. Frame systems and semantic networks allow describing the domain object structure and the relationship between them. Product systems (rules) are used to represent domain knowledge in the form of "if–that" statements. Based on the above-mentioned models, various knowledge representation languages were developed, which are the input languages for some universal envelopes and expert systems [20, 74, 87, 88, 109, 113].

The main methodological principles to determine the domain ontology are formulated in [20, 88].

Table 4.4   Cyber security risk assessment example

Levels of Risk	Power System Type/ Capacity	Characteristics of the Incident		Consequences of the Incident		Losses		
		Impact on the population	Impact on infrastructure	Impact on information resources	Legal effects	Human	Reputational	Financial
In the highest degree critical	International systems/ more than 10 GW/h	Suffered more than 50% of the country's population or more than 25% of the several countries' population	Suffered international critical infrastructure	Undefined	Closing the company or pledge	Presence of direct and casual victims as a result of incident	Irreversible trust loss to the company worldwide	Less than 50% from EBITDA
Critical	National systems / 1–10 GW/h	Suffered from 25 to 50% of the country's population	The national critical infrastructure suffered	Undefined	Temporary suspension of activities	Presence of casual victims as a result of the incident	Irreversible trust loss to the company inside the country	Less than 50% from EBITDA

High	Urban systems/ 0.1–1 GW/h	Suffered from 10 to 25% of the population	Suffered significant infrastructure	Unauthorized disclosure or modification of sensitive data	Fine to 10% from EBITDA *	Presence of direct victims as a result of incident	Temporary trust loss to the company within countries	Less than 33% from EBITDA
Average	Local systems/ 1–100 MWh	Suffered from 2 to 10% of the population	Other infrastructure	Unauthorized disclosure or modification of personal data	Penalty up to 10% of EBITDA	Presence of victims as a result of the incident	Temporary trust loss to the company inside region	Less than 10% from EBITDA
Low	Home systems/ less than 1 MW/h	Suffered less than 2% of the population	Infrastructure does not suffer	Personal and other sensitive data is not involved in the incident	Warning	Insignificant accidents	Short-term and insignificant loss of trust to the company	Less than 1% from EBITDA

Table 4.5 Known risk assessment techniques

Method Name	Type of Evaluation	The Percentage of Mentions in the Literature, from 100%	Country of Development
OCTAVE		22%	USA
CRAMM		16%	UK
CORAS	Qualitative	12%	Greece, Germany, Norway
FRAP		10%	Canada
COBRA		5%	UK
NIST		16%	USA
ISRAM		7%	Turkey
CORA		5%	USA
RiskWatch	Quantitative	5%	USA
IS		2%	South Korea

1. At the substantive level, ontology is understood to be an agreement set (domain terms definitions, their interpretation, statements that limit the possible meaning, as well as the interpretation of these statements). Unlike empirical knowledge, these agreements cannot be disproved by empirical observations.
2. Ontology, conceptualization, knowledge, and reality must be modeled by a single mathematical construction.
3. Between the domain features and the mathematical construction elements, a clear correspondence must be established.
4. Each domain ontology model should contain both formal elements and their meaningful interpretation in terms understandable to domain specialists.
5. The ontology and its model should be manageable even for complex domains that have a large number of concepts.

In ref [116][1], an ontology and possible multi-agent smart control cyber security mechanisms are considered in computer systems and networks that allow:

1. Data collection on the information system state and its analysis through mechanisms of processing and merging information from various sources;
2. Proactive cyber-attacks and their implementation prevention;

---

[1] http://comsec.spb.ru/ru/staff/kotenko/.

3. Detection of abnormal activity and evident cyber-attacks, as well as illegitimate actions and deviations of users work from the security policy, prediction of intentions, and possible malicious actions;
4. Active response to attempts to implement the malicious actions by automatic protection components reconfiguration to reply to the malicious actions in real time;
5. Attacker misinformation, concealment, and camouflage of important resources and processes, attacker "luring" into false (fraudulent) components to identify and clarify his purposes, reflexive control over the attacker behavior;
6. Monitoring the network functioning and controlling the correctness of the current security policy and network configuration;
7. Support for decision-making on the security policies management, including adaptation to subsequent intrusions and critical defense mechanisms strengthening.

The ontology and systemic image of intellectual cyber security systems with anticipation feature, in particular, a new systems class to prevent the cyber-attacks, which are self-learning intellectual systems of self-organizing gyromates, are explained in ref [122]. It is shown that the application of the proposed intellectual systems in practice makes it possible to more successfully solve the problems associated with the risks prevention of the cyberthreats implementation.

In ref [122][2], the authors studied the Smart Grid cyber security ontology, which was developed as a part of the grants at the Institute of Power Systems named after L. A. Melentieva of the Siberian Branch of the Russian Academy of Sciences (ISEM SB RAS):

- RFBR No. 13-07000140 "Development and integration methodology of intelligent, agent and cloud computing in Smart Grid (smart energy systems)";
- RAS Presidium Program "Methods and tools to support decision-making in research and ensuring energy security based on intellectual calculations".

To develop this cyber security ontology, the thesaurus of the following normative documents was used:

- GOST R 53114-2008 "Organization information security";

---

[2]http://proceedings.spiiras.nw.ru/ojs/index.php/sp/article/view/3096.

- GOST R 50922-2006 "Information security. Basic terms and definitions";
- ISO/IEC 27032: 2012 standard "Cybersecurity Guidelines";
- ISO/IEC 27000 standard "Information technology. Security methods. Information security management systems. General overview and terminology".

According to the developer [87], the proposed cyber security ontology reflects only one of the initial stages of addressing the cyber security problem, since it does not affect such areas as information security, security of applications, networks, information systems, etc. Therefore, it was decided to continue the research to further specify the concepts considered in their connection to the features of specific information systems and energy object management systems.

The paper [74] is based on a more general Smart Grid ontology, Gridpedia.[3] To existing Gridpedia classes and features were added classes and properties of the T.N. Vorozhtsov cyber security ontology [87]. At the same time, for better clarity, the main classes and features of the above-mentioned initial cyber security ontology were regrouped in accordance with the requirements to develop advanced Smart Grid systems.

However, under an information confrontation, a more sophisticated Smart Grid cyber security ontology is required, which would prevent the power systems reduction to catastrophic consequences.

This problem formulation is required to significantly reconsider the well-known concept of Smart Grid information security provision. The matter is that modern power systems that are complex distributed heterogeneous systems do not possess the required stability for the intended operation under the current and possible information confrontation because of the high construction complexity and the potential danger of the undeclared equipment functioning and system-wide software, including attacker hypervisors. The identification and full neutralization means of information and technical impacts, combining the capabilities of simultaneous integrated technologies use of unauthorized access obtainment, hardware-program bookmarks, and malicious software, are still insufficiently effective [56, 18].

Moreover, neither the traditional information security means at Level 4 – ERP; Level 3 – MES; Level 2 – SCADA; Level 1 – PLC/PZA; Level 0

---

[3]http://gridpedia.org/wiki/Main_Page/.

that are field devices that include traditional means (protection from unauthorized access, firewalling, filtering, cyber-attacks detection and prevention, antivirus protection, cryptographic information security, security analysis, integrity monitoring, and cyber security management generally based on SCIRT/CERT/SOC) nor the known means of ensuring the power systems stability, using the capabilities of redundancy, standardization, and reconfiguration, are no longer suitable to provide the required future Smart Grid systems efficiency under information confrontation.

To develop a new cyber security ontology, the likely scenarios of a targeted information impact implementation on the future Smart Grids were analyzed, a typical power systems structure was considered, and their vulnerabilities were described. The security threats implementation specifics and the possible performance disruption risks of a typical power system and the information and technical implementation impacts specifics on critical elements of future power systems are revealed [7, 11, 13, 57, 88].

A critical analysis of existing methods and means to detect and neutralize information and technical impacts, including intended or targeted attacks, was carried out. The traditional information protection systems computability to warn, detect, and neutralize the information and technical impacts on power systems was assessed. The organization means of weaknesses of ensuring and monitoring the cyber security policy based on IEC 62351-8 [5, 21, 22, 39, 56] are shown.

The traditional means of imperfection of Smart Grid performance control and restoration are shown. The ways to ensure the functioning stability of the power systems under malicious mass information and technical impacts are studied. A critical analysis of approaches and methods to ensure the power system performance stability under their destabilization was performed. The Smart Grid performance maintenance ideology was developed based on immunity. The goal and objectives of the future power system stability under information confrontation were formalized [19, 20, 24–26, 40].

As a result, the cyber security ontology of self-recovering Smart Grid was proposed, which allows describing the future energy systems organization under information confrontation based on the immunity to disturbances by analogy with the immune protection system of a living organism.

The relevance of the new Smart Grid cyber security ontology is confirmed by the requirements (Figure 4.16) of the following normative documents:

- "Russian Information Security Doctrine", 2016;
- Federal Law "On the Critical Information Infrastructure Security";

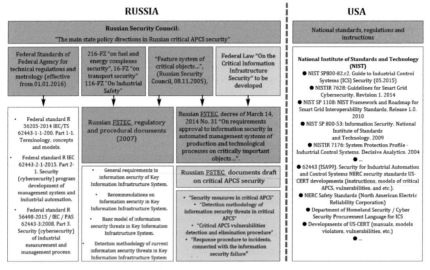

Figure 4.16   Regulatory requirements.

- "The main state policy directions in Russian critical APCS security";
- Russian FSTEC Decree of March 14, 2014 No. 31 "On requirements approval to information security in automated management systems of production and technological processes on critically important objects, potentially hazardous objects, and objects of increased danger to people life and health and the environment";
- Russian FSTEC documents on critical APCS security, 2016:
    - "Security measures in critical APCS";
    - "Detection methodology of information security threats in critical APCS";
    - "Critical APCS vulnerabilities detection and elimination procedure";
    - "Response procedure to incidents, connected with the information security failure";
- Federal Standards of Federal Agency for technical regulations and metrology "Industrial communication networks. Network and system security (cyber security) (effective from January 16, 2016)":
    - Federal standard R 56205-2014 IEC/TS 62443-1-1-200. Part 1-1. Terminology, concepts, and models.

○ Federal Law R IEC 62443-2-1-2015. Part 2-1. Security (cyber security) program development of management system and industrial automation.

○ Federal standard R 56498-2015/IEC/PAS 62443-3:2008. Part 3. Security (cyber security) of industrial measurement and management process.

4.3.3 Proposed Cyber Security Ontology

In this section, the cyber security ontology of self-recovering Smart Grid (hereinafter, the cyber security ontology) is understood as the basis for a reusable knowledge of a special kind, or the "conceptualization specification" of such hard-formalized domain as ensuring the functioning sustainability of future energy systems under information confrontation. This means that in this domain, based on the basic cyber security terms classification, first, it is necessary to separate the basic terms (concepts) and then to determine the connections between them (conceptualization). At the same time, the cyber security ontology can be represented both graphically and analytically (e.g., by a formal grammar and programming language or some mathematical model).

Two methodological approaches were used to develop the cyber security ontology. For the graphical cyber security ontology representation, the IDEF5 Schematic Language is used, and for the analytical description, the text language IDEF5 Elaboration Language is used (Tables 4.6 and 4.7).

In order to automate the cyber security ontology simulation, an SBONT tool demonstration prototype of Knowledge Based Systems, Inc. is used. The first methodological approach Implementation took 6 years (2000–2005). Currently, the cyber security ontology contains an 800-term description from the information security domain (two volumes of 1284 pages with text and graphic schemes were prepared) and is constantly updated.

For the current cyber security ontology version, the following terms and definitions of the following normative documents and best practice recommendations are used as input data:

1. Thesaurus of Russian Security Council normative documents "Russian Information Security Doctrine" (2016), "The main state policy directions in Russian critical APCS security" and "Feature system of critical objects".

Table 4.6 Correspondence between UML and IDEF5 graphical elements

IDEF5		UML	
Item name	Picture	Item name	Picture
Type (class)	Label type	Class	Class name
Individual	Label individual	Instance	*Instance name*
Double relations first-order *part-of*	⟶	Aggregate association	◇—
Double relations first order *subkind-of*	—○⟶ —○▶▶	The generalization relation, is-a	⟶▷
Transition between states		The transition between states	*Event*
Process	Process label	Action state	*Act*
Instant marker transition	Δ	Nontransitive transitions	⟶

Table 4.7 Correspondence between OWL tags and UML constructs

OWL Tag	Element Name in UML
<owl: Class rdf: ID = "...">	Class
<owl: Thing rdf: ID = "...">	Instance
<owl: ObjectProperty rdf: ID = "...">	Attitude
<owl: DatatypeProperty rdf: ID = "...">	Attribute

2. The thesaurus of the Federal Law No. 149-FZ of July 27, 2006 "On Information, Information Technologies and Information Security", Federal Law No. 16-FZ of February 9, 2007, "On Transport Security", Federal Law No. 256-FZ of July 21, 2011 "On Fuel and Energy Complex Security", Federal Law No. 116-FZ of July 21, 1997 "On Industrial Security of Hazardous Production Facilities", Federal Law No. 170-FZ of November 21, 1995 "On Atomic Energy Use", the Federal Law "On critical information infrastructure security".

3. Russian FSTEC Documents: Decree of March 14, 2014 No. 31 "On requirements approval to information security in automated management systems of production and technological processes on critically important objects, potentially hazardous objects, and objects of increased danger to people life and health and the environment"; Russian FSTEC documents in 2007: "Basic model of threats to information security in key information infrastructure systems", "Detection

methodology of current information security threats in key information infrastructure systems", "General requirements to information security in key information infrastructure systems", "Recommendations to information security in key information infrastructure systems", "Regulations on the registry of key information infrastructure systems"; draft documents 2016: Security measures in critical APCS"; "Detection methodology of information security threats in critical APCS"; "Critical APCS vulnerabilities detection and elimination procedure"; "Response procedure to incidents, connected with the information security failure".

4. Federal standard R 53114-2008: "information security in the organization" and Federal standard R 50922-2006 "Information security. Basic terms and definitions"; Federal Standards of Federal Agency for technical regulations and metrology on industrial communication networks. Network and system security (cyber security): Federal standard R 56205-2014 IEC/TS 62443-1-1-200. Part 1-1. Terminology, concepts and models. Federal Law R IEC 62443-2-1-2015. Part 2-1. Security (cyber security) program development of management system and industrial automation. Federal standard R 56498-2015/IEC/PAS 62443-3:2008. Part 3. Security (cyber security) of industrial measurement and management process. Federal Law R 56545-2015 "Information security. Information systems vulnerabilities". Vulnerability description rules (define the content of vulnerability information that security control providers should include in their solution database, while the document takes into account existing practices and vulnerability description tools such as Common Weakness Enumeration (CWE), the formal language Open Vulnerability and Assessment Language (OVAL), and the Common Vulnerability Scoring System (CVSS) vulnerability assessment methodology); Federal Law R 56546-2015 "Information security. Information systems vulnerabilities. Vulnerabilities classification" (defines the most common vulnerability types that allow unifying the terminology used by penetration testers).

5. Best practice: ISO/IEC 27000 standards in terms of general principles for the digital control systems security, including ISO/IEC 27032: 2012 "Cybersecurity Guidelines" and ISO/IEC 27000 "Information technology. Security methods. Information management systems security. General overview and terminology"; IEC TC57 standards: IEC 61850, IEC60870, IEC 62351 regarding the communication protocols security; INL Cyber Security Procurement Language 2008 standard.

6. Recommendations: NIST-800-82 r.2 "Guide to Industrial Control Systems (ICS Security) is a guide to the technological objects management system security" of May 2015, Control Systems Security Program/National Cyber Security Division (Measures list to ensure management system security. Recommendations for standard developers), IEC 62443 and ISA 62443 (documents of the International Electrotechnical Commission (IEC) and 99th Committee for the Security Standards development of the critical APCS of the International Automation Society (ISA), NERC CIP (Critical Infrastructure Protection) security standards is a set of requirements to security of the critically important objects of the North American Council for the Reliability of Electrical Networks (NERC), Department of Homeland Security: Cyber Security Procurement for ICS, US-CERT developments (manuals, models of threats and intruders, regulations on response to cyber incidents, vulnerabilities database, etc.).

The cyber security ontology development was carried out step by step [14, 15, 24, 57, 88]:

1) Cyber security ontology context definition;
2) Data collection is a terms sources identification and terms selection for the cyber security ontology;
3) Data analysis is a definition of the main terms and elements terms, relationships, and verbal terms description;
4) Cyber security ontology development is a development of a schematic and analytical ontology description;
5) Cyber security ontology validation is a validation of the ontology completeness and correctness, the compliance with the original requirements.

Cyber security ontology is represented by graphical schemes in IDEF5 Schematic Language (524 schemes) and corresponding analytical descriptions in IDEF5 Elaboration Language.

The above analytical cyber security ontology descriptions are performed in accordance with the previously developed methodology:

1) Entering the basic and auxiliary cyber security terms notation;
2) Terms-elements explanation by unrelated types;
3) Assigning a unique identifier to each term-element;
4) Input and output connections identification for each term;
5) Fixing elements connections;

6) Descriptions correctness verification;
7) If necessary, updating and clarifying the descriptions.

In the second methodological approach, the W3C consortium (The World Wide Web Consortium) methodical recommendations were used to represent the cyber security ontology in the semantic Web context (Web 3.0) [39]. The second approach implementation took 5 years (2006–2010). OWL is used to describe the hierarchy of possible Smart Grid cyber security ontologies with memory, OWL provides a detailed description of ontology classes, individuals belonging to these classes, and the existing relationships between them. This language extends the RDF language capabilities, providing an opportunity to operate with the basic "subject–predicate–object" structures, as well as the RDFS language that defines the basic structures and relationships between classes and individuals. At the same time, to provide the description capability of the complex connections between Smart Grid cyber security ontology individuals, the OWL DL variant is used (Table 4.7).

This allowed using the listed types to describe fixed vocabulary structures of the domain knowledge base, to define multiple connections to set "the many-to-many" relationships, and to apply logical (Boolean) classes combinations to define the complex structure connections of the Smart Grid cyber security ontology with memory. It was shown that the OWL allows specifying different cyber security ontology representations.

It was decided to use the OWL representation in the XML syntax as the most common and convenient for automatic processing and analysis of the cyber security ontologies texts by relevant software tools. In Inset 4.1, a cyber security ontology fragment example is described using this syntax.

Ontologies placing into each other at the language level was used to integrate separate cyber security ontology parts (the owl: imports design). This allowed describing the basic concepts, connections, and individuals related to the domain.

A rules description to build connections in the SWRL, which is integrated into ontologies formed in OWL, is used to dynamically expand and modify the knowledge base. The rules are used to describe the dynamic relationships between individuals of the ontologies that occur when certain conditions exist. Thus, such relationships can describe the method applicability to solve a problem of ensuring a required Smart Grid stability under information confrontation, depending on the input data characteristics. Using the dynamic relationships construction together with the ontology inclusion makes it possible to implement a partial logical inference already at the ontological structure interpretation level. To do so, the active facts set, formed in the interaction process with the user, is formalized as a separate ontology using the basic ontological structure inclusion. The obtained structure interpretation makes it possible to analyze the basic ontological structure taking into account the introduced facts.

Inset 4.1. The ontology fragment
An example of the representation of the ontology of cyber security <?xml version = "1.0"?> <!DOCTYPE rdf: RDF [<!ENTITY dl-safe "http://owldl.com/ontologies/dl-safe.owl#"> <!ENTITY swrl ≪http://www.w3.org/2003/11/swrl#≫> <!ENTITY owl ≪http://www.w3.org/2002/07/owl#≫> <!ENTITY xsd "http://www.w3.org/2001/XMLSchema#"> <!ENTITY rdfs "http://www.w3.org/2000/01/rdf-schema#"> <!ENTITY rdf "http://www.w3.org/1999/02/22-rdf-syntax-ns#"> <!ENTITY ruleml ≪http://www.w3.org/2003/11/ruleml#≫> <!ENTITY escience ≪http://escience.sec.ru/escience.owl#≫>]> <rdf: RDF xml: base = ≫http://escience.sec.ru/escience.owl#≫ xmlns = ≫http://escience.sec.ru/escience.owl#≫ xmlns: owl = ≫http://www.w3.org/2002/07/owl#≫ xmlns: rdf = ≫http://www.w3.org/1999/02/22-rdf-syntax-ns#≫ xmlns: rdfs = "http://www.w3.org/2000/01/rdf-schema#" xmlns: xsd = "http://www.w3.org/2001/XMLSchema#" xmlns: swrl = "http://www.w3.org/2003/11/swrl#" xmlns: ruleml = ≫http://www.w3.org/2003/11/ruleml#≫> <! – Knowledge domain -> <owl: Class rdf: ID = ≫FieldOfKnowledge≫ /> <! - Solution method -> <owl: Class rdf: ID = "Method" /> <! - Problem -> <owl: Class rdf: ID = "Problem" /> <! - Data set (input or output) -> <owl: Class rdf: ID = "DataSet" /> <! – Method generalization -> <owl: ObjectProperty rdf: ID = ≫generalizedBy≫> <rdf: type rdf: resource = "& owl; TransitiveProperty" />

Inset 4.1. Continued

```
<rdfs: domain rdf: resource = "#Method" />
<rdfs: range rdf: resource = "#Method" />
</ owl: ObjectProperty>
<! - Method parameterization ->
<owl: ObjectProperty rdf: ID = "hasParameter">
<rdfs: domain>
<owl: Class>
<owl: unionOf rdf: parseType = ≫Collection≫>
<owl: Class rdf: about = ≫# Method≫ />
<owl: Class rdf: about = "#Problem" />
</ owl: unionOf>
</ owl: Class>
</ rdfs: domain>
<rdfs: range rdf: resource = "# DataSet" />
</ owl: ObjectProperty>
<! - Input parameter ->
<owl: ObjectProperty rdf: ID = "hasInput">
<rdfs: subPropertyOf rdf: resource = "# hasParameter" />
</ owl: ObjectProperty>
<! - Output parameter ->
<owl: ObjectProperty rdf: ID = "hasOutput">
<rdfs: subPropertyOf rdf: resource = "# hasParameter" />
</ owl: ObjectProperty>
</ rdf: RDF>
```

To run queries on the ontological structure, the SPARQL is used, which allows using the existing ontological interpretation tools, to analyze the Smart Grid cyber security ontology with memory (including the construction of dynamic rule relationships). A query example is shown in Inset 4.2.

4.3.4 Ontology Structure Example

Here is a possible cyber security ontology structure to describe the knowledge set used in the Smart Grid self-recovery organization under information confrontation. This structure was tested in 2012 in the joint research of the scientific cyber security schools of ETU LETI, ITMO University, and the faculty of CMC of Lomonosov Moscow State University.

Inset 4.2. The example of a query
Example of a query for an ontological structure
PREFIX nano: <http://escience.ru/sec.owl#>
PREFIX escience: <http://escience.ru/escience.owl#>
PREFIX rdf: <http://www.org/1999/02/22-rdf-syntax-ns#>
PREFIX rdfs: <http://www. org/2000/01/rdf-schema#>
SELECT ?E ?L ?C WHERE {
?E rdf:type escience:DataSet .
?E rdfs:label ?L .
OPTIONAL {?E rdfs:comment ?C} .
nano:Hf escience:hasInput ?E .
?E escience:isValue ?V .
?V rdf:type escience:SelectedValue
}

In the ontology, two main layers are distinguished: the description of concepts (classes) and individuals that implement concepts. Thus, individuals can be connected by the relations defined at concepts level. Moreover, the relationship between separate concepts is acceptable (e.g., the generalization ratio). In the simplest case, the relation set can be limited by two-dimensional relations. Another ontology element is the individuals attributes (characteristics), detailing their description. In addition, one of the possible extensions is the characteristic association not only with individuals (as class implementations), but also with the relationships between them (as implementations of admissible connections classes).

Formally, the ontology class layer is defined as a graph $O = <C, R>$, where C is a class set and R is a set of the abstract relations connecting classes. Similarly, the ontology individual layer is defined as a graph $\tilde{O} = <\tilde{C}, \tilde{R}>$, where $\tilde{C}$ is an individual set and $\tilde{R}$ is a relation set between individuals. For each individual layer element, the following functions are defined:

a) Generalization ratio

$$gn^{(C)}\tilde{C} \rightarrow C, gn^{(R)}\tilde{R} \rightarrow R,$$

identifying the individuals relationship and connections between them with the corresponding classes and class relationships;

b) "Guard condition", which determines the elements' applicability under the given conditions

$$gc^{(C)}(F) : \tilde{C} \rightarrow \{0, 1\}, gc^{(R)}(F) : \tilde{R} \rightarrow \{0, 1\}$$

where F is a set of active facts defined for the current task;

c) Criterion estimation function

$$k^{(C)}(F) : \{\tilde{c}\epsilon\tilde{C}|gc^{(C)}(\tilde{c}) = 1\} \to \Psi^{(C)}$$
$$k^{(R)}(F) : \{\tilde{r}\epsilon\tilde{R}|gc^{(R)}(\tilde{r}) = 1\} \to \Psi^{(R)}$$

where $\Psi^{(C)}$ and $\Psi^{(R)}$ are, respectively, the individual evaluation criteria space and the relationships between them.

The logical inference block allows determining the method to solve the problem as a tuple $S = (s_1, s_2, \dots, s_N)$ of the fixed structure whose i-th element is a set in the form

$$s_i = \{\tilde{c}\epsilon\tilde{C}|gc^{(c)}(\tilde{c}) = c_i\},$$

where the sequence of classes $c_i\epsilon C$ and the requirements of the sets s_i determines the general solution structure. To evaluate the solution constructed by the criteria system, graph analysis is used

$$\tilde{O}' = \langle \tilde{C}', \tilde{R}' \rangle : \tilde{C}' = \cup_i s_i \cup \tilde{C}_s,$$

where $\tilde{C}_s$ is an associated system of classes $\tilde{C}_s = \{\tilde{c}_s|\tilde{c}_s \notin \cup_i s_i, \exists \tilde{c}_1 \in \cup_i s_i : rch(\tilde{c}_s, \tilde{c}_1)\}$ and $rch(\tilde{c}_s, \tilde{c}_1)$ is the attainability relation on the graph $\tilde{O}'$. The estimate $\tilde{O}'$ is implemented in the Ψ criteria space defined by the intersection of the criteria sets describing the $\Psi^{(C)}$ and $\Psi^{(R)}$ spaces.

A possible scheme of the immunity formation to disturbances is shown in Figures 4.17 and 4.18.

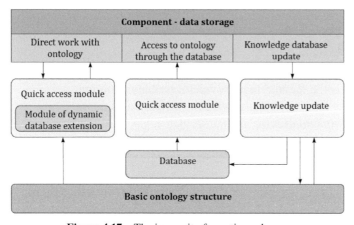

Figure 4.17 The immunity formation scheme.

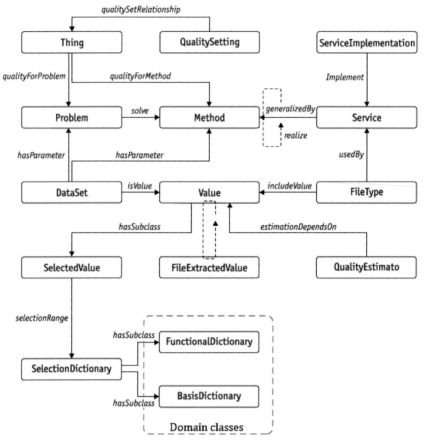

Figure 4.18   The class structure of cyber security ontology.

- **Problem** is the problem solved within the domain.
- **Method** is the method that provides a problem solution.
- **Service** is the computing service that implements the method.
- **ServiceImplementation** is the service copy, available as part of the software package.
- **DataSet** is a set of input or output data for a given method or task.
- **Value** is the domain values used as input and output data to solve problems. Two specific value classes are distinguished, differing in the manner of their tasks:
- **FileExtractedValue** is the values retrieved from the files (the extraction method is described as the class (in the source component code) that implements the unified IfileValueExtractor interface).

- *SelectedValue* is the value selected from the list of available ones (the list of available values is specified in the ontology by individuals belonging to the subclasses of the SelectionDictionary class).
- *FileType* is the file containing the values available for extraction.

The structure of the accumulated immunity database is specified by the ADO.NET Entity Framework model. To organize access to the database, a library that provides access to the entity instances stored in the database through the ADO.NET Entity Framework is built. This approach provided the possibility to access the database as a set of interrelated collections storing instances of classes equivalent to the database entities. The implementation of direct access to the ontological structure by the Pellet API (*RunLib variant*) is proposed. The interface implemented by this module includes the following basic methods of work with an ontological structure:

- **CreateSession** () creates a session, returns the session string identifier;
- **AddOWLModel** (<**session id**>, <**ontology**>) is an ontological structure extension that is specified in OWL in the form of a separate ontology with possible references to existing elements;
- **ExecuteQuery** (<**session ID**>, <**query**>) is a query for an ontological structure extended within the current session (the query is specified in SPARQL, the result of which is a string in XML format). A general interaction scheme of the RunLib implementation with the ontology interpreter is shown in Figure 4.19.

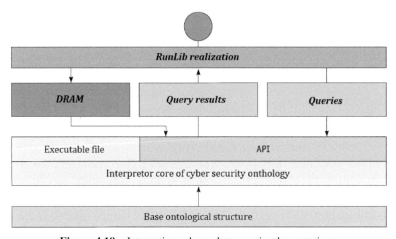

Figure 4.19   Interaction scheme between implementations.

***

As a result of the work completed [1, 3, 5, 8, 9, 11, 13, 20–22, 24, 25, 39], the conceptual foundations of the future power systems self-recovery under information confrontation were put forward and proved, and a new, more advanced cyber security ontology of self-recovering Smart Grid was developed. Basic notions, which are a part of self-recovery, were defined. The contents of elementary, complex, and disturbed processes of power system functioning were determined. The characteristic features of single, group and mass disturbances are shown. A model of future energy system self-recovery involving the formal apparatus of R.E. Kalman dynamical systems was proposed. The Smart Grid system operation ideology with memory to transfer the immunity to disturbances was introduced and developed. A general analysis of the immune system to ensure the Smart Grid systems operation stability under information confrontation was considered [7, 13].

The stratification of future energy systems self-recovery under mass and group information and technical attacker impacts is offered (Figure 4.20). The organization of destabilization facts monitoring and the disturbances type and nature identification were considered. The disturbance dynamics was analyzed and solutions of returning to balanced state were developed. An abstract self-recovery plan was developed. The immune system implementation that provides the power system recovery after disturbances was proposed. In accordance with the proposed concept, based on the identification theory and complex technical systems control, a generalized problem to develop the relevant intellectual subsystem to ensure the Smart Grid cyber security was set.

The scientific and methodical apparatus relevant to solve the problems of the Smart Grid self-recovery organization was selected. The multilevel hierarchical systems theory application to design the mentioned intellectual cyber security subsystem is explained. The formal languages and grammar theory application to generate and recognize the possible mass disturbance structure types is proposed. The feasibility of the catastrophe theory application to analyze the dynamics behavior of the disturbed energy systems operation processes was explained by analogy with disturbance simulation in living nature. The immunity formation to destructive disturbances is proposed by using the control and restoration theory results of the Smart Grid systems functioning [1, 3, 5, 8, 9, 11, 13, 20–22, 24, 25, 39].

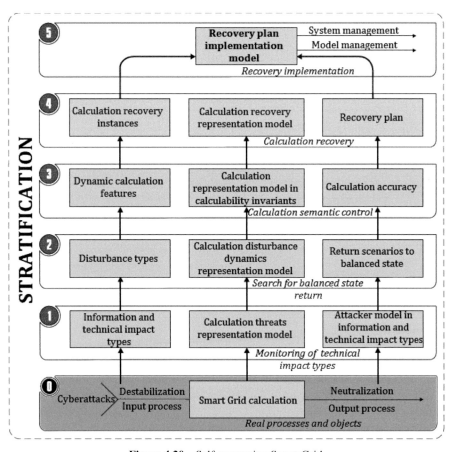

Figure 4.20   Self-recovering Smart Grid.

All of the above allows developing a system image of the future intelligent subsystem providing the Smart Grid cyber security in the model types of those theories that allow synthesizing the desired self-recovery structure, as well as constructing and implementing a relevant self-recovery plan under growing threats to information security.

4.4 Development of Cyber Security Platforms

A feasible-type design of evolutionary "von Neumann architecture" modifications is presented to choose an advanced hardware platform for the Russian early-warning cyber security system. The overall readiness assessment of the

Russian Program of supercomputer technologies development until 2025 is given to solve the problem of preventing and anticipating cyber-attacks. An approach is proposed to create an early-warning cyber security system based on the so-called "computational cognitivism". In this relatively new scientific research trend, cognitive processes are a kind of symbolic computation. It is shown that the cognitive approach makes it possible to create systems that are fundamentally different from the traditional monitoring systems of information security threats, a unique ability to independently associate and synthesize new knowledge about qualitative characteristics and quantitative patterns of information confrontation. The feasible architecture of the Russian early-warning cyber security system against the Russian information resources based on the analysis and processing of extremely large amounts of structured and unstructured information from various Internet/Intranet and IoT/IIoT sources (Big Data and Big Data Analytics) is proposed.

4.4.1 Principles of Designing Special Computing Systems

The functional model of John von Neumann's electronic computer, called "von Neumann's architecture" (1945), for many years ahead, determined the main development ways of general-purpose and special-purpose computers. There are five main generations of computers (from classical computers to high-performance super-systems). In this case, it is often said about generations of computers in certain families. At the beginning of the 21st century, the approach based on the canonical "von Neumann architecture" came into conflict with the requirements imposed on the developed and advanced information security means. It involved the requirements of productivity and the trust of the advanced monitoring centers for the security state of the Russian critical infrastructure.

Background

In 1945, a talented American mathematician of Hungarian origin, John (Janos) von Neumann, for the first time, described the technical implementation's model of the "universal calculator" by Alan Turing (1936) in line with W-670-ORD-4926 contract between the US Army and the Moore School of the Electrical Engineering University of Pennsylvania.

The corresponding 101-page scientific report, dated June 30, 1945, contained a description of the main architectural principles for building the advanced computer EDVAC (Electronic Discrete Variable Automatic Computer), created by research engineers John Mokley and John Eckert.

In the above-mentioned scientific and technical report, John von Neumann formulated the following principles of computer design:

1) Principle of program management: the computer operation is carried out automatically by executing a certain program (here, the program is a nonempty set of machine instructions, each of which carries out a single act of information transformation, and all the variety of commands forms the machine language or its command system);
2) Principle of sequential operation – the commands are performed sequentially until a special stop command (or emergency exit) is executed; the command choice is determined by the stacking order in the memory or by the program control command;
3) Principle of the stored program, predetermining the program storing with the original data in memory; as a consequence, the program may change during execution;
4) Use of the binary system for the information presentation in a computer that significantly simplified the technical computer implementation;
5) Principle of the hierarchical memory construction made possible developing both expensive but fast-acting memory of relatively small capacity and less expensive external memory with higher capacity and less speed on external data storage media (magnetic tapes, log drums, disks, etc.).

In 1947, in the Soviet Union academician Lebedev, S. A., independently of John von Neumann works, proposed similar architectural principles for the design of the first Russian computers (Small Electronic Calculating Machine named MESM, Big Electronic Calculating Machines named BESM-1 and BESM-2):

1) Principle of architectural construction – the computer should include an arithmetic device, memory, control device, and input–output device;
2) Principle of the stored program – the program in the machine codes should be stored in the same memory as the numbers;
3) Use of the binary system – the binary system should be applied for representation of numbers and commands;
4) Principle of sequential operation – calculations must be performed automatically in accordance with the program stored in memory;
5) Principle of performing logical operations – logical operations must be performed along with arithmetic operations;
6) Principle of hierarchical memory construction – the machine memory must be organized according to a hierarchical principle.

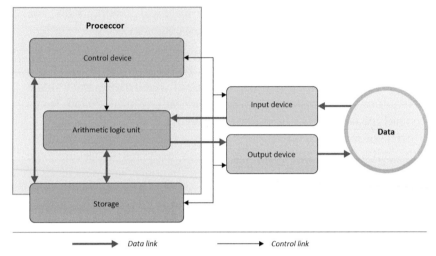

Figure 4.21   von Neumann Architecture.

The canonical "von Neumann architecture" consisted of the following key components (Figure 4.21):

- Single control device (CU);
- Arithmetic logic unit (ALU) for performing arithmetic and logical operations on data (operands are numbers or words), including conditional and unconditional transitions;
- Storage or memory for storing programs and data;
- Devices for programs and data input (ID) and result output (OD).

Over time, the composition of the CU, ALU, and part of the memory (random access memory, RAM) was called "the processor", and in the case of performance on one or several large-scale integrated circuits (LSI), it was called "the microprocessor" [123].

Based on the canonical "von Neumann architecture", the first three generations of computers were developed [123–127]; the first generation (1949) was based on the lamp element base, for example:

- BINAC, EDVAC, SEAC, ORDVAC, IAS-machine, MANIAC I, ORACLE, etc. were created in the USA;
- BESM-1 (BESM-2 and BESM-4), MESM, M-20 (M-40, M-50) and 5E92, E926, etc. were created in the Soviet Union;
- EDSAC, Mark I, WITCH, ICT 1301 / Flossie, ICL 2966, etc. were created in Great Britain;

- CSIR Mk 1, CSIRAC, SILLIAC were created in Australia;
- WEIZAC, etc. were created in Israel.

Today, the listed models and their operating models can be seen in the largest computer museums in the world.

In the second (1955) and third (1963) generations, computers gradually moved away from the principle of sequential processing of information and used a different element base (transistors and integrated circuits).

Among the popular examples of these computers generations were IBMS/360 and S/370 in the USA and their alternatives "Single system of computers" ("Series 1" and "Series 2") in the Soviet Union, as well as the mini-computer and microcomputer of the HP family (Hewlett Packard), PDP (Digital Equipment Corp.) and ASVT-M, SM computers, "Electronics", respectively.

The evolution limit of the canonical "von Neumann architecture" was the conveyor way of processing information in combination with the data vectorization, which made possible performing in practice the parallel operation of a small operations number [124, 126, 123].

In the early 1960s, the classical approach to computer design based on "von Neumann architecture" came into conflict with the requirements for high-performance and fault-tolerant means of collecting and processing information [123, 124]. The solution of this contradiction required the development of new principles for designing a computer based on the model of "collaborative computing" (or "collaborative calculators"):

- Principle of parallelism (parallelism, concurrency) – parallel execution of operations on a set of computers interacting with each other;
- Principle of programmability (programmability, adaptability) – programmability of the computer system's structure for performing certain calculations;
- Principle of homogeneity (homogeneity) – homogeneity of key components' computer system.

New principles of architectural computer design differed significantly from John von Neumann's initial principles and provided developers with more opportunities. So, the principle of programmability allowed "optimizing" the structure of computing systems for performing certain calculations in an optimal way, and the principle of homogeneity – to achieve optimal values of performance indicators, fault tolerance, and safety.

Technical implementations of "collaborative computing" model based on these principles of architectural computer design were called "computing systems" and were assigned to the fourth and fifth computers generations [21, 45, 46, 68, 75, 123, 124, 126]. Named systems were architecturally close to third-generation computers on macrolevel. However, at the microlevel – the level of the functional microprocessor structure – revolutionary changes took place and the possibilities of pipelining calculations and parallel data processing were achieved. As a result, the computing systems began to represent super-systems that combined the sets of pipeline microprocessors and matrix processors.

In the Soviet Union, many outstanding Russian researchers made a fundamental contribution to the theory and practice of architectural computer design.

4.4.2 Feasible Computing Classifications

At the present time, various classifications of computer architecture are known [68, 75, 90, 125, 126]. Among them, the classification of Stanford University professor M. J. Flynn, which in 1966 proposed distributing all known computer architectures in four main classes (Table 4.8), was most widely spread.

According to Flynn's classification, the first sequential computers can be classified as SISD class, the multiprocessor computing systems with a single control device, for example, ILLIACIV or CM-1 from Thinking Machines, as SIMD class, the systolic computing systems or pipelining systems as MISD class, the most known parallel multiprocessor computing systems as MIMD class.

In 1972, a number of researchers [68, 75, 90, 125, 126] proposed some refinements to Flynn's classification. In particular, subclasses of multiprocessors or systems with shared memory, as well as multi-computers or systems with distributed memory were branded in the MIMD class (Figure 4.22).

In this case, among the multiprocessors began to distinguish the system with a uniform memory access (UMA) and nonuniform memory access (NUMA), the first ones (Figure 4.23(a) became the basis for the building of parallel vector processors (PVP) and symmetric multiprocessors (SMP), for example, the supercomputer Cray T90 and IBM eServer, Sun StarFire,

Table 4.8   M. J. Flynn computer architectures classification

	Single instruction	Multiple instructions
Single data	SISD	MISD
Multiple data	SIMD	MIMD

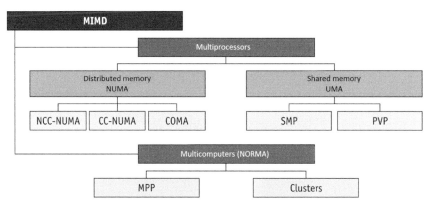

Figure 4.22 Classification of multiprocessor computing systems (MIMD).

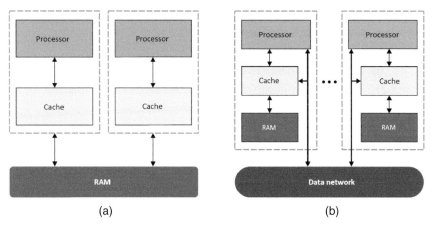

(a)   (b)

Figure 4.23 The architecture of multiprocessor systems with shared memory: systems with uniform (a) and nonuniform (b) memory access.

HP Superdome, and SGI Origin. The second ones (Figure 4.23(b) became the basis for building following systems:

- With shared access to the local memory with noncache coherent NUMA (NCC-NUMA) of different processors, for example, Cray T3E system;
- With cache coherent NUMA (cache-coherent NUMA or CC-NUMA) of different processors, including SGI Origin 2000, Sun HPC 10000, and IBM/Sequent NUMA-Q 2000;
- With cache-only memory architecture (COMA) of processors for data representation, in particular, the KSR-1 and DDM system.

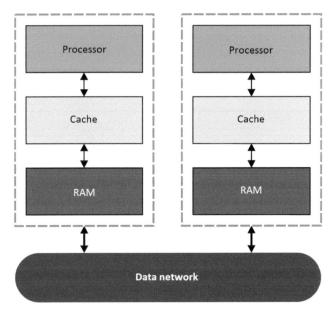

Figure 4.24 The architecture of multiprocessor systems with distributed memory.

Multicomputers began to distinguish massively parallel processors (MPP) and clusters – systems of typical computers connected to a network, for example, IBM RS/6000 SP2, Intel PARAGON, ASCI Red, transputer systems Parsytec and AC3 Velocity, and NCSA NT Supercluster. In these systems, memory access was no longer provided (no-remote memory access or NORMA) (Figure 4.24). Here, each processor could only use local memory, and it was necessary to explicitly perform message-passing operations to access data from other processors.

In the literature [21, 45, 46, 68, 86, 90], other classifications of computing systems are known (R. Hockney, T. Feng, L. Schneider, and others). For example, in the works of Khoroshevskii, V. G., and his student Kurnosov, M. G. [124, 125], all the main types of computer systems (CS) are listed in chronological order:

- Pipeline computing systems (CS);
- Matrix CS;
- Multiprocessor CS;
- Distributed CS;
- CS with programmable structure;
- Cluster SC;
- Spatially distributed multicluster CS.

Here, the pipeline CSs are the simplest technical implementations of "collaborative computing" models. Such systems are based on the pipeline way of information processing, and their functional structure is represented as a "sequence" of related elementary blocks of information processing. In this case, the blocks work in parallel, but each of them implements only its operation on the data of the same flow. Pipeline CSs are classed as MISD systems. In matrix CSs, it is possible to simultaneously implement a large number of operations on elementary processors (EP), "combined" into a matrix. Here, each EP is a composition from an arithmetic logic unit (ALU), a control unit (CU), and local memory (LP) for storing instructions and data. Matrix CSs belong to the class of SIMD systems.

Multiprocessor CS are systems from a set of unconnected processors and a common (possibly, sectioned, modular) memory with connections between them. The interaction between processors and memory is carried out through the switch (the common bus, etc.), and between the processors is carried out through memory. Multiprocessor CSs are classified as the class of MIMD systems.

Distributed CSs are multiprocessor CSs with no single resource (shared memory). The main components of a distributed CS (such as a switch, control device, arithmetic logic unit or processor, memory) allow representation as a composition of the same elements (local switches and control devices, local processors and memory modules). Distributed CSs are classified as the class of MIMD systems.

Computing systems with a programmable structure are completely based on the collective calculator model and are a composition of interconnected elementary machines. Each elementary machine (EM) in its composition necessarily has a local switch (LS), a processor, memory, and, possibly, external devices. The local EM memory is intended for storing both data part and, most importantly, the parallel program's branch. CS with a programmable structure architecture belongs to the class MIMD. Such CSs for their potential architectural capabilities are not inferior to any of the above systems classes. In this case, spatially concentrated and distributed CSs are distinguished.

A characteristic feature of the concentrated CSs is the compact spatial arrangement of information processing and storage facilities, in which the average word transfer time between functional modules (processors, memory modules, computers, etc.) is commensurate with the average time of one operation execution in the processor. Spatially distributed CSs include macrosystems – the complex configuration's systems in which spatially dispersed computing facilities based on the calculator models and collective calculators, as well as communication networks that provide mutual tele-access between information processing facilities, act as functional elements. Spatially distributed CSs, in which the program compatibility condition of the computer is fulfilled, are classified as homogeneous CSs.

At the end of the 20th century, cluster computing systems became widespread. As a rule, a cluster CS or cluster is a composition of a set of calculators, a network of connections between them, as well as software for parallel information processing (the parallel algorithm implementation for solving complex problems). When forming a cluster CS, standard industrial components and special computing facilities can be used. As a rule, mass hardware and software are used, which ensures high technical and economic efficiency of the systems mentioned. At the same time, MISD, SIMD, and MIMD architectures are implemented, as well as various functional structures and design solutions. Here, the main functional and structural unit is EM, its configuration allows for variation from the processor core to the functionally developed computer. As a rule, modern clusters are configured from multiprocessor nodes and multicore processors, and the corresponding communication environments are built based on Gigabit Ethernet, InfiniBand, and Myrinet technologies (Figure 4.25). To combine the computing nodes in the system, switches are used with a fixed number of input ports for connecting end devices (computing nodes, switches, storage systems, etc.). Various schemes are used to build the component switches, for example, Fat tree or Clos.

Computing clusters belong to the class of distributed memory systems. As a rule, multicluster computing systems are spatially distributed systems,

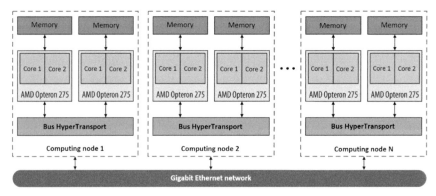

Figure 4.25   An example of a hierarchical organization's communication environment of a cluster CS.

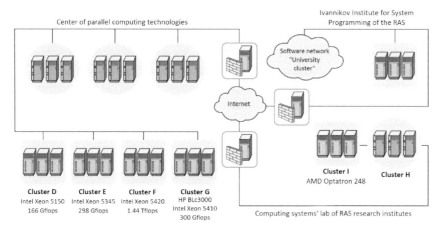

Figure 4.26 An example of a spatially distributed multicluster computing system.

which are built by combining the distributed cluster resources (Figure 4.26). At the same time, the development of appropriate parallel processing programs is based on the Message Passing Interface (MPI) standard, the Virtual Machine (PVM), the IBM Message Passing Library (MPL), P4, etc.

In the early 21st century, GRID technologies (Global Resource Information Distribution) (2008), followed by "cloud" (2012) and "foggy" (2016) technologies have become widespread. Main features of these technologies were the implementation of "service" collaborative computing model based on the composition of physical and virtual components of data centers, supercomputers, and canonical and modern distributed computing systems. Hypervisors, virtual machines, integration buses, and special protocol stacks are in demand for distributed (composite) collaborative computing.

4.4.3 Characteristics of the Known Computing Systems

It should be noted that the architecture of modern computers of the fourth and fifth generations differs significantly from the original John von Neumann' canons.

In this case, the dominant majority of the mentioned computing systems is multiarchitecture and is classified as MISD, SIMD, and MIMD systems. The functional structure of such systems inherits the properties of pipeline, matrix, and multiprocessor systems, for example, as it was implemented in systems with mass parallelism or MPP systems (Massive Parallel Processing

Systems) Cray TZD, Cray TZE, Cray TZE-900, Cray TZE-1200, Cray TZE-1350. At one time, these systems became an alternative to the well-known vector-parallel systems and were characterized by high-performance values (from dozens of Gflops to hundreds of Tflops), as well as a larger memory capacity (from gigabytes to hundreds of terabytes). The development of the pipeline computing systems' architecture led to the transformation of these systems into distributed multiprocessor systems with a MIMD architecture and mass parallelism. In this case, the canonical pipeline structure (the basis of the supercomputer architecture (1980) was adopted in the microprocessors of Intel Pentium, IBM Power PC, DEC Alpha, AMD Opteron, Intel Xeon, etc. As a result, "pipeline" microprocessors were created and outperformed vector supercomputers (1980).

Matrix systems or systems with mass parallelism have surpassed the pipeline systems. In this case, the architecture of the matrix systems, depending on the analysis "depth", can be attributed to MIMD and SIMD simultaneously. For example, the system could have MIMD architecture and its subsystems could have SIMD architecture.

Known examples of such families are Connection Machine (CM-1, CM-2, and CM-5) (1985), as well as nCube (nCube-1, nCube-2, and nCube-3) (1995).

Most of the known parallel systems can be referred to the class of "multiprocessor computing systems", consisting of multiple processors and a single shared resource (usually with shared memory). For example, a C.mmp (Carnegie-Mellon Multi-Processor) computer by the Carnegie Mellon University, Elbrus-1 and Elbrus-2 by the Lebedev Institute of Precise Mechanics and Computer Engineering (IPMCE), a family of modular-scalable multiprocessor systems with a programmable architecture by the Kaliayev Research Institute of Multiprocessors and others.

Modern distributed computing systems can vary from several processor units to hundreds of thousands (MVS-15000VM has 1148 processors, IBM Roadrunner – 122,400, and IBM BlueGene systems of the second generation – 884,736). Note that the first generation of IBM BlueGene (2008) had 212,992 processors with a performance of 2.8 Gflops. The system was characterized by a record peak performance of 596,378 Tflops, and IBM Blue Gene/P and IBM Blue Gene/Q-versions had 1 and 3 Pflops, respectively. These systems were characterized by a hierarchical organization, heterogeneous communication channels with different bandwidths.

Over time, developers of functional "collaborative computing" models independently from each other came to understand the need to create distributed systems with a programmable structure. They use the MIMD architecture and the possibility of software migration to MISD or SIMD architecture. The distinctive characteristics of such systems were the lack of a single common resource, the failure of which would lead to the system failure as a whole, as well as the ability of automatic reconfiguration with respect to the structure and parameters of the solved problem. Bright examples of Russian computing systems family with a programmable structure were Minsk-222 (1965), MINIMAKS (1975), SUMMA (1976), and three models of the MICROS family (1986, 1992 and 1996) by the Siberian Branch of the USSR Academy of Sciences.

In ref [124], the following requirements to the structures of modern CS have been singled out:

1. Simplicity of embedding a parallel algorithm for solving a complex problem in the CS structure. The CS structure should be adequate to a sufficiently wide class of solved problems, and the configuration of problem-oriented virtual configurations should not be associated with significant overhead costs.
2. The convenience of addressing elementary machines and "transferring" subsystems within the computing system. The computing system should allow implementing the simplest "mechanism" for converting virtual EM addresses into real (physical) addresses of the system machines. The need for simultaneous solution of several problems on the CS (i.e. the need to separate the space of elementary machines between tasks) justifies the requirement of subsystems' relative "movement ease" within the system (while retention of their topological properties).
3. Feasibility of the short-range interaction and minimum delays principle in inter-machine information transmissions in the CS. The short-range interaction principle predetermines the information exchanges implementation between EM's "remote" from each other through the intermediate system machines. Consequently, in the conditions of limited connection number for each EM, the structure should provide a minimum of delays in "transit" information transfers.
4. Scalability of the SC structure. For the CS configurations with a given efficiency, it is required that the structure possess the ability to increase and reduce the number of vertices (machines). At the same time, the change in the EMs number should not lead to connectivity disruption and a significant increase in the connections number between EMs.

5. CS structure commutability. The computer system should be adapted to the implementation of group inter-machine information exchanges. Consequently, the CS structure should have the ability to carry out a given number of simultaneous disjoint interactions between elementary machines.

6. Vitality (stability and reliability) and safety of SC structure. An important CS property is the ability to ensure the system's operability both in the case of abruption or failure of hardware and software components and in the context of targeted information and technical impact on the system.

7. CS structures legacy. The network structure of inter-machine CS communications should not impose special requirements on the element base and on the manufacturing microprocessor LSI technology. The systems are required to be susceptible to mass technology, and their "computing core" should be formed from mass microprocessor LSI. The latter makes it possible to achieve high values of technical and economic CS indices.

According to Russian computer developers [123, 124], the homogeneous structures satisfy the most fully listed requirements. Such structures are, in particular, advanced for building the large computer systems, CS with a programmable structure. The advanced structures include n-dimensional or circulant structures. For the first time, they were identified and investigated by the Computing Systems Department of the Mathematics Institute of the Siberian Branch of the USSR Academy of Sciences in the early 1970s and were originally called *Dn* graphs.

4.4.4 Development of the Supercomputer Technologies

In 1966, in the Soviet Union, Kartsev, M. A., was one of the first to put forward the idea of creating a multi-computer complex M-9 with a capacity of about 1 billion operations per second (M-9 did not receive industrial development). In the 1990s, together with a number of RAS research institutes and industrial enterprises, a well-known line of Russian supercomputers MVS-1000 (1999) was created. They consisted of multiprocessor arrays combined with external disk storage, information input/output devices, and a control computer. In these systems, microprocessors Alpha 21164 (*DECCompaq*) with a capacity of up to 1–2 billion operations per second and RAM of 0.1–2 Gbytes were used.

At present, supercomputers are created in a small number based on typical universal components of a large-scale industrial production. It is believed that this provides high speed, operational reliability, cost-effectiveness, etc. Also, a significant contribution is made by problem orientation (specialization) in the form of various coprocessors and reconfigurable structures (field-programmable logic devices – FPLD). Along with FPLD, the custom ASIC (application-specific integrated circuits) and SoC ("system on crystal") are applied, if mass production is required [68, 75, 90]. The integral evaluation of supercomputers characterizing the achievements in the field of modern supercomputer technologies (SCT) is the magnitude of peak performance demonstrated on special tests, for example, Linpack (HPL), which today has the value of several dozen petaflops. The plan of the technologically advanced countries of the world (Russia, USA, China, Germany, France, Great Britain, Japan, etc.) is to achieve an exaflops level (1000 Pflops) by 2018–2020. Because of its importance, this task was given a special name ExaScale.

In the Russian Federation, in order to achieve the set goals, the Concept of the federal program "Development of high-performance computing technology based on the exaflop class supercomputer (2012–2020)" was developed.

More recently, on November 14, 2016, at the international conference on supercomputer technologies SC 16 in Salt Lake City (USA), the next 48th rating of the 500 most productive supercomputers in the world was presented. The performance calculation of these supercomputers was carried out traditionally on the Linpack test (HPL). It is a test problem for solving a system of linear equations with a large number of variables (Inset 4.3).

Inset 4.3. Performance units

To assess the nominal computer speed and performance by Gibson, when fixed-point operations come into account only, the following units of measurement are applied:
- MIPS (million of instructions per second), 1 MIPS = 106 operations/s;
- GIPS, 1 GIPS = 10^9 operations/s.

Performance measurement on test task sets is performed in the following units:
- 1 flop (one FLoating-point Operations Per Second);
- 1 Mflops = 10^6 operations/s = 1 million floating-point operations per second;
- 1 Gflops = 10^9 operations/s = 1 billion operations/s;
- 1 Tflops = 10^{12} operations/s = 1 trillion operations/s;
- 1 Pflops = 10^{15} operations/s = 1 quadrillion operations/s.

Ahead was Sunway TaihuLight by the China supercomputer center running Sunway Raise OS 2.0.5, which showed a performance value of 93.015 Pflops (peak performance – 125.436 Pflops). It is 2.7 times more than the performance of the previous winner – Chinese Tianhe-2 33,863 Pflops), 5.2 times more than the best American supercomputer Titan, and 44.2 times more than the best Russian supercomputer "Lomonosov-2".

The Sunway TaihuLight supercomputer (Figure 4.27) uses more than 10.5 million processor cores and operates a sub-system of its own Sun way Raise OS 2.0.5 based on Linux. The system also includes its own OpenACC 2.0

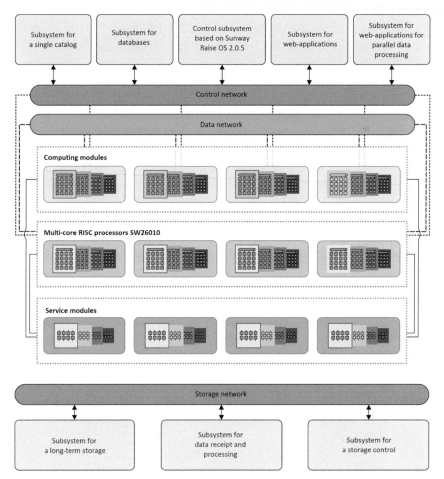

Figure 4.27   Sunway TaihuLight supercomputer architecture, China.

implementation, intended for parallelizing the code. This supercomputer uses multi-core 64-bit RISC processors SW26010, based on the ShenWei architecture. The total number of processors in the system is 40,960, and each processor contains 256 general-purpose computing cores and 4 control cores, which together yields 10,649,600 cores. The processor cores contain 64 KB of internal data memory and 16 KB for instructions and are communicated via a network on the chip instead of using the traditional cache memory hierarchy (Figures 4.28 and 4.29).

It is interesting to note that for the first time in the Chinese history of information technology, the national processors were used in the Sunway TaihuLight supercomputer. The previous most productive Chinese supercomputers Tianhe-1A and Tianhe-2 were built on the American company Intel processors. Previously, it was going to double the performance of Tianhe-2 based on the American microprocessors. However, in April 2015, the USA was toughened by the export control of supplies to the NUDT and supercomputer centers NSCC-CS, NSCC-GZ, NSCCTJ due to a suspicion that China is using US computer technology for scientific research in the field of new weapons (including cyber weapons), and the planned update has not yet taken place.

The closest competitor from the US supercomputer Titan was put into operation in October 2012 and almost immediately took the first line in the TOP500 ranking. However, already in July 2013, it was moved by Tianhe-2

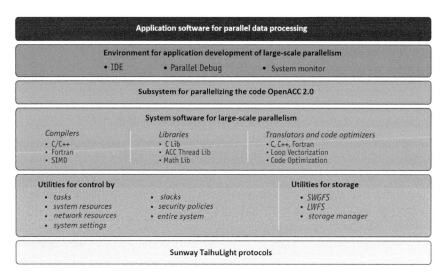

Figure 4.28 Sunway TaihuLight supercomputer software stack architecture, China.

to the second place, and in June 2016, it was in the third place. Titan is a massively parallel supercomputer by Cray Inc., located in the Oak Ridge National Laboratory (ORNL is US DOE National Laboratories). It is an update of the famous Jaguar supercomputer, which increased the number of CPUs and added the Nvidia Tesla K20x GPU. The Titan is built on a Cray XK7 platform with hybrid architecture: in addition to the 16-core AMD Opteron 6274 processors, the NVIDIA Tesla K20x general-purpose graphics processor (Kepler Architecture) is installed in each of the 18,688 nodes of the supercomputer system.

Thus, the total number of supercomputer cores was 299,008. The compiler, specially developed for Titan, automatically parallelizes code execution between the central and graphics processors.

The results of the last two TOP500 ratings indicate that the world leaders in the total number of supercomputers are the USA – 171 and China – 170, followed by Germany with 31, Japan – 27, France – 20, the United Kingdom – 14, Poland – 7, Italy – 6, Russia – 5, Saudi Arabia – 5, India – 5, South Korea – 4, Switzerland – 4, Netherlands – 3, Sweden – 3, Austria – 3, Australia – 3, New Zealand – 3, Brazil – 3, Denmark – 2, Finland – 2, Canada – 1, Norway – 1, South Africa – 1, Belgium – 1, Ireland – 1, Spain – 1, Greece – 1, and Singapore – 1.

The interesting trends in the supercomputer development in the world include the following:

- Minimum threshold of peak performance to enter into the TOP500 has increased over the 6 months from 285.9 to 349.3 Tflops, and to enter the TOP100, from 958.7 Tflops to 1.073 Pflops;
- Total performance of all systems in the rating for 6 months increased from 566.7 to 672 Pflops (3 years ago, it was 223 Pflops), and 117 modern supercomputers demonstrate a performance of more than 1 Pflops (in the previous rating – 94);
- Total distribution by supercomputers number in different parts of the world looks as follows: 217 supercomputers are in Asia (180 in the previous list), 171 in America (formerly 212), and 112 in Europe (formerly 108);
- Intel CPUs are leading as a processor core – 91% (it was 89%), followed by IBM Power – 4.6% (5.2%) and AMD – 2.6% (4.2%);
- 10% (in the previous rating was 8.6%) of all processors used have 16 cores, 30.4% (23%) – 12 cores, 15% (16.2%) – 10 cores, 12.4% (there were 30.4%) – 8 cores, 13.2% (14.2%) – 6 cores (dual- and single-core systems are not included in the rating);

- 93 out of 500 supercomputers (in the previous rating - 104) additionally use accelerators or co-processors, while in 67, the NVIDIA chips were used (there were 68), in 26 – Intel Xeon Phi (29), in 3 – AMD GPU (unchanged), in 3 were used hybrid solutions (there were 4);
- Hewlett-Packard is the leader among the supercomputer manufacturers – 25.4% (was 31%), followed by Lenovo with 16.8% (5%), Cray with 12% (13.8%), Sugon with 10.2% (9.8%), IBM with 7.6% (9%), and SGI with 5.2% (6.2%).

Among TOP500 leaders, Russian supercomputers occupy the modest places (Inset 4.4). It should be noted that in terms of performance, Russian supercomputers are up to the best foreign analogues. However, their total number, as well as the total productivity, is much lower than that in China, the USA, Germany, France, the UK, and Japan. For example, the total number of supercomputers used in Russian science and education is no more than 20, systems focused on specific applied research are less than 10, systems used in industry are less than 7, and systems used in the credit and financial sphere are less than 5.

Inset 4.4. Three of the best Russian supercomputers in the TOP500 (11.2016)

- "Lomonosov-2" (Lomonosov Moscow State University) moved from the 41st place (06.2016) to the 52nd, Linpack performance is 2.102 Pflops.
- "Lomonosov" (Lomonosov Moscow State University) moved from the 108th place (06.2016) to 132nd; 0.902 Pflops.
- "Tornado" (Peter the Great St. Petersburg Polytechnic University) moved from 158th place (06.2016) to the 226th; 0.658 Pflops.

Apparently, this state of affairs is caused by the underdevelopment of the needs and practices of using super systems and underestimating the SCT capabilities, in general, certain difficulties in the industrial production of key hardware and software components, as well as problems with the import of missing foreign components.

The main architectural design problems of Russian supercomputers are:

- Excessive power consumption of microprocessors and microcircuits, power and cooling circuits;
- Insufficient capacity and memory bandwidth;
- Low performance and energy efficiency of interconnect;

- Insufficient scalability, stability, and security of system software;
- Overestimation of stability and computation correctness in the context of possible malfunctions, failures, and impacts of various kinds of opponents;
- Weak efficiency of known program algorithms for calculations on exaflop systems;
- Overestimation of manageability, stability, and security of control systems for structured and unstructured Big Data;
- Weak support for mass concurrency in languages and programming systems;
- Insufficient information support of scientific research and scientific and technical developments;
- Low labor efficiency of scientists performing engineering data calculations.

September 26, 2016 MSU SRCC and RAS JSC announced the next, 25th, edition of the TOP50 most powerful supercomputers of the CIS countries. According to the presented data, a slight increase in the supercomputer performance in the post-Soviet space was recorded. The total system performance on the Linpack benchmark for 6 months has grown from 7.8 quadrillion floating point operations per second (Pflops) to 8.1 Pflops. The total peak performance of the systems was 12.52 Pflops (12.2 Pflops in the previous list version). In just 6 months, one new supercomputer appeared on the list and three more systems were updated.

Hewlett-Packard Enterprise remained the leader in terms of the systems number on the list: 14 systems (15 in the previous version), followed by IBM Corporation – 11 systems (11), T-Platforms – 9 (9), and a group of companies RSC – 9 (8).

All 50 supercomputers are built on Intel processors. The number of hybrid supercomputers using graphics processors for computing has decreased from 19 to 18, and the number of systems using Intel Xeon Phi accelerators remains the same – 6. The number of supercomputers using InfiniBand remained at the same level – 32, and the number of supercomputers using for the nodes interaction only communication network Gigabit Ethernet, decreased from 15 to 14. In addition, in the list the first system based on Intel Omni-Path technology appeared.

***

The current level of information and communication technologies' development determines the high operational and technical requirements for the developed and advanced software and hardware systems for ensuring information security, including the requirements to the monitoring centers of the protection state of the Russian critical infrastructure [1, 5, 6, 8, 10, 11, 22, 24, 94, 111]. At the same time, the capabilities of the Russian industry do not allow the production of the required complexes to be fully implemented on the Russian electronic base and software. The situation is further dramatized by the sanctions of foreign countries for the supply of high-tech electronic components and related software to Russia.

This state of affairs is due to the limited nomenclature and incompleteness of the technical characteristics of the Russian component base and software. At the same time, the significant dependence of the Russian information security tools production from the supply of foreign hardware and software products, potentially containing undeclared capabilities and bookmarks, raises information security problems. Problems in this area under such conditions are solved by using additional components aimed at suppressing possible bookmarks and hidden control channels. This inevitably leads to a decrease in the functional reliability of the developed samples of software and hardware systems for ensuring information security, increasing the product cost and the terms of design, development, and implementation of those in general.

In turn, the import substitution implementation of the required components is complicated by:

- Insufficient performance, functionality, reliability, and security of required components;
- Imperfection and inadequacy of development, debugging, and verification means of the said complexes;
- High price for the Russian electronic component base (for some products, it exceeds the cost of foreign analogues);
- Insufficient level of developer awareness about existing hardware and software products and their technical characteristics due to the lack of information in open sources;
- Lack of normative legal regulation in the cataloging organization of the relevant components intended for creating software and hardware security systems.

Thus, in the conditions of the mass use of the foreign and developed in Russia (but produced abroad) electronic component base, the definition of the so-called "trusted component base" for the developed software and hardware security systems is urgent.

Apparently, there is a need to significantly expand the nomenclature of the corresponding hardware and software components. To solve this task, it is necessary to implement a set of measures for the period up to 2025 to organize their development and serial production, mainly various types of Russian microprocessors, as well as special and general-purpose microcircuits.

4.5 Security Software Development based on Agile Methodology

Today, the development of applications for the technological platforms of the leading Russian digital enterprises is carried out taking into account the new requirements of the Russian State Program "Digital Economy". Implementation of these requirements assumes the transition to modern Industry 4.0 technologies, including the creation of a trusted "cloud" data storage of high and ultrahigh performance, the application of technologies for collecting, analyzing, and processing Big Data, building intelligent forecast models, generating new knowledge, etc. Which techniques and practices of modern software technology can help to do it and achieve the desired effect? Let us look at the example of the well-known Agile software development methodology, adapted to the cyber security requirements.

4.5.1 Main Ideas and Principles of the Agile Methodology

At present, the agile ("live" or "flexible") software development methodology is widely known. The methodology relevance is due to the imperfection (insufficiency) of the known "heavy" and "slow" software development practices based on the traditional "waterfall" software engineering model. Prerequisites for its appearance was the similarly named "Agile Manifesto" (released in 2001, Utah, USA).

This manifesto has been adopted by members of other well-known methodologies of flexible software development [19, 21, 45–47, 49, 51, 53, 128, 129], such as extreme programming (Extreme Programming, XP), team developers' levelling (Crystal Clear), dynamic model (Dynamic Systems Development Model, DSDM), a methodology that considers at future changes (feature-driven development, FDD), the development of missed deadlines and ideas crisis (Scrum), adaptive software development, pragmatic programming, etc.

The Agile Manifesto contains four main ideas:

- People and interaction are more important than processes and tools;
- Working product is more important than exhaustive documentation;
- Collaboration with the customer is more important than contract negotiation;
- Readiness for change is more important than following the original plan, as well as 12 principles of the flexible software development.

4.5.2 Best Practices of Agile Methodology

Briefly, the Agile methodology essence (Figure 4.29) is formulated as follows:

- Development is carried out in short cycles (iterations), lasting 1–4 weeks;
- At the end of each iteration, the customer receives a specific result, ready to use;
- Development team collaborates with the customer throughout the entire project;
- Changes in the project are welcomed and promptly taken to be proceeded.

The main reasons for the increased interest of application developers in Agile's flexible software development methodology are [31, 48, 50, 52, 122, 130–133].

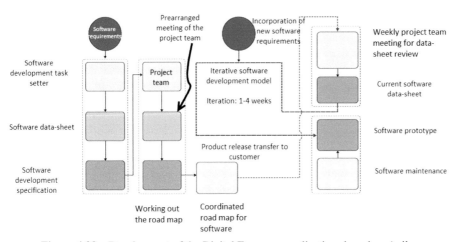

Figure 4.29   Development of the Digital Economy applications based on Agile.

The need to take into account the new requirements

As a rule, at the start of a software development project, the customer cannot formulate exhaustive requirements to the product.

- A customer can suggest an idea, but he does not know how to implement it in practice;
- Project teams cannot come to a consensus on the solution functionality;
- Project team cannot agree on the best course of action.

The Agile methodology suggests using prototypes and developing intermediate solution versions as often as possible, which makes it possible to remove the uncertainty in requirements and to test the solution functionality in practice. In the Agile methodology, the reaction to change is more important than following the plan. We welcome the introduction of new requirements to improve the solution competitiveness.

Increase the speed of accounting for changes. In response to changes, we need a fundamentally different flexibility and readiness for change, as well as speed of decision-making in the software development process. If we continue to use classical software development methods based on a rigid "waterfall" development model, we can inevitably get a result that is irrelevant and lagging behind the business requirements for several months, if not several years.

The Agile methodology allows us to quickly make changes and increase the solution functionality.

Improving communication opportunities

Agile approach focuses on the direct face-to-face communication of task managers and developers. Most Agile teams are located in the same office, sometimes called bullpen. The team includes the customer (also called product owner) or its authorized representative, who determines the requirements for the development result; this role can also be performed by the project manager, business analyst, or client. Also the team includes interface designers, testers, technical writers, and other professionals.

The Agile methodology declares that collaboration with the customer is more important than contract obligations. By giving preference to direct communication, the Agile methodology can significantly reduce the amount of written documentation besides other software development approaches.

What are the difficulties that developers face when using the Agile methodology?

The project team should transform into the so-called full-fledged tribe with its clusters and functional teams (squalls and chapters). As a tribe automation tool, we can choose, for example, the Atlassian JIRA solution, which can support hundreds and thousands of active tasks. We need to learn how to work in sprints and calculate the tasks for story points, which can be taken into account, including the developers' motivation system.

Finally, it is necessary to establish full-fledged change management using Agile practices, as well as an urgent task solution for Digital Economy applications development [52, 61, 73, 75, 83, 84, 85, 130, 134–136].

4.5.3 Adapting Agile for Secure Application Development

In practice, to develop security applications and technological platforms of the Digital Economy, a number of special methodical techniques had to be added to the classical Agile methodology (Figures 4.30 and 4.31). Today, the best practices for secure software development include requirements and recommendations: SDL PCI DSS, Microsoft SDL, Cisco SDL, 7.3.5 SRT BR IBBS-1.4-2018, etc. For example, Microsoft SDL includes in the development of secure software development the following a standard set of measures: training, configuration security requirements, designing, risk

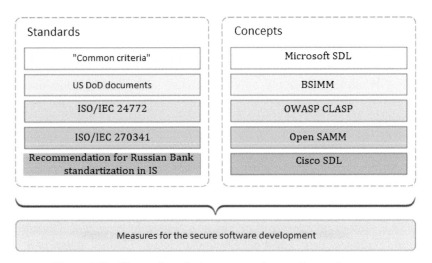

Figure 4.30   The need to take into account the security requirements.

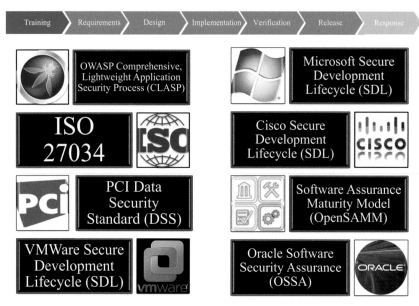

Figure 4.31   Practice of Microsoft SDL.

analysis of software architecture (modeling of information security threats), static and dynamic analysis of program source code, security testing, release, and support [21, 45–47, 50, 51, 134, 135]. However, the said practice does not contain recommendations for an independent assessment of the completeness and reliability of the implemented security measures. Note that the so-called "Common Criteria" (ISO/IEC 15408) requirements, which widely used to assess software for information security requirements, are also functionally limited. For example, they apply only to software with security functions, and the measures' nomenclature does not require static and dynamic analysis, training, etc.

Therefore, to remove these limitations, it was suggested to use the recommendations of national standards:

- GOST R 56939 "Information protection. Secure software development. General requirements". Has a number of general requirements for the measures implementation to develop secure software.
- GOST R "Information protection. Secure software development. Threats to information security when developing software". Defines the nomenclature of typical threats to information security.

Table 4.9 Accounting for security requirements and software functionality

Stages	Functionality	Security
Requirements development	Functionality, portability, technology	Security requirements
Design	Architecture, functions, data streams	Security functions modeling, threats modeling
Code development Applications test	Functional test, test data	Code analysis, security test
Migration to real environment	Data test delete, undo algorithm	
Support	Creating the new models	Secure development cycle observation

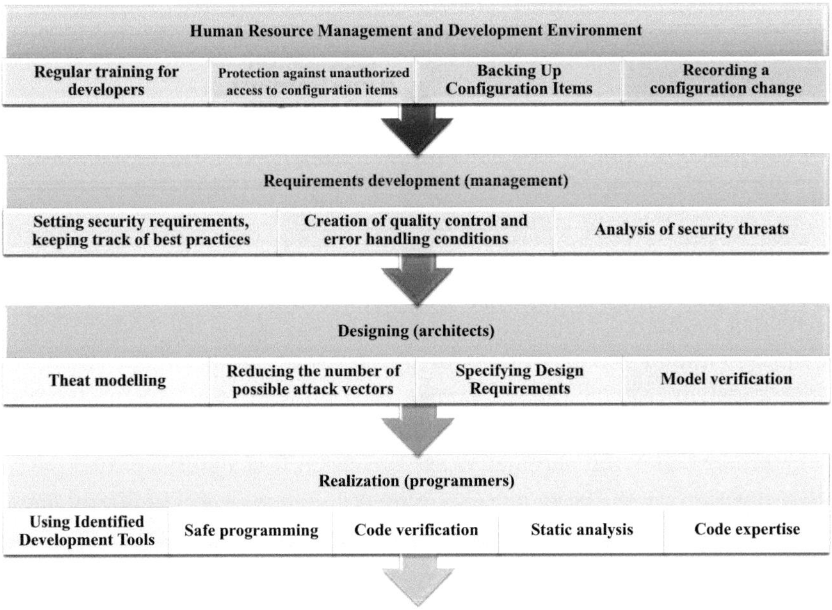

Human Resource Management and Development Environment

| Regular training for developers | Protection against unauthorized access to configuration items | Backing Up Configuration Items | Recording a configuration change |

Requirements development (management)

| Setting security requirements, keeping track of best practices | Creation of quality control and error handling conditions | Analysis of security threats |

Designing (architects)

| Theat modelling | Reducing the number of possible attack vectors | Specifying Design Requirements | Model verification |

Realization (programmers)

| Using Identified Development Tools | Safe programming | Code verification | Static analysis | Code expertise |

Figure 4.32 The role and place of the GOST R 56939 recommendations.

- GOST R "Information protection. Secure software development. A guide to software development". Includes a set of practical recommendations for the implementation of secure software development measures in accordance with GOST R 56939.

- GOST R "Information protection. Secure software development. Compliance assessment methodology of conformity assessment". Describes a number of standard procedures for verifying the compliance of software development organizations with the requirements of GOST R 56939.

In practice, the application of these recommendations has made possible solving the following particular problems [19, 21, 45–47, 51, 53, 83, 122, 128, 129, 131, 132, 135]:

- Assess the sufficiency of measures for reducing the number of vulnerabilities in the software being developed, and their applicability in conducting software conformity assessment;
- Generate a basic set of requirements for the secure software development, allowing to assess the processes conformity to these requirements;
- Develop a methodology for the reasonable generation of a measures set for the secure software development.

It should be noted that the recommendations of GOST R ISO/IEC 27034-1 were used to generate tools and means for monitoring and managing software security, and the recommendations of GOST R ISO/IEC 15408-3 are used to specify and expand the trust components.

Thus, a basic set of requirements for the secure Digital Economy software development based on GOST R 56939 was developed. At the same time, aspects related to the requirements generation for documentary evidence of requirements compliance, as well as to the assessor actions (e.g., testing laboratory expert) performed in the course of conformity assessment, were taken into account. To describe the requirements for the secure software development, the following parameters set was used. It is significant that the reliability and validity of the mentioned requirements set was confirmed by the Technical Committee for Standardization TK-362 "Information Protection" examination. As a result, a method was developed to justify the measures set generation for the secure Digital Economy software development, the main stages of which are presented in Figure 4.33.

This methodology was based on the recommendations of the "General Criteria" and GOST R 56939 regarding the definition of the basic requirements set for the secure software development, as well as GOST R ISO/IEC 12207 regarding the secure software development in accordance with the software life cycle.

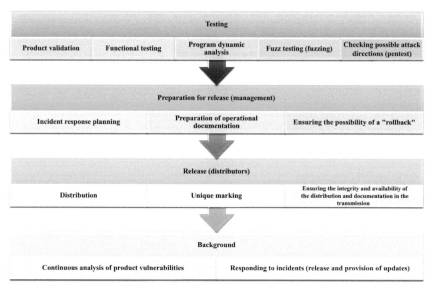

Figure 4.33 The main stages in the methodology for measures justification.

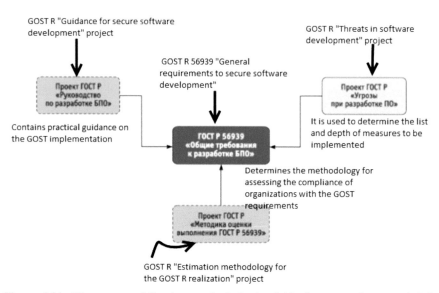

Figure 4.34 The system of Russian standards in the field of secure software and their interconnections.

In practice, the use of the Agile-adapted methodology for Digital Economy applications development allows:

- Determining the environment characteristics for the Digital Economy applications' development;
- Carrying out a reasoned choice of processes, tasks and works from the GOST R ISO/IEC 12207 nomenclature, taking into account the development environment characteristics, as well as requirements for Digital Economy applications;
- Specifying the list of tasks and works, taking into account the proposed nomenclature of measures to develop secure Digital Economy applications according to GOST R 56939;
- Documenting decisions on the implementation of selected processes, tasks, and work, as well as relevant justifications for selected solutions.

It is significant that this allowed us to implement two main scenarios:

a) Conformance statement: in this case, the developer of the Digital Economy applications must implement all measures proposed in the GOST R 56939, and appropriate certificates are created;

b) Use the GOST R 56939 as recommendations to improve the protection level of Digital Economy applications: in this case, a software developer can choose a subset of measures to be implemented.

At the same time, the security measures justification of the Digital Economy applications can be based on the risk assessment associated with information security threats that arise in the software development processes. To do this, it was necessary to create an up-to-date list of the information security threats when developing Digital Economy applications, as well as to develop a violator model and a taxonomy of information security threats related to software development processes.

4.6 Development of BI-platforms for Cyber Security Predictive Analytics

At present, security services of leading state companies and enterprises are increasingly paying their attention to the new analytical information systems, the so-called business intelligence platforms (BI-platforms). BI-systems allow the top-managers and analysts to work in real time with large information volumes (Big Data). At the same time, tools for secure access to business data sources that have advanced possibilities for consolidating, analyzing, and presenting information are especially relevant. The security trend of recent years is the software integration and the transition from specialized analytical solutions to multi-purpose BI platforms. What kind of BI platforms may be in demand for the Russian security services?

4.6.1 BI-security Platform Requirements

Currently, leading Russian state and military companies are mainstreaming the various business intelligence systems from the traditional ones of the Business Intelligence class (ETL + SQL + reports) to more advanced systems of Advanced Analytics class (Big Data, NoSQL storage, thread-specific data processing online).

According to TATA Consultancy Services [1, 2, 5–7, 9], which summarized data for 1,217 companies from 4 world regions (USA, Europe, Asia, and Latin America), about half of the companies experienced difficulties in obtaining and analyzing "Big Data". The investment volume in business analytics varies considerably among different companies: for half of them, it amounts to more than $ 10 million. In addition, companies that receive revenue through the Internet invest in the development of business intelligence systems even more financial resources.

The research revealed the following key problems in arrangement of the Big Data (in descending order of priority):

- Low computing speed for Big Data and its high variability;
- Time-consuming data sources identifying for analysis;
- Complexity of recruiting employees, able to analyze Big Data;
- High laboriousness of creating visual representations for the results of data processing.

In the Information Week study conducted in 2013 [1, 2, 5–7, 9], 248 companies from various fields, including the state financial sector, took part. The study results revealed the following key problems:

- Low quality of source data (59%);
- Software complexity for data analysis by employees who do not have special training (46%);
- Difficulties in deploying the solution to the entire organization (42%).

At the same time, BI-systems users expect the development of the system functional in the following aspects:

- Visualizations (sparklines, heatmaps, treemaps);
- Analysis (forecasting, statistical analysis, etc.);
- Applying interactive dashboards[4].

According to the well-known analytical company Gartner [1, 2, 5–7, 9], the basis for successful BI-projects is a qualitative and understandable presentation of the results and conclusions from BI-studies: 70–80% of projects that do not pay due attention, end in failure.

TDWI (The Data Warehousing Institute) estimated that poor quality and nontransparency of the presented results and conclusions, based on the gathered information, do not allow responsible executives to make high-quality decisions timely. According to Gartner, the best tool that allows quickly and clearly conveying useful information and conclusions from the BI system to the making decisions person in the company is dashboard: 94% of respondents said that dashboard has become an integral part of their BI project success.

Finally, the study [1, 2, 5–7, 9], which involved 752 companies from around the world, shows that even large companies face difficulties in analyzing large amounts of data. More than 41% of companies said they do not derive maximum benefit from continually collected data. At the same time, 67% of respondents noted that it is important for them to process data in real time. However, attempts to increase the processing data speed lead to a number of problems. Among the main difficulties, the respondents said the following:

- 41% – insufficient skills;
- 39% – delays related to cleaning up and data validation;
- 32% – lack of necessary technologies.
- Thus, if we summarize the results of the above-mentioned and other relevant studies, the main problems of implementing, operating, and maintaining BI-systems are as follows.
- *Existing BI systems are not able to perform calculations on large amounts of data quickly enough.* According to experts [1, 2, 5–7, 9], the amount of data generated around the world in 2013 was more than 7,000 petabytes, and the annual growth rate of this parameter exceeded 40%. More than 80% of BI experts believe that existing systems do not have enough computing speed. According to the Gartner forecast [104, 106–110], while maintaining the growth rates of data generation, 33%

---

[4]Dashboard is a graphical user interface, usually single-page, visualizing the current baseline data values of the information system.

of companies by 2017 will not be able to analyze data because of the inability of existing BI systems to quickly process such volumes.

- *Calculations in BI systems often produce incorrect results due to poor data quality.* The analysis is complicated by the fact that 60% of calculations must be made according to poor quality, unreliable data [1, 2, 5–7, 9], so IT-specialists put the data quality problem in the second place among all the complexities when using analytics. Almost 30% of respondents found the data of their companies unsuitable for analysis with the help of existing BI systems.
- *Developing procedures for converting and cleaning up the source data is time-consuming and costly* [1, 2, 5–7, 9]. In large companies, up to 70% of the study time is spent preparing data for analysis. More than 90% of companies use ETL solutions for transforming data. ETL costs take more than 65% of the data management budget and are up to $67,000 per terabyte per year.
- *Companies do not have specialists skilled to develop mathematically correct algorithms and data analysis methods.* About 41% of large companies named the lack of such specialists a key problem in the implementation of existing BI systems, and by 2018, only in the USA, a shortage of almost 200,000 specialists in this field is expected [1, 2, 5–7, 9].
- *Analysis takes a long time, and its results are received by business users too late.*
- More than 40% of the surveyed companies do not have BI systems capable of making calculations quickly enough. Among executives, more than 70% believe that faster data analysis would help them to improve the company's efficiency, while 26% of companies believe that speeding up payments is the key driver success of working with Big Data [1, 2, 5–7, 9].
- The analysis results are often incomprehensible and impractical. Almost 42% of users are dissatisfied with the conclusions they can see on their dashboards, while 28% of respondents admitted that the analysis results do not allow them to obtain maximum information from the collected data [1, 2, 5–7, 9].
- Existing BI systems do not provide users with the necessary visualization tools. Currently, almost 50% of existing BI systems display information in the form of reports, rather than dashboards, while 43% of users are dissatisfied with the use of visualization tools. Another 44% users lack the predictive dashboard function, and 61% are sure that

dashboards do not provide enough information to make decisions [1, 2, 5–7, 9].

- The analysis results and conclusions are presented in an uncomfortable form and therefore often remain unrequited. About 46% of the interviewed companies stated that the interface complexity is the main problem in using BI systems. Studies note three key disadvantages of existing interfaces:
  - o No complex visualization views;
  - o Interactive functions are not developed;
  - o Access from mobile devices is restricted [1, 2, 5–7, 9].

To solve the identified problems of implementation, operation, and maintenance of modern BI systems, the following technical solutions and corresponding technological trends for the development of advanced AA class BI systems are offered [1, 2, 5–7, 9].

First, most CIOs of large companies began to abandon ETL technology, arguing that 70–80% of the costs associated with BI are spent on permanent rewriting and adaptation of ETL systems.

The possible way to reduce costs is associated with the following new approach: data are directly uploaded to a distributed cluster of analytics and storage, where aggregation, sorting, transformation, and primary data analysis are implemented. Time delays can thus be reduced by a factor of 5–10, since the data at the conversion and download stage do not leave the analytic cluster; also, no additional costs are required for ETL systems software and hardware.

Second, the single storage database was replaced with a distributed NoSQL storage. Today, the typical infrastructure of a large company contains between 300 and 500 internal systems; therefore, it is rather difficult to store data in a single database. As a solution, 53% of companies made attempts to implement NoSQL storage and 43% of them expect to receive ROI[5] more than 25%.

Third, analytics should be performed in a distributed cluster. Due to the high cost of processing in BI systems, companies use on average not more than 12% of their data for analytics. Nevertheless, 12% of companies are already applying a new approach, during harmonization and data enrichment, as well as most of the analytics is performed not in a separate BI system, but

---

[5]Return on investment (ROI) is the ratio of the net profit to cost of investment resulting from an investment of some resource.

in a distributed analytical cluster that stores, processes, and provides data to users.

Fourth, a massive escape from static graphs to interactive online dashboards on a variety of output devices began. Static reports and graphs do not allow us to use up-to-date information, so 75% of companies preferred to use dashboards continually. But dashboards by themselves are not a panacea, interactive solutions are needed to investigate the causes that led to result-based parameters. In addition, 70% of users prefer to see analytics on mobile devices.

4.6.2 BI Security Platform Startup

Using the example of the Liberty Grant BI platform [1, 2, 5–7, 9], we summarize the requirements and specification of a possible multipurpose BI security platform. Doing so, we will take into account that this platform should allow performing high-speed calculations for solving the problems of Russian security services based on a functional library of algorithms and analytical applications (such library has a wide range of the expressive capabilities), for example:

- Analysis of the information technology protection effectiveness;
- Analysis of the computer systems' performance;
- Monitoring of IS threats;
- Analysis of information security tools and CIPF;
- IS analysis, etc.

Typically, BI platforms are based on the following technologies:

- Mixed processing of streaming and packet data;
- Creation of complex analytical procedures from the completed algorithmic blocks;
- Multilevel parallelization of calculations with the provision of high data locality;
- Distributed data storage, parameters and calculation results, parameters and metadata;
- Disclosure of deviations and nonstandard behavior of the analyzed parameters;
- Disclosure of formats, structure and correlations in the source data, etc.

The typical BI platform content can consist of data loading and visualization agents (Figure 4.35), which go to the high-speed distributed storage directly from external systems; in this storage, processing is carried out using

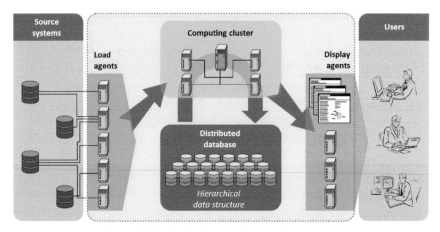

Figure 4.35 Typical BI security platform content.

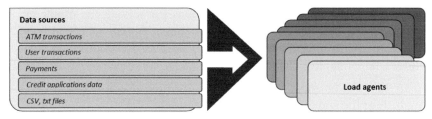

Figure 4.36 Typical data loading agents.

computing cluster procedures. In this case, during the periods of the peak load absence in the repository, there is a data harmonization process, as well as other procedures of the computing BAM (bidirectional associative memory) core. Analytical shell tools request data prepared for more advanced analytics, the results of which are passed to the display agents. Dashboard, as one of the most convenient types of up-to-date information visualization, allows the user to interactively monitor parameter changes and to investigate the causes of such changes.

BAM can integrate applications that are designed to perform applied information security tasks for both state companies and enterprises in various fields of activity.

As components (load agents) (Figure 4.35 and Table 4.10), we can use software adapters to connect to different data sources, for example:

- Connectors for various databases, such as MySql, MsSql, Oracle, MongoDB, MsAccess, and PostgreSQL;

Table 4.10 Typical formats of data loading agents

Databases	CRM and SIEM	Files	Data Streams and Data Buses	Web Services Protocols
MySql	*SAP*	*csv*	*TCP*	*SOAP*
MsSql	*Oracle*	*xls*	*UDP*	*XML-RPC*
Oracle	*HP*	*txt*	*Oracle*	*REST*
PostgreSQL	*IBM*	*XML*	*Tibco*	*WSDL*
MsAccess	*EMC*	*dif*		*UDDI*
MongoDB	*etc.*			

- Some CRMs (such as SAP, Oracle, etc.) and SIEM (such as HP, IBM, EMC, etc.);
- Files and data streams (such as UDP, TCP, xls, txt, csv, dif);
- Popular Web interfaces (such as SOAP, XML-RPC, REST, WSDL, UDDI, etc.).

Depending on the source, the connector device (Figure 4.37) can be very different and include:

- Connection driver;
- Source structure description;
- Data format description;
- Query pool;
- Multiplexer (makes it possible to group some requests into one and then to parse the overall response into several).

A specialized database developed for high-speed computations over multidimensional data can be used in BI platform. For example, as a data model and storage format, HDF5 can be used (Figure 4.38), which:

- Allows processing a wide range of multidimensional data, storing a rich metadata history;

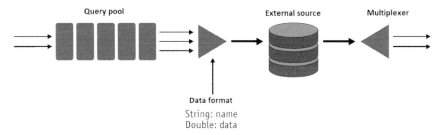

Figure 4.37 Functional connector diagram.

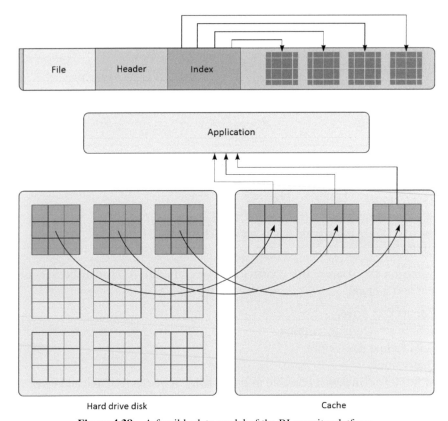

Figure 4.38   A feasible data model of the BI security platform.

- Has a wide selection of predefined data formats, as well as the feasibility to create custom formats;
- Supports parallelization at all work stages without exception: from uploading to data transfer.

In this case, the system allows storing various types of information (Figure 4.40):

- Raw harmonized data;
- Aggregates;
- Time series;
- User data;
- Metadata;
- Safety information.

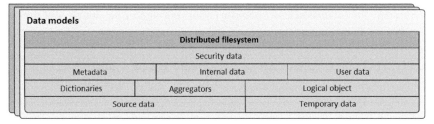

Figure 4.39 Feasible data types of the BI security platform.

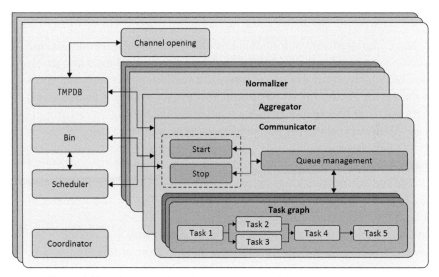

Figure 4.40 The computing cluster's structure of the BI security platform.

Various structures are used for storage:

- B-trees;
- Hash-tables;
- Heaps;
- Arrays;
- Graphs.

The computing cluster's structure of a typical BI platform is shown in Figure 4.40.

In this case, the coordinating function in the system is performed by a gateway (Figure 4.41):

- Evenly distributes user computing tasks to the nodes of the computing cluster;

- Performs the functions of routing data between the local storage databases of processing nodes.
- In this case, the computing cluster itself can consist of a number of nodes. Here, each computing node is controlled by a micro-hub component, which is responsible for starting the server and managing its operation. Each micro-hub can launch several analytical procedures – the pipelines, consisting of several separate computing tasks. In addition, each noted task is a standard analytical block (clustering, classification, identification of main components and outliers, data normalization, etc.), and the output of one block is the input for the next.

The system can perform automatic asynchronous parallelization of the pipelines to servers (Figures 4.41, 4.42, and 4.43), as well as the same parallelization of separate blocks of a single pipeline between processors and cores within the server. Micro-hubs create buffer stores for greater parallelization efficiency between blocks of pipelines.

In addition to parallelizing the pipelines, the system can achieve acceleration in data processing due to the special organization of the on-line server space. So, each of the computing nodes has a local database (cache) containing data with whom the pipeline blocks will operate. The distribution gateway uses the semantic analysis to determine in advance and transmit to the cache the data required for calculations.

Note that the main visualization tool is a dashboard – a flexibly customizable operational panel for monitoring the situation in real-time mode. Access to the dashboard is via the Web interface or with native applications.

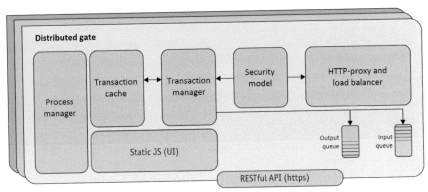

Figure 4.41 The structure of the BI security platform gateway.

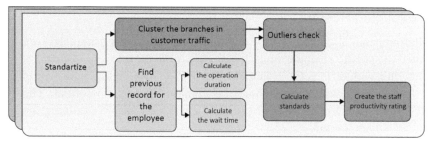

Figure 4.42 An example of a computing algorithm.

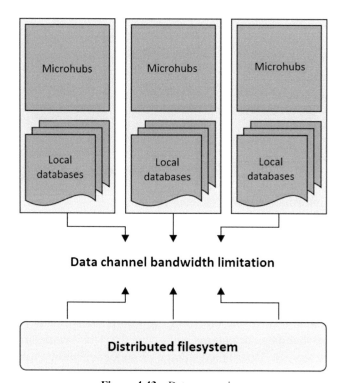

Figure 4.43 Data processing.

For different types of users, the system allows creating several dashboard configurations, consisting of a filtering panel and a set of widgets, whose layout is described in the layout JSON file, which determines the location of graphic elements on the dashboard.

4.6.3 Expected Results

The filtering panel allows narrowing the data set on which the parameters are calculated to specific attribute values.

Here, the widget is a GUI element that allows displaying information useful to the user and system-defined information in a given format. The widget contains a set of parameter values, a chart, and allows changing the section for values' output.

A large set of standard and specialized diagrams can be integrated into the system, such as:

- Histogram;
- Bar chart;
- Bar chart with quartiles;
- Bullet chart;
- Compound duration of the process stages;
- Diagram of the process routes;
- Gantt chart;
- Process graph, etc.

The dashboard view is fully described using the layout. Layout is a universal tool for customizing the dashboard view and allows users to flexibly personalize the informative space for their needs.

Dashboards can have their own hierarchy. For example, from a top-level dashboard with a set of aggregated parameters, we can go to the lowest level and see what the parameter values are.

To connect to an external system, we need to install a module of the corresponding connector on the system side, configure the data transmission channel, and set the update rate (or configure triggers).

When the system receives an incoming request to receive data, Gateway checks the addressant data correctness and opens the data download session to the distributed storage. Distributed storage provides scalability, reliability, and fault tolerance of the system.

Different data from the storage can be processed by different system blocks with different intensities; therefore, the system contains mechanisms that allow reducing the time of request and data transfer:

- Requests to the repository are saved in order to avoid repeated unloading of the same data;
- User request parameters and the results of their processing are saved;
- Multilevel caching is used, which allows us to make a compromise between the speed and the data volume.

The system can contain a wide variety of customizable triggers for various events to launch processing. Therefore, after the data were saved in the repository, the insertion trigger can launch a number of procedures from the computing core, for example, advanced algorithms of the analytic shell.

Here, the computing core (built-in handlers) can consist of procedures (Figures 4.45, 4.46, and 4.47), which help to put the data in a more convenient form for further study (Figure 4.47).

1. Quality estimation of data incoming from heterogeneous sources through multiple connectors puts in correspondence to the data received a set of permissible analytical operations on them, depending on the availability of the necessary fields and attributes.
2. Compiling metadata, that is, information about the data itself. Here is an automatic recognition of types and data formats (structural metadata). Metadata is also collected about individual records for subsequent cataloging and quick access.

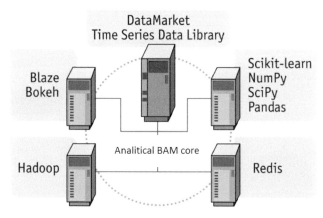

Figure 4.44 Analytical core's composition of the BI security platform.

Figure 4.45 Adapting data models.

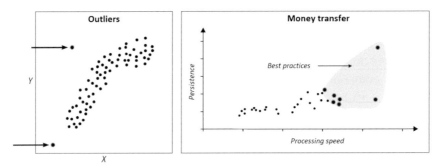

Figure 4.46   Representation of calculation results.

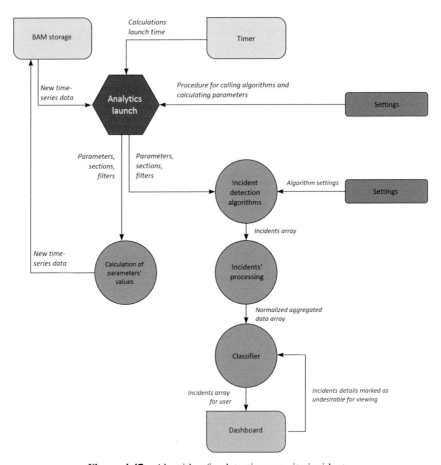

Figure 4.47   Algorithm for detecting security incidents.

3. Harmonization is the data transformation to a unified view while storing the semantic significance by reducing the dimension and put to the final set of types and formats detected automatically. The harmonization procedure can reduce the number of problems associated with heterogeneity of sources, but unlike ETL, it does not set the goal of put all the data to previously known formats.

4. Forming the time series – sequences of data values at equal discrete time intervals. A such time series creation allows further application of a set of analytical procedures, including regression analysis, prediction of future values on a retrospective basis, disclosure of seasonality, autocorrelation analysis, and autoregressive prediction (e.g., the ARIMA algorithm).

5. Searching and cleaning up from the outlayers, which can significantly distort the real parameter value (e.g., the average). On the other hand, the search for outliers often reveals best practices or potential problem areas and failures.

The system uses data that have passed by the harmonization in an easily scalable analytical shell. Thus, the system allows going beyond the standard "what happened?" and getting answers to the questions "why did it happen?", "what happens if the trend continues?", "what will happen in the future?", and "what is the best output?". For this purpose, the following software packages are integrated into the system:

- Mathematical modeling;
- Analysis and optimization;
- Calculating the number and developing strategies;
- Predicting development trends;
- Searching for "hidden" patterns, data mining;
- Machine learning;
- Detecting nonstandard behavior;
- Detecting security incidents, etc.

Here, the main system interface is a dashboard with diagrams for the most important business parameters. Dashboard is an interactive tool that allows the user to choose a data section, filter, and decompose the parameter. The user can also look at problematic areas using regulations, benchmarking or incident reporting.

Choosing a section (Figure 4.48) allows user to see the same parameter from a different point of view: he selects "by RDD" in the combo box of the dashboard section selection for monitoring threats, the widget displays all the important information for each of the RDDs.

For a more detailed exposure in a particular parameter, in order to study it from different points of view and in dynamics, the dashboard performs decomposition, allowing to "fall" in the parameter to an each employee, process, a specific document, and each operation. With this detail, we can see all the problems on the dashboard, identify the reasons for their occurrence, and understand how to fix them.

It is possible to create specialized applications (based on the system) to solve specific security services tasks, first of all, to assess the effectiveness of monitoring IS threats.

Details of the composition, structure, and functionality of the discussed BI security platform can be found in refs [1, 2, 5–7, 9].

***

The largest implementations of BI systems fall on state companies and enterprises. As a rule, customers use several analytical systems simultaneously. For example, Wells Fargo introduced the system Tableau, SAS BI, Terradata, in Royal Bank of Canada – SAS, Terradata, Esperant. Barclays introduced the systems Tableau, SAS BI, and Microsoft Technologies.

The Russian market volume in 2012 amounted to about $1.3 billion and continues to grow at 15% per year. In 2012, more than 250 projects were implemented, the total number of projects was more than 1,400. In 2013, only in the field of public administration, more than 450 projects were implemented, which is 35% of all projects implemented in Russia. The growth rate of investments in BI is more than 100% per year.

Sberbank is actively implementing business intelligence systems (67 projects), VTB24 (39), Uralsib (24), and public control authorities: the Russian Ministry of Finance (13 projects), the Russian Ministry of Health (4), and the Federal Property Management Agency (4). More than 20 companies-users of BI systems own 100 or more licenses.

In general, the dynamics of the BI systems market is positive: from $14 billion in 2013 to $20 billion in 2018. At the same time, the growth BI market rate of Advanced Analytics (AA) class significantly exceeds the growth market rates for traditional BI systems and IT in general and account for more than 33% annually. In particular, significant growth is observed in the segments Big Data, NoSQL, and Real-time analysis. It is projected that by 2020 the volume of the Big Data systems market will amount to $20 billion against $4 billion in 2013. The growth rate (27%) is four times higher than the growth BI market rate. In 2013, the share of companies that planned investments in Big Data increased from 58% to 64%. Also, the NoSQL storage market is growing: by 2017, it will amount to $1.2 billion, compared to $0.3 billion in 2013, which corresponds to an increase of 40% per year.

Today, customers increasingly prefer cloud solutions: 63% of companies see significant advantages in cloud analysis computing technologies. As a result, by 2017, the volume of the cloud analytics market is expected to amount 3.65 billion dollars against 1.2 billion in 2013, which corresponds to a 30% growth annually. Moreover, by 2017, 50% of analytics will account for streaming data; 67% of the companies surveyed consider it important to process and analyze data in real time. As a result, many companies that work in advanced analytics show very high business growth rates, for example, QlikTech (23%), Pitney Bowes (38%), Tibco (50%), and Tableau (76%).

With regard to Russian security services that provide adequate security for banking transactions and electronic commerce, it is also necessary to consider the following. According to the approved "Fundamentals of the Russian Federation state policy in the field of international information security for the period until 2020", much attention is now given to the information technology security and import substitution in technological platforms.

It implicitly means that critical and sensitive information should be processed by software and hardware complexes exclusively from Russian manufacturers and suppliers.

To reduce the dependence level on foreign technologies, the public sector should completely switch to work with Russian software and hardware platform developers. In turn, the commercial segment should also facilitate the transition to Russian developments and more actively use them in their activities. Thus, the most popular in the short term will be Russian BI systems for Big Data work, independent of foreign hardware and software.

For the first time, such systems began to appear on the BI systems market. For example, the hardware and software complex "INNA" for Big Data work under the control of the "Elbrus" OS based on the Linux 2.6.33 kernel and the Russian processor "Elbrus 2C +" by MCST. The mentioned BI system undergoes certification tests on safety requirements of Russian FSTEC on the 2nd level of undocumented features control and the 2nd class of unauthorized access security.

Conclusion

Dear Reader,

We hope that our book was interesting and beneficial to you!

Nowadays, the relevant areas of technological development of digital enterprises include:

- *Artificial intelligence* (speech analytics, computer vision, text analysis, and decision-making systems, NLP). Algorithms based on the models and techniques of artificial intelligence (AI) and machine learning are embedded in the key decision-making processes, first of all, in terms of risk management and offers for customers, as well as in the development of new products and services with the maximum personalization. Voice and text assistants, based on speech recognition and text recognition algorithms, allow creating fundamentally new interfaces for customers.
- *Virtual and Augmented Reality (VR/AR)*. VR/AR technologies are used in the field of creating a multi-faceted client experience. For example, customer service with real-time holograms instead of client managers allows increasing customer satisfaction, and simulations of the complex client situations increase the effectiveness of corporate training programs.
- *Cloud technologies*. Cloud technologies are at the heart of the modern technology platforms and can significantly reduce costs and increase the speed of computing processes.
- *Robotics*. Software robots (bots) allow the user to automate the simple banking processes; they function 24/7 with a small number of errors and cost 66% less than outsourcing. Mechanical robots in the banking sector are also being developed by replacing the physical labor, increasing the accuracy and replacing the work of employees associated with the conditions of an increased danger. The predicted automation potential of robots will reach, for example, 50% in services.

- *Blockchain*. The technology of blocking allows creating decentralized online services on the basis of "smart contracts". At the same time, Blockchain accelerates the decision-making process in the conditions of interaction with a large number of counterparties and also improves the security of transactions.
- *Biometric identification*. Technologies for recognizing the client face, fingerprint, palm, voice, retina, etc. allow adjusting the communication under the psychological profile and is used in personnel management.
- *Internet of Things (IoT)*. The implementation of IoT technologies allows bringing a number of supporting processes in the bank to a new level. For example, tracking the cash movement, optimizing the purchasing process, and operation of ATMs and payment terminals.
- *Gamification*. Application of the gaming principles allows increasing the customer involvement, making the service interesting and exciting, increasing the motivation of employees in the learning and evaluation processes.
- *Cyber security*. The use of innovative, the so-called "end-to-end" cyber security technologies, is an indispensable condition for the sustainable operation of Digital enterprises. Today, the world is actively introducing the advanced solutions in the field of personal data protection and countering cyber fraud based on cognomorphic, neural-like, and genetic algorithms.

The evidence that the proper organization and innovative research conduction in the field of the digital economy is one of the most important tasks today is no longer in a doubt. In this regard, the creation of an effective system of organization and implementation of innovative activities in practice is an urgent and timely scientific and technical problem and requires finding the solution. The approach of the authors of this monograph to the solution of the above-mentioned problem yielded the following results.

In the critical sustainability analysis of the Digital Economy technological platforms operation:

1) System analysis of the computing methods in software and technical influences and the requirements for the stability of the Digital Economy technological platforms functioning are defined, the high vulnerability and insufficient stability of the named platforms in the conditions of destructive effects, the inconsistency of the known methods of information security, control and restoration of computing processes for the sustainable platform operation with destructive intruders impact are revealed;

2) It is established that the solution of the self-restoration problem of the Digital Economy technological platforms in the mass and group nature of cyber-attacks requires the search for new ways of organizing the stable functioning of the named platforms;

3) The expediency of the developing models and methods for organizing sustainable computing under cyber-attack conditions by giving the Digital Economy technology platform possibility to develop immunity to the perturbations of computational processes under conditions of mass attacker intrusions by analogy with the immune system of protecting a living organism, as well as resist the cyberneviation during, not after a hostile impact.

In terms of ensuring the required stability of the Digital Economy technological platforms operation based on self-healing:

1) The concepts of computing with memory for countering cyber-attacks and self-healing of disturbed states, and the technological platforms of the Digital Economy are introduced;

2) The principle of structuring models of disturbance states of the Digital Economy technological platforms and the corresponding stratification of the system organization of platform stable behavior under group and mass cyber-attacks is proposed;

3) The scientific-methodical device suitable for the stable functioning problem solution of Digital Economy technological platforms is chosen and justified:

- Abstract calculator of self-repair model of the Digital Economic platforms under cyber-attack based on the theory of dynamic systems;
- Architecture of the organization system of sustainable operation of the Digital Economic platforms based on the theory of multi-level hierarchical systems;
- Language description of previously unknown types of mass perturbation structures based on the formal languages and grammar theory;
- Self-applicable translational model for the synthesis of abstract self-recovery programs for disturbance states of the Digital Economy technological platforms under mass and group disturbances based on theoretical and system programming methods;
- Architecture of the technological environment for the synthesis of abstracted self-healing programs;

- Specification of the behavior dynamics of perturbed calculations by analogy with the modeling of perturbations in living nature on the basis of catastrophe theory;
- Methods of proving computability properties of reconstructed computations on the basis of similarity theory;
- Method of immune development to destructive disturbances with the use of the results of control and computation restoration theory.

As part of the early detection and prevention of group and mass cyber-attacks on the Digital Economic platform:

1) Ontological model of a structured and context-dependent presentation of knowledge concerning processes of cyberspace conflict in Russian Federation on the basis of the Arnold catastrophe theory was developed;

2) Functional model of the cognitive early-warning cyber security system for critically important governmental information assets of Russian Federation was created;

3) Experimental prototypes of software hardware complexes of the cognitive early-warning cyber security system based on the theory of multilevel hierarchical systems and methods of "cognitive computations" were designed;

4) Common techniques for early-warning cyber security system for Digital Economy technological platforms implemented through the following operations:

   - Input and initial data and knowledge processing used to form the specifications of early-warning cyber-attack in the context of rising cyber threat;
   - Laying down specifications for early-warning cyber-attack operations in information technical conflict based on "cognitive calculations" and logical inference by analogy;
   - Data and knowledge preparation and presentation for the prediction of possible outcomes of information conflicts in state cyberspace and selection of process specifications potentially leading to the prevention of cyber-attacks on critical infrastructure as a whole

The scientific novelty of the results, conclusions, and recommendations lies in the fact that for the first time, the following were offered:

1) Concept of ensuring the functioning stability of the Digital Economy technological platforms with the immunity system based on self-healing calculations;

2) Concept of building specifications for early warning processes of a cyber-attack on the Digital Economy technological platforms in the context of rising cyber threat, contributing to the prevention of destructive impact on these objects.

At the same time, the corresponding scientific and methodical devices were developed:

a) Organization of sustainable computing under group and mass cyber-attacks, including:

- Computing model, resistant to destabilization in a hierarchical multilevel control environment with feedback;
- Models of the simplest perturbations for the synthesis of scenarios for the return of computational processes to equilibrium using the dynamic equations of catastrophe theory;
- Model of semantics of correct calculations representation based on the static and dynamic similarity invariants;
- Self-healing methods with memory using permissive standards;
- Technique for detecting and neutralizing distributed cyber-attacks such as "denial of service" (DDoS) applying the immunity system;

b) Semantic-syntactic modeling of processes, which allows realization of the synthesis of early warning scenarios about cyber-attack and proactive behavior, while the implementation of the cognitive approach to the creation of an early warning system on cyber threat for the critical state infrastructure, allowing managing the depth processing of special knowledge on information security, is fundamentally new.

The theoretical significance and scientific value of the obtained results are:

- Formalization of a structured knowledge system of the early-warning cognitive system for cyber-attacks using the complementary formal semantic models: denotational semantics of computable structures, axiomatic semantics of computability properties, and operational semantics of computational actions;
- Synthesis of the intensional expansion of cyber security ontologies for an early-warning cyber security system for constructing an arbitrarily complex multilayered ontological construction;
- Development of a consistent intensional theory of partially ordered structured models of functioning for formalizing the semantics of context-dependent knowledge concerning the conflict subject in information space;

- Construction of a formal axiomatized solvable interpreted theory for modeling the data and context-dependent knowledge hierarchical representation and generation of truth-based reasoning, applying the semantics of roles in the mentioned multilayered cyber security ontologies;
- Construction of a hierarchical level-by-level coordinated early-warning system for cyber-attacks, based on the theory of hierarchical systems and methods of "cognitive calculations" used to search for potential solution algorithms;
- Proof of model completeness, consistency, and the named theory solvability.

Practical significance of the obtained results is the outcome of work that has been brought to a level suitable for practical testing and used for:

- Design and development of prototypes and an open segment range of the national early-warning cyber security system for critically important government information assets, ensuring a general increase in the completeness of the parameters monitored;
- Construction of formalized models for cyber security threat, providing increased and detailed depth of the forecasts for multilayered potentially implemented attacks on critically important state information assets;
- Generation of process specifications to be potentially implemented in information conflict;
- Development of private methods for giving early warning of and preventing potential attacks;
- Setting up and conducting national and international cyber-lessons, as well as various types of training sessions, seminars, and practices aimed at improving and developing models and methods for an early-warning cyber security system for Digital Economy technological platforms.

It is significant that here the use of the immunity system allows developing and accumulating measures to counteract previously unknown cyber-attacks, to detect group and mass impacts leading to borderline with catastrophic states, partially restore computing processes that support aiming the targets based on the Digital Economy technological platforms, preventing their degradation, and draw unrestorable or difficult-to-recover perturbations against intruders.

In conclusion, we would like to note that the current book is a first attempt to share with beginners and experienced specialists the experience available in research and development in the field as well as to develop some practical recommendations. We intend to continue our work in this direction, and for this, your feedback, dear readers, is required. Taking this into account, we will be very grateful for any comments and suggestions on expanding and improving the quality of the book's material. You may contact the authors at this email address:

S.Petrenko@innopolis.ru

Germany – Russia

August 2018

References

[1] S. A. Petrenko, A. S. Petrenko, Creation of a cognitive supercomputer for the cyber - attack prevention, Protection of information. Inside, No. 3 (75), pp. 14–22, Russia, 2017.

[2] S. A. Petrenko, A. Ya. Asadullin, A. S. Petrenko, Evolution of the von Neumann architecture, Protection of information. Inside. No. 2 (74), pp. 8–28, Russia, 2017.

[3] A. S. Petrenko, S. A. Petrenko, Super-productive monitoring centers for security threats, Part 1, Protection of information. Inside, No. 2 (74), pp. 29–36, Russia, 2017.

[4] A. S. Petrenko, S. A. Petrenko, Profile of the security of the mobile operating system, Tizen, Information security. Inside, No. 4 (76), pp. 33–42, Russia, 2017.

[5] S. A. Petrenko, A. S. Petrenko, New Doctrine of Information Security of the Russian Federation, Information Protection, Inside. No. 1 (73), pp. 33–39, Russia, 2017.

[6] S. A. Petrenko, T. I. Shamsutdinov, A. S. Petrenko, Scientific and technical problems of development of situational centers in the Russian Federation, Information protection, Inside, No. 6 (72), pp. 37–43, Russia, 2016.

[7] S. A. Petrenko, V. A. Kurbatov, I. A. Bugaev, A. S. Petrenko, Cognitive system of early cyber-attack warning, Protection of information, Inside, No. 3 (69), pp. 74–82, Russia, 2016.

[8] A. S. Petrenko, S. A. Petrenko, Big data technologies in the field of information security, Protection of information, Inside, No. 4 (70), pp. 82–88, Russia, 2016.

[9] A. S. Petrenko, I. A. Bugaev, S.A. Petrenko, Master data management system SOPKA, Information protection, Inside. No. 5 (71), pp. 37–43, Russia, 2016.

[10] A. S. Petrenko, S. A. Petrenko, Designing the corporate segment SOPKA, Protection of information, Inside. No. 6 (72), pp. 47–52, Russia, 2016.

[11] A. S. Petrenko, S. A. Petrenko, The first interstate cyber-training of the CIS countries: "Cyber-Antiterror-2016", Information protection, Inside, No. 5 (71), pp. 57–63, Russia, 2016.

[12] S. A. Petrenko, D. D. Stupin, National Early Warning System on Cyber-attack: a scientific monograph [under the general editorship of SF Boev] "Publishing House" Athena, University of Innopolis; Innopolis, Russia, p. 440, 2017.

[13] S. A. Petrenko, A. A. Petrenko, Ontology of the cyber-security of self-healing SmartGrid, Protection of information, Inside, No. 2 (68), pp. 12–24, Russia, 2016.

[14] A. A. Petrenko, S. A. Petrenko, The way to increase the stability of LTE-network in the conditions of destructive cyber – attacks, Questions of cybersecurity, No. 2 (10), pp. 36–42, Russia, 2015.

[15] A. A. Petrenko, S. A. Petrenko, Cyberunits: methodical recommendations of ENISA, Questions of cybersecurity, No. 3 (11), pp. 2–14, Russia, 2015.

[16] A. A. Petrenko, S. A. Petrenko, Research and Development Agency DARPA in the field of cybersecurity, Questions of cybersecurity, No. 4 (12), Russia, pp. 2–22, 2015.

[17] S. A. Petrenko, Methods of Information and Technical Impact on Cyber Systems and Possible Countermeasures, Proceedings of ISA RAS, Risk Management and Security, pp. 104–146, Russia, 2009.

[18] A. A. Petrenko, S. A. Petrenko, Intranet Security audit (Information technologies for engineers), DMK Press, p. 416, Moscow, Russia, 2002.

[19] S. A. Petrenko, S. V. Simonov, Management of Information Risks, Economically justified safety (Information technology for engineers), DMK-Press, Moscow, Russia, p. 384, 2004.

[20] S. A. Petrenko, V. A. Kurbatov, Information Security Policies (Information Technologies for Engineers), DMK Press, p. 400, Russia, Moscow, 2005.

[21] S. A. Petrenko, A. S. Petrenko Lecture 12, Perspective tasks of information security, Intelligent Information Radiophysical Systems, MSTU, N. E Bauman; [ed. SF Boev, DD Stupin, AA Kochkarova], Moscow, Russia, pp. 155–166, 2016.

[22] A. S. Petrenko, S. A. Petrenko, Super-productive monitoring centers for security threats, Part 2, Protection of information, Inside, No. 3 (75), pp. 48–57, Russia, 2017.

[23] S. A. Petrenko, The concept of maintaining the efficiency of cybersystem in the context of information and technical impacts, Proceedings of the ISA RAS, Risk management and safety, Vol. 41, pp. 175–193, Russia, 2009.

[24] S. A. Petrenko, Methods of detecting intrusions and anomalies of the functioning of cybersystem, Risk management and safety, Vol. 41, pp. 194–202. Russia, 2009.

[25] S. A. Petrenko, The Cyber Threat model on innovation analytics DARPA, Trudy SPII RAN, Issue. 39, pp. 26–41, Russia, 2015.

[26] S. A. Petrenko, Methods of ensuring the stability of the functioning of cybersystems under conditions of destructive effects, Proceedings of the ISA RAS, Risk management and security Vol. 52, pp. 106–151, Russia, 2010.

[27] A. S. Nashekin, "Rostelecom" As a global service provider, Telecommunications. No. 9, Russia, 2011.

[28] R. V. Kravtsov, "Rostelecom" builds an information society: from the electronic government to the electronic state, Telecommunications. No. 4, Russia, 2013.

[29] The Order of FSTEC of Russia of 11 February 2013 No. 17 "On Approving the Requirements for the Protection of Information that Is not a State Secret Contained in State Information Systems".

[30] Order FSTEC of Russia of February 18, 2013 No. 21 "On the approval of the composition and content of organizational and technical measures to ensure the safety of personal data when processing them in personal data information systems."

[31] D. N. Biryukov, A. S. Petrenko, S. A. Petrenko, Method for synthesizing the structure of the self-healing program for computations with memory: in the collection. "Distance educational technologies", Proceedings of the II All-Russian Scientific and Practical Internet Conference, pp. 188–192, Russia, 2017.

[32] A. G. Lomako, S. A. Petrenko, A. S. Petrenko, Model of the Immune System of Stable Computations, In: Information Systems and Technologies in Modeling and Control. Materials of the all-Russian scientific-practical conference, pp. 250–254, Russia, 2017.

[33] A. G. Lomako, A. S. Petrenko, S. A. Petrenko, Representation of perturbation dynamics for the organization of computations with memory, In: Remote educational technologies, Materials of the II All-Russian Scientific and Practical Internet Conference, pp. 355–359, 2017.

[34] A. G. Lomako, S. A. Petrenko, A. S. Petrenko, Realization of the immune system of the stable computations organization, In: Information systems and technologies in modelling and management, Materials of the All-Russian scientific and practical conference, pp. 255–259, Russia, 2017.

[35] K. A. Makoveychuk, S. A. Petrenko, A. S. Petrenko, Organization of calculations with memory, Information Systems and Technologies in Modeling and Control. Materials of the all-Russian scientific-practical conference, pp. 260–266, Russia, 2017.

[36] K. A. Makoveychuk, S. A. Petrenko, A. S. Petrenko, Modeling the recognition of destructive effects on computer calculations, Information Systems and Technologies in Modeling and Control. Materials of the all-Russian scientific-practical conference, pp. 155–161, Russia, 2017.

[37] K. A. Makoveychuk, S. A. Petrenko, A. S. Petrenko, Modeling of self-recovery of computations under perturbation conditions, Information Systems and Technologies in Modeling and Control. Materials of the all-Russian scientific-practical conference, pp. 162–166, 2017.

[38] S. A. Petrenko, A. S. Petrenko, The task of semantics of partially correct calculations in similarity invariants, Remote educational technologies, Materials of the II All-Russian Scientific and Practical Internet Conference, pp. 365–371, Russia, 2017.

[39] M. A. Mamaev, S. A. Petrenko, Technologies of information protection on the Internet. - St. Petersburg.: publishing house "Peter", p. 848, Russia, St. Petersburg, 2002.

[40] S. A. Petrenko, Stability problem of the cybersystem functioning under the conditions of destructive effects, Proceedings of the ISA RAS, Risk management and security, Vol. 52. pp. 68–105, Russia, 2010.

[41] S. A. Petrenko, A. A. Petrenko Information Security Audit Internet/Intranet (Information Technologies for Engineers), 2 nd ed, DMK-Press, p. 314, Moscow, Russia, 2012.

[42] Federal Law No. 149-FZ of July 27, 2006 (edition of July 6, 2016) "On Information, Information Technologies and Information Protection".

[43] E. G. Vorobiev, S. A. Petrenko, I. V. Kovaleva, I. K. Abrosimov, Analysis of computer security incidents using fuzzy logic, In Proceedings of the 20th IEEE International Conference on Soft Computing and Measurements (24–26 May 2017), SCM 2017, pp. 349–352, St. Petersburg, Russia, 2017.

[44] E. G. Vorobiev, S. A. Petrenko, I. V. Kovaleva, I. K. Abrosimov, Organization of the entrusted calculations in crucial objects of informatization under uncertainty, In Proceedings of the 20th IEEE International Conference on Soft Computing and Measurements (24–26 May 2017). SCM, pp. 299–300. DOI: 10.1109/SCM.2017.7970566, St. Petersburg, Russia, 2017.

[45] S. M. Abramov, Research in the field of supercomputer technologies of the IPS RAS: a retrospective and perspective. Proc, Proceedings of the International Conference "Software Systems: Theory and Applications", Publishing house "University of Pereslavl", vol. 1. pp. 153–192. Russia, Pereslavl, 2009.

[46] S. M. Abramov, E. P. Lilitko, State and prospects of ultra-high performance computing systems development, Information Technologies and Computing Systems, No. 2. pp. 6–22, Russia, 2013.

[47] A. G. Arbatov, Real and imaginary threats: Military power in world politics in the beginning of the XXI century. [Electronic resource], Russia in global politics, Access mode: http: //www.global-affairs.ru/number/Ugrozy-realnye-i-mnimye-15863, Russia, March 3, 2013.

[48] Aristotle. Comp. in 4 volumes (Series "Philosophical heritage"). - M.: Thought. - 1975–1983.

[49] D. N. Biryukov, A. G. Lomako, S. A. Petrenko, Generating scenarios for preventing cyber – attacks, Protecting information, Inside, No. 4 (76), 2017.

[50] D. N. Biryukov, A. P. Glukhov, S. V. Pilkevich, T. R. Sabirov, Approach to the processing of knowledge in the memory of an intellectual system, Natural and technical sciences, No. 11, pp. 455–466, Russia, 2015.

[51] V. V. Andreev, K. B. Zdiruk, IV Jupiter: implementation of corporate security policy in computer networks, Open systems, No. 7–8, pp. 43–46, Russia, 2003.

[52] E. V. Batueva, American concept of threats to information security and its international political component: doctoral thesis of polit. Sciences, MGIMO (U) Ministry of Foreign Affairs of the Russian Federation, p. 207, Moscow, Russia, 2014.

[53] P. A. Baranov, Detection of anomalies based on the application of the criterion of the dispersion degree, Proceedings of the XIV All-Russian Scientific Conference "Information Security Problems in the Higher

School System", Publishing house department of the St. Petersburg State Polytechnic University, pp. 25–27, St. Petersburg, Russia, 2007.

[54] A. V. Bedritsky American policy of cyber space control, Problems of national strategy, No. 2 (3), pp. 25–40, Russia, 2010.

[55] S. M. Abramov, History of development and implementation of a series of Russian supercomputers with cluster architecture, History of domestic electronic computers, 2nd ed., Rev. and additional; color. Ill, Publishing house "Capital Encyclopedia", Moscow, Russia, 2016.

[56] S. A. Petrenko, A. S. Petrenko, From Detection to Prevention: Trends and Prospects of Development of Situational Centers in the Russian Federation, Intellect & Technology, No. 1 (12), pp. 68–71, Russia, 2017.

[57] S. A. Petrenko, A. S. Petrenko, Practice of application the GOST R IEC 61508, Information protection, Insider, No. 2 (68), pp. 42–49, Russia, 2016.

[58] L. V. Massel, Problems of Smart Grid Creation in Russia from the Perspective of Information Technologies and Cyber Security, Proceedings of the All-Russian Seminar with International Participation, Methodological Issues of Research into the Reliability of Large Energy Systems, Reliability of energy systems: achievements, problems, prospects, ISEM SB RAS, vol. 64, pp. 171–181, Irkutsk, Russia, 2014.

[59] V. B. Tarasov From multiagent systems to intellectual organizations, A series of "Sciences about the artificial", Editorial URSS, p. 352 Moscow, Russia, 2002.

[60] Report on the implementation of the project for the technological platform "Intellectual Energy System of Russia" (TPII) in 2014 implementation and the action plan for the TP IES for 2015 - M. - 2015.

[61] B. M. Velichkovsky Cognitive technical systems, Computers, brain, cognition: successes of cognitive sciences, Nauka, pp. 273–292, Moscow, Russia, 2008.

[62] GOST R 51583. Information protection. Sequence of automated operational system formation in protected mode. Basic provisions.

[63] O. M. Ovdei, G. Yu. Proskudina, Review of Ontology Engineering Tools - Institute of Software Systems, National Academy of Sciences of Ukraine, Vol. 7. - Vol. 4, Kiev, Ukraine, 2004.

[64] GOST 22.0.05–97. Safety in emergencies. Technogenic emergencies. Terms and definitions.

[65] GOST R 51624. Information protection. Protected automated systems. General requirements.

[66] E. S. Zinovieva, International Internet Governance: Conflict and Cooperation: Textbook, - MGIMO-University. p. 169, 2011.

[67] I. N. Panarin, L. G. Panarina, Information War and Peace, OLMA-PRESS, p. 384, Moscow, Russia, 2003.

[68] I. I. Levin, A. I. Dordopulo, I. A. Kalyaev, Yu. I. Doronchenko, M. K. Razkladkin, Modern and promising high-performance computing systems with reconfigurable architecture, Proceedings of the international scientific conference "Parallel Computing Technologies (PaVT'2015)", Ekaterinburg, March 31 – April 2, 2015, Publishing Center of SUSU, pp. 188–199, Chelyabinsk, Russia, 2015.

[69] K. B. Zdiruk, The organization of a secure system for storing and processing data in heterogeneous computer networks, Questions of information protection, No. 3 (78), pp. 6–9, Russia, 2007.

[70] Appliance of information and communication technologies for development. Resolution of the General Assembly of the UN. Document A / RES / 65/141 dated December 20, 2010 [Electronic resource]. - Access mode: http://www.un.org/en/ga/search/view_doc.asp?symbol=A/RES/65/141.

[71] Advances in the field of information and telecommunications in the context of international security. Report of the UN Secretary-General. Document A / 66/152 of 15 July 2011 [Electronic resource]. - Access mode: http://www.un.org/en/documents/ods.asp?m=A/66/152.

[72] S. A. Petrenko, A. S. Petrenko New Doctrine as an Impulse for the Development of Domestic Information Security Technologies // Intellect & Technology, No. 2 (13), pp. 70–75, Russia, 2010.

[73] V. Pashkov,: US information security. Foreign Military Rev. 10, pp. 3–13, Russia, 2010.

[74] I. N. Pashchenko, V. I. Vasiliev, M. B. Guzairov, Protecting Information in Smart Grid Networks Based on Intelligent Technologies: Designing the Rules Base, YuFU. Technical science, Izvestia, pp. 28–37, Russia, 2015.

[75] S. F. Boev, A. A. Kochkarov, D. D. Stupin, The role and possibilities of pre-university training in the problem of the formation of highly qualified specialists for high-tech branches of the real economy and the experience of the RTI Systems Concern: materials of the International Scientific Conference "Forming the Identity of Finno-Ugric world

and Russian education", Mordovian state publishing house University, pp. 330–333. Saransk, Russia, 2011.

[76] GOST 15.000–94. System of product development and launching into manufacture. Basic provisions.

[77] S. F. Boev, A. A. Kochkarov, D. D. Stupin, The role and possibilities of pre-university training in the problem of the formation of highly qualified specialists for high-tech branches of the real economy and the experience of the RTI Systems Concern: materials of the International Scientific Conference "Forming the Identity of Finno-Ugric world and Russian education", Mordovian state publishing house University, pp. 330–333. Saransk, 2011.

[78] GOST 15.000-94. System of product development and launching into manufacture. Basic provisions.

[79] GOST 15.000-94. System of product development and launching into manufacture. Basic provisions.

[80] S. F. Boev, A. A. Kochkarov, D. D. Stupin, The role and possibilities of pre-university training in the problem of the formation of highly qualified specialists for high-tech branches of the real economy and the experience of the RTI Systems Concern: materials of the International Scientific Conference "Forming the Identity of Finno-Ugric world and Russian education", Mordovian state publishing house University, pp. 330–333. Saransk, 2011.

[81] GOST 15.000-94. System of product development and launching into manufacture. Basic provisions.

[82] GOST R MEK 61508–2012. Functional safety of electrical electronic programmable electronic safety-related systems. Part 1–7. Standartinform, Moscow (2014).

[83] N. Wiener, Cybernetics, or Control and Communication in Animal and Machine. 2nd edn, Science, The main edition of publications for foreign countries, p. 344 Moscow, Russia, 1983.

[84] V. M. Vishnevsky, A. I. Lyakhov, S. L. Portnoy, I. V. Shakhnovich, Broadband Wireless Information Transmission Networks. The technosphere, Moscow, Russia, 2005.

[85] V. V. Voevodin, V. L. B. Voevodin, Parallel Computing, BHV-Petersburg p. 609 St. Petersburg, Russia, 2002.

[86] V. E. Wolfengagen, Categorical abstract machine. Lecture Notes: An Introduction to Computing. 2nd edn, JSC "Center YurInfo", p. 96 Moscow, Russia, 2002.

[87] T. N. Vorozhtsova: Ontology as a basis for the development of an intellectual system for ensuring cybersecurity. Ontol. Des. 4(14), pp. 69–77, Russia, 2014.

[88] T. A. Gavrilova, V. F. Khoroshevsky, Bases of Knowledge of Intellectual Systems, A Textbook for High Schools p. 384 Peter, St. Petersburg, Russia, 2000.

[89] S. N. Grinyaev, The battlefield - cyberspace: theory, techniques, means, methods and systems of information warfare, Mn Harvest. p. 448, 2004.

[90] V. F. Guzik, I. A. Kalyaev, I. I. Levin, Reconfigurable computing systems; [under the Society. ed. I.A. Kalyayeva], Publishing house SFU, p. 472, Rostov-on-Don, 2016.

[91] Russia-US bilateral project on cybersecurity. Fundamentals of critical terminology. Ed. 1; [main. Ed .: Carl Frederick Rauscher, V. V. Yashchenko]. - 2011 [Electronic resource]. - Access mode: http://iisi. msu.ru/UserFiles/File/Terminology% 20IISI% 20EWI/Russia-U% 20S% 20% 20bilateral% 20on% 20terminology% 20RUS.pdf.

[92] Report of government experts on achievements in the field of information and telecommunications in the context of international security. Document A / 65/201 of 30 July 2010 [Electronic resource]. - Access mode: http://www.un.org/disarmament/HomePage/ODAPublications/ DisarmamentStudySeries/PDF/DSS_33_Russian.pdf.

[93] Achievements in the field of information and telecommunications in the context of international security. Report of the UN Secretary-General. Document A / 66/152 of 15 July 2011 [Electronic resource]. - Access mode: http://www.un.org/en/documents/ods.asp?m=A/66/152.

[94] Doctrine of Information Security of the Russian Federation (approved by the Decree of the President of the Russian Federation of December 5, 2016 No. 646).

[95] S. M. Ermakov, Transformation of NATO after the Lisbon Summit in 2010: from the defense of the territory to the protection of the public domain, Probl. Natl. Strateg. 4(9), pp. 107–128, Russia, 2011.

[96] L. Y. Zhilyakova,: The associative memory model based on a dynamic resource network. In:Proceedings of the conference "Management in technical, ergatic, organizational and network systems (UTEOSS2012)", State Scientific Center RF, JSC Concern CSRI, Elektropribor, pp. 1160–1163. St. Petersburg, Russia, 2012.

[97] V. Zhukov, The views of the US military leadership on the information warfare. Foreign Military Rev. 1, pp. 2–8, Russia, 2001.

[98] K. B. Zdiruk, A. V. Astrakhov, A. V. Lonsky, The model of information protection in heterogeneous computer networks based on the architecture of built-in "protected circuits". Proceedings of the Xth Russian Scientific and Technical Conference "New Information Technologies in Communication Systems and management", 1–2 June 2011, pp. 543–545. Kaluga, Russia, 2011.

[99] K. B. Zdiruk, The organization of a secure system for storing and processing data in heterogeneous computer networks, Questions of information protection, No. 3 (78), pp. 6–9, 2007.

[100] E. S. Zinovieva, International Internet Governance: Conflict and Cooperation: Textbook, MGIMO-University, p. 169, Moscow, Russia, 2011.

[101] I. Yu. Alekseeva, et al. Information Challenges of National and International Security; [under the Society. ed. A. V. Fedorova, VN Tsigichko], PIR Center, p. 328, Moscow, Russia, 2001.

[102] V. N. Tsygichko, D. S. Votrin, A. V. Krutskikh, G. L. Smolyan, D. S. Chereshkin, Information weapons are a new challenge to international security, Institute of System Analysis of the Russian Academy of Sciences, p. 52, Moscow, 2000.

[103] Use of information and communication technologies for development. UNGA Resolution. Document A / RES / 65/141 dated December 20, 2010 [Electronic resource]. - Access mode: http://www.un.org/en/ga/search/view_doc.asp?symbol=A/RES/65/141.

[104] I. D. Klabukov, M. D. Alekhin, A. A. Nekhina, The DARPA research program for 2015, Moscow, 2014.

[105] I. A. Kalyaev, I. I. Levin, E. A. Semernikov, V. I. Shmoilov, Reconfigurable Multicopy Computing Structures; [under the Society. ed. I. A. Kaliayev] 2nd edn, Pub. House of the Southern Scientific Center RAS, p. 344, Rostov-on-Don, Russia, 2009.

[106] E. Kaspersky, Computer Malignity, Peter, St. Petersburg, p. 208, Russia, 2008.

[107] I. D. Klabukov, M. D. Alekhin, S. V. Musienko, The sum of the national security and development technologies, Moscow, 2014.

[108] Cl. Richard, N. Robert, The Third World War. What will it be like?, Publishing house "Peter", St. Petersburg, 2011.

[109] A. S. Kleschev, I. L. Artemieva,: Mathematical models of ontologies of subject domains. Part 2. Components of the model. STI. Ser. 2. 3, pp. 19–29, 2001.

[110] A. N. Kolmogorov, Automats and life, In: Berg, A.I., Kolman, E. (eds.) Cybernetics: Expected and Cybernetics Unexpected, Science, pp. 12–30. Moscow, 1968.

[111] The concept of the state system for detecting, preventing and eliminating the consequences of cyber - attacks on the information resources of the Russian Federation (approved by the President of the Russian Federation on December 12, 2014, No. K 1274).

[112] The concept of foreign policy of the Russian Federation (approved by the Decree of the President of the Russian Federation of November 30, 2016, No. 640

[113] The concept of the development of an intelligent electric power system in Russia with an actively adaptive network. OJSC "FGC UES" OJSC "Scientific and technological center of electric power industry". Moscow, 2011.

[114] G. Korsakov,: Information weapons of the superpower. Ways. Peace. Secur. 1(42), pp. 34–60, Moscow, 2012.

[115] I. V. Kotenko,: Intellectual mechanisms of cybersecurity management. Proceedings of ISA RAS. Risk Manag. Safety, 41, pp. 74–103, Moscow, Russia, 2009.

[116] International Information Security: World Diplomacy: Sat. materials; [under the Society. Ed. S. A. Komov], p. 272, Moscow, Russia, 2009.

[117] International information security: problems and solutions; [under the Society. Ed. S. A. Komov], p. 264, Moscow, Russia, 2009.

[118] N. Marz, J. Warren, Big data. Principles and practice of building scalable data processing systems in real time, Williams, p. 292, Moscow, Russia, 2016.

[119] Common internal security space in the EU: political aspects; [responsible. Ed. - S. V. Utkin], IMEMO RAS, p.146, Moscow, Russia, 2011.

[120] Report on the implementation of the project for the implementation of the technological platform "Intellectual Energy System of Russia" (TPII) in 2014 and the action plan for the TP IES for 2015, Moscow, Russia, 2015.

[121] I. Panarin Information war and power, OLMA-PRESS, p. 224, Moscow, Russia, 2001.

[122] D. N. Biryukov, A. G Lomako, Yu. G. Rostovtsev, The appearance of anticipatory systems to prevent the risks of cyber threat realization, Proceedings of SPIIRAS, Issue. 2 (39), pp. 5–25, Russia, 2015.

[123] V. G. Khoroshevsky, Architecture of Computing Systems. MSTU Them, p. 520 N. E. Bauman, Moscow, Russia 2008.

[124] M. G. Kurnosov, Models and algorithms for embedding parallel programs in distributed computing systems: Doctoral thesis in Technical. Science, Siberian State University of Telecommunications and Informatics, p. 177, Novosibirsk, Russia, 2008.

[125] V. K. Levin, Communication network MVS-express, Inf. Technol. Comput. Syst. 1C, pp. 10–24, Russia, 2014.

[126] A. A. Sidnev, A. V. Gorshkov, A. V. Linev, A. V. Sysoev, V. P. Gergel, E. A. Kozinov, I. B. Meerov, S. I. Bastrakov, Introduction to the principles of functioning and application of modern multinuclear architectures (by the example of Intel Xeon Phi). INTUIT, Moscow (2008) [Electronic resource]. Access mode: http://www.intuit.ru/goods_store/ebooks/9709/

[127] R. L. Smelyansky, Program-Configurable Networks, Open Systems. 5 (2012) [Electronic resource]. Access mode: http://www.osp.ru/os/2012/09/13032491/

[128] A. V. Bedritsky, Information War: Concepts and Their Implementation in the US, RISI, p.183, Moscow, Russia, 2008.

[129] A. V. Bedritsky, The Evolution of the American Concept of Information War, RISI, Analytical Rev. (3), p. 26, Moscow, Russia, 2003.

[130] D. N. Biryukov, Cognitive-functional memory specification for simulation of purposeful behavior of cyber systems. Proc. SPIIRAS. 3(40), pp. 55–76 Russia, 2015.

[131] D. N. Biryukov, A. G. Lomako, T. R. Sabirov, Multilevel Modeling of Pre-Emptive Behavior Scenarios. Problems of Information Security. Computer systems, Publishing house of Polytechnic University, vol. 4, pp. 41–50. St. Petersburg, Russia, 2014.

[132] D. N. Biryukov, Y. G. Rostovtsev, Approach to constructing a consistent theory of synthesis of scenarios of anticipatory behavior in a conflict. Proc. SPIIRAS. 1(38), pp. 94–111, Russia, 2015.

[133] D. N. Biryukov, A. G. Lomako, Approach to Building a Cyber Threat Prevention System. Problems of Information Security. Computer systems, Publishing house of Polytechnic University, vol. 2, pp. 13–19, St. Petersburg, Russia, 2013.

[134] M. M. Bongard, The Problem of Recognition, Fizmatgiz, Moscow, Russia, 1967.

[135] V. A. Bocharov, V. I. Markin, Fundamentals of Logic. Moscow State University, Moscow, 2008.

[136] S. V. Vasyutin, S. S. Zavyalov, Neural network method for analyzing the sequence of system calls for the detection of computer attacks and the classification of application modes. Methods and Means of Information Processing: Proceedings of the Second All-Russian Scientific Conference; [ed. member corr. RAS L. N. Koroleva], Pub. Department of the Factor of Computational Mathematics and Cybernetics of the Moscow State University. M. V. Lomonosov, pp. 142–147, Russia, Moscow, 2005.

Index

A

Agent 58, 115, 228, 420
Alert 151, 152, 337
Anomaly Detection 159, 283, 355
Antivirus Software 123, 147, 356
Application-Based Intrusion Detection and Prevention System 176, 180
Artificial Intelligence Models and Methods 68, 156

B

Big Data 24, 157, 211, 431
Black list 176
Blindly 77, 80

C

Cognitive Computing 436
Console 105, 176, 180, 291

D

Database Server 300, 317
Digital Enterprise 4, 118, 217, 406
Digital Transformation 7
Distributed ledger technologies (block chain) 24

F

False Positive 248, 282, 297, 299
Flow 24, 100, 216, 393

H

Host-Based IDs 283

I

IIoT/IoT Internet Things 1, 29, 126, 161
Incident 12, 136, 292, 429
Industry 4.0 2, 29, 126, 406
Internet of things (IIoT/ Industri 4.0) 24, 124, 345, 434
Intrusion Detection 283
Intrusion Detection and Prevention 176, 180
Intrusion Detection System 283
Intrusion Prevention 163
Intrusion Prevention System 180

J

Jamming 49

M

Malware 136, 182, 216, 237
Management Network 31, 46
Management Services 172, 173

N

Network-Based Intrusion
Detection and Prevention
System 176
Neuro and cognomorphic
computations 37

P

Promiscuous Mode 69

Q

Quantum technologies
(Q-computing) 23, 38

S

Security Software
Development 354, 406

Sensor 64, 219, 285, 296
Signature 143, 180, 288
Signature-Based Detection 240

T

Threshold 160, 241, 402
Tuning 153

V

Virtual and augmented reality
technologies 37, 433

W

White list 178
Wireless technology
(5G) 67, 110

About the Author

Prof. Sergei Petrenko
Innopolis University, Russia

He was born in 1968 in Kaliningrad (the Baltic). In 1991 he graduated with honors from the Leningrad State University with a degree in mathematics and engineering. In 1997 - adjuncture and 2003 doctorate.

The designer of information security systems of critical information objects:

- Three national Centers for Monitoring Information Security Threats and two Situational-Crisis Centers (RCCs) of domestic state;
- Three operators of special information security services MSSP (Managed Security Service Provider) and MDR (Managed Detection and Response Services) and two virtual trusted communication operators MVNO;
- More than 10 State and corporate segments of the System for Detection, Prevention and Elimination of the Effects of Computer Attacks (SOPCA) and the System for Detection and Prevention of Computer Attacks (SPOCA);
- Five monitoring centers for information security threats and responding to information security incidents CERT (Computer Emergency Response Team) and CSIRT (Computer Security Incident Response Team) and two industrial CERT industrial Internet IIoT/IoT.

Head of the State Scientific School "Mathematical and Software Support of Critical Objects of the Russian Federation".

Expert of the Section on Information Security Problems of the Scientific Council under the Security Council of the Russian Federation.

Scientific editor of the magazine "Inside. Data protection".

Doctor of Technical Sciences, Professor.

It is part of the management of the Interregional Public Organization Association of Heads of Information Security Services (ARSIB), an independent non-profit organization Russian Union of IT Directors (SODIT).

Author and co-author of 12 monographs and more than 350 articles on information security issues (Proceedings of ISA RAS and SPIIRAS, journals "Cybersecurity issues", "Information security problems", "Open systems", "Inside: Information protection", "Security systems", "Electronics", "Communication Bulletin", "Network Journal", "Connect World of Connect", etc.). Including, monographs and practical manuals of publishing houses "Peter", "New Athena" and "DMK-Press": "Methods of information protection in the Internet", "Methods and technologies of information security of critical objects of the national infrastructure", "Methods and technologies of cloud security", "Audit of corporate Internet/Internet security", "Information Risk Management", "Information Security Policies" and others.

Awarded the "Big ZUBR" and "Golden ZUBR" in 2014 for the national projects of the Russian Federation in the field of information security.